Zero Waste

Zero Waste
Management Practices for Environmental Sustainability

Edited by
Ashok K. Rathoure

CRC Press
Taylor & Francis Group
Boca Raton London New York Leiden

CRC Press is an imprint of the
Taylor & Francis Group, an **informa** business

CRC Press
Taylor & Francis Group
6000 Broken Sound Parkway NW, Suite 300
Boca Raton, FL 33487-2742

© 2020 by Taylor & Francis Group, LLC

CRC Press is an imprint of Taylor & Francis Group, an informa business

No claim to original U.S. Government works

Printed on acid-free paper

International Standard Book Number-13: 978-0-367-18039-3 (Hardback)

This book contains information obtained from authentic and highly regarded sources. Reasonable efforts have been made to publish reliable data and information, but the author and publisher cannot assume responsibility for the validity of all materials or the consequences of their use. The authors and publishers have attempted to trace the copyright holders of all material reproduced in this publication and apologize to copyright holders if permission to publish in this form has not been obtained. If any copyright material has not been acknowledged, please write and let us know so we may rectify in any future reprint.

Except as permitted under U.S. Copyright Law, no part of this book may be reprinted, reproduced, transmitted, or utilized in any form by any electronic, mechanical, or other means, now known or hereafter invented, including photocopying, microfilming, and recording, or in any information storage or retrieval system, without written permission from the publishers.

For permission to photocopy or use material electronically from this work, please access www.copyright.com (www.copyright.com/) or contact the Copyright Clearance Center, Inc. (CCC), 222 Rosewood Drive, Danvers, MA 01923, 978-750-8400. CCC is a not-for-profit organization that provides licenses and registration for a variety of users. For organizations that have been granted a photocopy license by the CCC, a separate system of payment has been arranged.

Trademark Notice: Product or corporate names may be trademarks or registered trademarks, and are used only for identification and explanation without intent to infringe.

Library of Congress Cataloging-in-Publication Data

Names: Rathoure, Ashok K., 1983– author.
Title: Zero waste : management practices for environmental sustainability / Ashok K. Rathoure.
Description: Boca Raton : Taylor & Francis, a CRC title, part of the Taylor & Francis imprint, a member of the Taylor & Francis Group, the academic division of T&F Informa, plc, 2020. | Includes bibliographical references.
Identifiers: LCCN 2019020870 | ISBN 9780367180393 (hardback : acid-free paper)
Subjects: LCSH: Waste minimization.
Classification: LCC TD793.9 .R38 2020 | DDC 628.4—dc23
LC record available at https://lccn.loc.gov/2019020870

Visit the Taylor & Francis Web site at
www.taylorandfrancis.com

and the CRC Press Web site at
www.crcpress.com

Printed and bound in Great Britain by
TJ International Ltd, Padstow, Cornwall

Contents

Foreword ix
Preface...xi
Acknowledgements xiii
Editor xv
Contributors xvii
List of Abbreviations...xix

Chapter 1 Introduction to Zero Waste: Management Practices 1

Ashok K. Rathoure

Chapter 2 Zero Liquid Discharge: Industrial Effluent Management 13

Ashok K. Rathoure, Tinkal Patel and Devyani Bagrecha

Chapter 3 Zero Noise Pollution: Green Belt Development 25

Ashok K. Rathoure and Jahanvi Modi

Chapter 4 Zero Defects in Woven Shirt Manufacturing: Application of Six
Sigma Methodology Based on DMAIC Tools 51

R. Rathinamoorthy

Chapter 5 Biomedical Waste Disposal and Treatment.. 75

Rakesh K. Sindhu, Gagandeep Kaur and Arashmeet Kaur

Chapter 6 Best Practices in Construction and Demolition
Waste Management 91

Ashok K. Rathoure and Hani Patel

Chapter 7 Plastic Waste Management Practices 105

Savita Sharma and Sharada Mallubhotla

Chapter 8 Industrial Waste Management System 115

Rakesh K. Sindhu, Gagandeep Kaur and Arashmeet Kaur

v

vi   Contents

Chapter 9   Role of Microbes in Solid Waste Management:
An Insight View 131

Debajit Borah and Kaushal Sood

Chapter 10   Management of Solid and Hazardous Waste as per
Indian Legislation.. 151

Ashok K. Rathoure and Unnati Patel

Chapter 11   Good Practices of Hazardous Waste Management 175

*Vidushi Abrol, Manoj Kushwaha, Nisha Sharma,
Sundeep Jaglan and Sharada Mallubhotla*

Chapter 12   Hydroponic Treatment System Plant for Canteen Wastewater
Treatment in Park College of Technology....................................... 187

Shailendra Kumar Yadav and Kanagaraj Rajagopal

Chapter 13   Ecological and Economic Importance of Wetlands and
Their Vulnerability: A Review..203

Sudipto Bhowmik

Chapter 14   Methylene Blue Dye Degradation by Silver Nanoparticles
Biosynthesized Using Seed Extracts of *Foeniculum vulgare* 219

S. Anjum Mobeen and K. Riazunnisa

Chapter 15   A Green and Sustainable Tool for Physicochemical Analysis
of Liquid Solutions: Survismeter...227

Parth Malik and Rakesh Kumar Ameta

Chapter 16   Recent Advances in Bioremediation of Wastewater for
Sustainable Energy Products...247

Dolly Kumari and Radhika Singh

Chapter 17   Effect of Manure on the Metal Efficiency of
Coriandrum sativum 277

*Deepshekha Punetha, Geeta Tewari, Chitra Pande,
Lata Rana and Sonal Tripathi*

Contents vii

Chapter 18 Fly Ash: A Potential By-Product Waste 301

Bhavana Sethi and Saurabh Ahalawat

Chapter 19 Fugitive Dust Control in Cement Industries 317

Ashok K. Rathoure

Index 329

Foreword

By 2050, the world population is projected to reach 9.1 billion. Ninety-nine percent of global population growth is projected to occur in developing nations and by 2050, 68.7% of the world population is projected to live in urban areas. Due to this rapid urbanization, waste generation is increasing day by day. Waste is traditionally thought of having no value. The main focus is more on downstream or end-of-pipe solutions. Local governments spend significant amounts of money on waste collection and disposal without adequate consideration on resource-saving measures. Their economic return or input and up-stream solutions provide the opportunities for resource reduction (increased resource efficiency/minimize raw material input), waste prevention/ minimization of environmental risks through eco-friendly designs, products and structured or reorganized production processes. Therefore, waste of one industry is a valued input to another. Over-reliance on conventional waste management such as landfills and incineration are not sustainable nowadays due to increasing pollution load. So, major opportunities for win-win solutions through partnership with the informal sector are building recycling rates, moving towards zero waste, improving livelihoods, working conditions and saving money.

Zero waste is the springboard to sustainability. It is a new direction to the back end of waste management and the front end of the industrial design. Zero waste continues with a series of simple steps, which are practical, cost-effective and politically acceptable. Zero waste refers to management practices like waste minimization, cleaner production technologies, minimum use of natural resources, pollution control or reduction in use of renewable resources or green belt development. It creates a closed loop system; no waste or liquid will go outside from industrial premises.

In order to write a foreword to the present book focused on zero waste, I would like to appreciate the sincere efforts of my colleague Dr. Ashok K. Rathoure, environmental scientist, to publish this book which contains the latest information on the various aspects of management practices to achieve zero waste in the industrial sector to reduce the pollution load.

I hope this book will serve as a strong reference material and milestone to the new and current researchers and also help the scientist working in identifying and discovering the gaps in this research area. I consider this book a valued addition to the scientific knowledge on waste minimization practices focusing zero waste.

Dr. Pawan Kumar Bharti
(Scientist, Shri Ram Institute of Industrial Research, New Delhi, India)

Preface

Industrial sector is growing day by day with an increasing amount of industrial waste (solid/liquid/gas). Zero waste management provides significant environmental and economic benefits, but with all eyes on climate change, it's time to prioritize zero waste as a proven, cost-effective climate solution. By taking action now, we can take a significant chunk out of our climate impact and can buy ourselves the much-needed time to address our longer-term challenges around energy and transportation. A total of 75% of construction demolition waste is recycled by facilities and services. Recovered plastic has returned to campuses as benches and recycled wood has become cabinetry. Zero waste is a whole system approach to resource management centred on reducing, reusing and recycling. To make recycling work for everyone, we need to buy products made from the materials we recycle. This reduces the need to utilize non-renewable resources by reusing materials that have already been consumed. Producing recycled materials uses less energy and saves more trees than producing virgin materials.

The goal of zero waste is to:

- Maximize recycling
- Minimize waste
- Reduce consumption
- Ensure products are made to be reused, repaired or recycled
- Purchase sustainable products.

Zero waste systems reduce greenhouse gases by:

- Saving energy, especially by reducing energy consumption associated with extracting, processing and transporting raw materials and waste;
- Reducing and eventually eliminating the need for landfills and incinerators.

The waste comes in many alternative forms and will be classified in a variety of ways. Waste management can be challenging for industrial, commercial and institutional sectors because waste management is closely linked to any organization's performance. However, a strategic waste management planning will help to define solid solutions. In many cases, the most efficient and cost-effective way is waste diversion and waste minimization. New and improved technologies are emerging which can help manage waste in a more efficient way, which is more beneficial in the long run as well. Thus with these techniques, nowadays the focus has moved upstream, addressing the problem from the beginning; this starts at the point of designing waste, preventing it, reducing both the quantities and the use of hazardous substances, minimizing and reusing resources and, where residuals still occur, keeping them concentrated and separated to preserve their potential value for recycling and recovery and prevent them from contaminating anything else with economic value after recovery. The main idea is to move away from waste disposal to waste management and from waste to zero waste. The book is written keeping in mind the requirement of industrial sector today for pollution control and best management practices.

Ashok K. Rathoure

Acknowledgements

This book is the result of the dedicated effort of numerous individuals. We would like to acknowledge the help of all the people involved in this project and, more specifically, to the authors and reviewers who took part in the review process. Without their support, this book would not have become a reality.

First, we would like to thank each one of the authors for their contributions. Our sincere gratitude goes to the chapter authors who contributed their time and expertise to this book.

Second, we wish to acknowledge the valuable contributions of the reviewers regarding the improvement of quality, coherence and content presentation of chapters.

Additionally, we would like to thank our editorial board.

The generous participation of Biohm researchers and trainees to eliminate errors in the text and to refine the presentation is greatly acknowledged.

In addition to this, Dr. Debajit Borah is thankful to DBT-Delcon facilities for providing access to more than 900 e-journals to the Centre for Biotechnology and Bioinformatics, Dibrugarh University. S. A. Mobeen is grateful to the Maulana Azad National Fellowship (MANF), New Delhi, for awarding the MANF-JRF fellowship to carry out this study. Dr. Geeta Tewari is thankful to UGC, New Delhi, for financial support and the head of the Department of Chemistry, D.S.B. Campus Nainital, for providing necessary laboratory facilities.

Last but not least, we beg forgiveness of all those who have been with us over the course of the years and whose names we have failed to mention.

Ashok K. Rathoure

Editor

Ashok K. Rathoure holds a doctoral degree in Bioremediation from Central University (HNBGU Srinagar UK). He is working as Independent Researcher in the field of Environment protection. Currently, he is working as Managing Director at Biohm Consultare Pvt. Ltd. Surat. Previously, he was associated with M/s Eco Group of Companies Surat; M/s Vardan EnviroNet Gurgaon and En-vision Group Surat (En-vision Environmental Services and En-vision Enviro Engineers Pvt. Ltd.) for EIA studies; Himachal Institute of Life Sciences Paonta and Beehive College of Ad. Studies, Dehradun for teaching to Biotechnology, Microbiology, Biochemistry and other biosciences subjects. He has 12 years of working experience in various domains. His area of research is environmental biotechnology and publication includes 70 full length research papers in international and national journals of repute and published more than 40 books from reputed publishers in India and abroad. He had reviewed more than 100 research manuscript for many international journals. He is member of APCBEES (Hong Kong), IACSIT (Singapore), EFB (Spain), Society for Conservation Biology (Washington) and founder member of Scientific Planet Society (Dehradun). Dr. Rathoure is also associated as Editor-in-Chief for Octa Journal of Environmental Research, Managing Editor (Octa Journal of Biosciences) and Executive Editor (Scientific India). He is working as expert for Ecology & Biodiversity and Water pollution Monitoring, Prevention & Control along with working as EIA coordinator for various sectors for EIA studies as per NABET (QCI, New Delhi) guidelines. Dr. Rathoure has worked for various industrial sectors viz. Synthetic organic chemicals industry, Mining of minerals, Isolated storage facility, Metallurgical industries (ferrous & non-ferrous), Thermal power, Nuclear Power, Coal Washeries, Cement plants, Pesticides industry, Chlor Alkali industry, Chemical Fertilizers, Paint Industry, Sugar Industry, Oil & gas exploration/transportation, Manmade fibers manufacturing, Distilleries, Common hazardous waste treatment, storage & disposal facilities (TSDFs), CETP, Port & Harbor, Highways, Waterways, Township & Area development, Building & Construction projects for environmental clearance, CRZ clearance, Forest clearance, Wildlife clearance, EC Compliance, ECBC compliance, Replenishment Study, Wildlife Conservation Plan, Mangrove Management Plan, Authorization/permission/objections, etc.

Contributors

Vidushi Abrol
Indian Institute of Integrative Medicine
Jammu, India

and

Shri Mata Vaishno Devi University
Katra, India

Saurabh Ahalawat
Vikram Cement Works
UltraTech Cement Ltd
Khor, Madhya Pradesh, India

Rakesh Kumar Ameta
Central University of Gujarat
Gandhinagar, Gujarat, India

Devyani Bagrecha
Biohm Consultare Pvt. Ltd.
Surat, Gujarat, India

Sudipto Bhowmik
University of Calcutta
Kolkata, India

Debajit Borah
Dibrugarh University
Dibrugarh, Assam, India

Sundeep Jaglan
Indian Institute of Integrative Medicine
Jammu, India

Arashmeet Kaur
Chitkara University
Rajpura, Patiala, Punjab, India

Gagandeep Kaur
Chitkara University
Rajpura, Patiala, Punjab, India

Dolly Kumari
Dayalbagh Educational Institute
Dayalbagh, Agra, India

Manoj Kushwaha
Indian Institute of Integrative
  Medicine
Jammu, India

Parth Malik
Central University of Gujarat
Gandhinagar, Gujarat, India

Sharada Mallubhotla
Shri Mata Vaishno Devi University
Katra, India

S. Anjum Mobeen
Vemana University
Kadapa, Andhra Pradesh, India

Jahanvi Modi
Biohm Consultare Pvt. Ltd.
Surat, Gujarat, India

Chitra Pande
Kumaun University
Nainital, Uttarakhand, India

Hani Patel
Biohm Consultare Pvt. Ltd.
Surat, Gujarat, India

Tinkal Patel
Biohm Consultare Pvt. Ltd.
Surat, Gujarat, India

Unnati Patel
Biohm Consultare Pvt. Ltd.
Surat, Gujarat, India

Deepshekha Punetha
Kumaun University
Nainital, Uttarakhand, India

Kanagaraj Rajagopal
IIT Roorkee
Roorkee, Uttarakhand, India

Lata Rana
Kumaun University
Nainital, Uttarakhand, India

R. Rathinamoorthy
Department of Fashion Technology
PSG College of Technology
Coimbatore, India

Ashok K. Rathoure
Biohm Consultare Pvt. Ltd.
Surat, Gujarat, India

K. Riazunnisa
Vemana University
Kadapa, Andhra Pradesh, India

Bhavana Sethi
Academy of Business and Engineering
    Sciences
Ghaziabad (UP), India

Nisha Sharma
Indian Institute of Integrative
    Medicine

and

Academy of Scientific and Innovative
    Research (AcSIR), Jammu
Jammu, India

Savita Sharma
Shri Mata Vaishno Devi University
Katra, India

Rakesh K. Sindhu
Chitkara University
Rajpura, Patiala, Punjab, India

Radhika Singh
Dayalbagh Educational Institute
Dayalbagh, Agra, India

Kaushal Sood
Dibrugarh University
Dibrugarh, Assam, India

Geeta Tewari
Kumaun University
Nainital, Uttarakhand, India

Sonal Tripathi
Navsari Agriculture University
Navsari, Gujarat, India

Shailendra Kumar Yadav
IIT Roorkee
Roorkee, Uttarakhand, India

List of Abbreviations

AC	Alternating current
AD	Anaerobic digestion
AEPC	Apparel Export Promotion Council
AIDS	Acquired immune deficiency syndrome
APC	Air pollution control
BMW	Biomedical waste
BOD	Biochemical oxygen demand
BTEX	Benzene, toluene, ethylbenzene and xylenes
BTX	Benzene, toluene and xylenes
CEMS	Continuous Emission Monitoring System
CO	Carbon monoxide
COD	Chemical oxygen demand
CP	Cleaner production
CPCB	Central Pollution Control Board
CSTR	Continuous stirred-tank reactor
DAF	Dissolved air flotation
DC	Direct current
DCS	Distributed control system
DFSS	Design for Six Sigma
DHU	Defects per hundred unit
DMADV	Define, Measure, Analyze, Design and Verify
DMAIC	Define, Measure, Analyze, Improve and Control
DPMO	Defect Per Million Opportunities
E(P)A	Environmental Protection Act
ESP	Electrostatic precipitator
ETP	Effluent treatment plant
e-wastes	electronic wastes
FAD	Flavin adenine dinucleotide
FAO	Food and Agriculture Organization
GBD	Green Belt Development
HF	High Frequency
HIV	Human Immunodeficiency Virus
ISLM	Integrating Sound Level Meter
KLD	Kilolitre per day
LF	Low frequency
LU/LC	Land use/land cover
MEE	Multiple effect evaporator
MSW	Municipal solid waste
MVR	Mechanical vapour recompression
NAD	Nicotinamide adenine dinucleotide
NADPH	Nicotinamide adenine dinucleotide phosphate
NIHL	Noise-induced hearing loss

NMS	Noise monitoring system
NO$_x$	Oxides of nitrogen
OPC	Operator performance card
OPD	Outpatient department
OSIR	Operator Self-Inspection Report
OWC	Organic waste composting
PAH	Polycyclic aromatic hydrocarbon
PM	Particulate matter
PPE	Personal protective equipment
PUC	Pollution under control
PVC	Polyvinyl chloride
RMG	Ready-made garments
RO	Reverse osmosis
RWH	Rainwater harvesting
SDA	Spray dryer absorption
SLM	Sound level meter
SO$_2$	Sulphur dioxide
SPCB	State Pollution Control Board
SPLs	Sound pressure levels
SPM	Suspended particulate matter
SW	Solid waste
SWM	Solid waste management
UASB	Upflow anaerobic sludge blanket
UV	Ultraviolet
VLF	Very low frequency
VOCs	Volatile organic compounds
WHO	World Health Organization
ZLD	Zero liquid discharge

1 Introduction to Zero Waste
Management Practices

Ashok K. Rathoure

CONTENTS

1.1 Introduction 1
    1.1.1 Types of Waste 2
        1.1.1.1 Gaseous Waste 2
        1.1.1.2 Liquid Waste 2
        1.1.1.3 Solid Waste 2
    1.1.2 Sources of Waste 3
        1.1.2.1 Domestic Waste 3
        1.1.2.2 Industrial Waste 3
        1.1.2.3 Agricultural Waste 4
        1.1.2.4 Municipal Waste 4
        1.1.2.5 Biomedical Waste 5
        1.1.2.6 Nuclear Waste 5
    1.1.3 What is Zero Waste? 5
    1.1.4 Why Zero Waste? 5
1.2 Zero Waste Management Practices 6
    1.2.1 Cleaner Production (CP) 7
    1.2.2 Minimum Use of Natural Resources 7
    1.2.3 Rainwater Harvesting (RWH) 7
    1.2.4 Organic Waste Composting (OWC) 8
    1.2.5 Zero Liquid Discharge (ZLD) 8
    1.2.6 Green Belt Development (GBD) 9
1.3 Conclusion 10
References 11

1.1 INTRODUCTION

Wastes are substances which are disposed or intended to be disposed or required to be disposed of by the provisions of national laws. Wastes are unnecessary products for which the generator has no further use for the production, transformation or consumption, and which it will dispose. Wastes are generated from the extraction and the processing of raw materials into intermediate and final products, the consumption of final products and other human activities. Residuals which will be recycled or reused at the place of generation are excluded. Of the many different types of wastes, several are hazardous in nature, causing many diseases (UNEP, 2004).

1.1.1 TYPES OF WASTE

Waste comes in many alternative forms and will be classified in a variety of ways. Here, waste can be classified into three types based on physical composition.

1.1.1.1 Gaseous Waste

Gaseous waste is a waste material in gas form which ensues from varied human activities, such as manufacturing, processing, material consumption or biological processes. Gaseous waste that commands in exceedingly closed instrumentality falls into the class of solid waste for disposal functions.

1.1.1.2 Liquid Waste

Liquid waste means wastewater containing fats, oils or grease, liquids, solids, gases, or sludge and hazardous household liquids. These liquids are hazardous or harmful for human health and/or the environment. They are also discarded products classified as liquid industrial waste such as cleaning fluids or pesticides or the by-products of manufacturing processes (Abercrombie, 2013).

1.1.1.3 Solid Waste

Solid waste means garbage, refuse materials or sludge from a wastewater treatment plant, water treatment plant, sewage treatment plant or air pollution control facility and other discarded materials which include solid, liquid, semi-solid or gaseous form and which is resulting from industrial, commercial, mining, agricultural operations and from community activities, but it does not include solid or dissolved materials in domestic manure or solid or dissolved materials in irrigation back flows or industrial discharges (DEC, 2014).

Solid waste can be classified into different types depending on their source:

1. *Household waste* is generally classified as municipal solid waste. This waste includes domestic waste, sanitation waste, waste from streets, construction and demolition detritus that arises through the construction and demolition of buildings and alternative construction activities. With the increase in urbanization, municipal solid waste is forming the bulk of solid waste. The growth of metropolitan cities is even resulting in a huge quantity of municipal solid waste.
2. *Industrial waste* or hazardous waste is dangerous because it contains toxic substances that are chemical in nature. This type of waste is very dangerous to humans, plants, animals and the overall environment. Improper disposal of the industrial solid waste might cause death, illness and sometimes environmental damage which will continue to future generations. For example, any oil spill in the seas, release of poisonous gases or chemicals into the

Introduction to Zero Waste

air and improper disposal of industrial effluents into the soil will lead to destruction of all living species and additionally to environmental harm.

3. *Biomedical waste* is an infectious waste being generated day in and day out by various hospitals, clinics, research centres, pharmaceutical companies and health care centres. This type of solid waste is most infectious and can spread diseases, viral and bacterial infections among humans and animals if not managed properly in a scientific way. Hospital waste includes solid waste such as disposable syringes, bandages, cotton swabs, body fluids, human excreta, anatomical waste, expired medicines and other types of chemical and biological waste. Hospital waste is equally hazardous and dangerous, just as industrial waste, if not disposed of or managed properly (Prokerala, 2012).

1.1.2 Sources of Waste

1.1.2.1 Domestic Waste

Domestic waste is generated due to various household activities. It includes paper, glass bottles and broken pieces of glass, plastics, cloth rags, kitchen waste, garden litter, cans and so forth. Kitchen waste includes waste food, peels of fruits and vegetables and ashes due to the burning of wood, coal or cow-dung cakes. Domestic waste is classified into subsequent types:

1. Garbage includes peels of fruit and vegetables, leftover food articles, garden litter and so forth. It is organic in nature and might biodegrade quickly.
2. Rubbish includes waste paper, glass and pottery pieces, plastic goods, rubber goods, polythene bags, cloth rags and so forth. These are mostly inorganic in nature.
3. Ash is the main source of solid domestic waste, which is generated due to the burning of wood, coal and dung cakes in the kitchen.
4. Sewage includes the wastewater from kitchen and bathrooms.

Domestic waste is often seen lying in the streets and along the roads in heaps. This makes the environment unhygienic and it may be the breeding ground for mosquitoes and other different harmful organisms.

1.1.2.2 Industrial Waste

Almost all kinds of industries use raw materials to manufacture finished expendable goods. In this method any leftover material, which is of no use, is called industrial waste. It includes general plant rubbish, packaging waste, demolition and construction waste, damaged parts of machines and tools and so forth. This can cause toxicity within the air, water and soil. It is harmful to human beings and the environment. The waste generated by a number of industries is given below:

1. The mining and other metallurgical type industries generate ash from coal, rocks of no value, furnace slag, metallic waste and so forth.

2. The chemical industries generate harmful chemicals like acids, toxic gases, oils, alkalis, and many types of synthetic materials.
3. The waste material from oil refineries and petrochemical industries are petroleum gases, hydrocarbons, oils and other toxic organic chemicals.
4. The cement factories generate suspended particulate matter (SPM) in the form of coarse and fine particles which can pollute the air and cause respiratory disorders.
5. The waste from the industries of cellulose fibre, paper scraps, bleaching powder, alkalis and so forth. About 40% wood or bamboo used in the production of paper generally goes to waste.
6. The construction companies use a variety of construction material like cement, sand, bricks, stone, wood, limestone and so forth for construction of buildings. The waste is discharged within the form of debris.

1.1.2.3 Agricultural Waste

The waste generated throughout the varied processes of crop production and livestock rearing is called agricultural waste. It is generally of the following types:

1. *Plant remains* include husk and straw, wood and rubber waste, cotton and tobacco waste, coconut waste products, nutshells and so forth. These are common plant remains which are produced during agricultural practices.
2. *Animal waste* includes heaps of animal waste that are left unattended, which emit a foul smell and are the breeding ground for many harmful microorganisms. Managing the massive quantity of animal waste generated in rural areas is a major task.
3. *Processing waste* includes food crops processed for the preparation of rice from paddies, flour from wheat and jowar, dal from pulses, edible oils from oilseeds and so forth. A huge quantity of husks are produced as waste. Improper handling might degrade the environment and can harm the health of people involved in the process. Agricultural activities include ploughing, sowing, harvesting, threshing, winnowing, poultry farming, dairy farming and so forth. Threshing and shifting are waste generating activities of agriculture. Modern agricultural practices use chemical fertilizers, pesticides, insecticides, weedicides, fungicides and herbicides on a large scale to enhance crop production. Excessive use of agrochemicals is harmful for crops and the environment.

1.1.2.4 Municipal Waste

Municipal waste is waste from municipal areas which is managed by the municipal corporation of the concerned area. It includes domestic waste, community waste and commercial waste. The waste collected from educational institutes and hospitals sweeping of the streets, lanes and roads are called community waste. Sewage is the foul-smelling, grey liquid which contains waste organic matter from towns or cities and is carried off in underground drains.

Introduction to Zero Waste

Commercial waste is generated from the business institutions like shops, offices, godowns (storage area), stores, markets and so forth. They include packaging materials, paper, spoiled or discarded goods and so on. The waste can be organic or inorganic. Fish and vegetable markets additionally generate a huge amount of waste.

1.1.2.5 Biomedical Waste

Waste generated from hospitals, pathological laboratories, clinics and so forth are called biomedical waste. These wastes are hazardous for human beings and the environment. They include syringes, needles, blades, scalpels, empty plastic bottles, polythene bags, gloves, tubes, expired medicines, pathological wastes and waste from surgeries and autopsies. They are toxic and hazardous. Biomedical waste falls underneath two main classes: infectious and non-infectious wastes. The infectious waste contains large number of pathogens, which are dangerous. Most biomedical waste is non-infectious, which gets mixed with infectious waste. Biomedical wastes are more hazardous than most of the other wastes.

1.1.2.6 Nuclear Waste

Radioactive wastes from the nuclear power plant and weapons industries are a matter of great concern. The waste from the spent nuclear fuel comprises unconverted uranium, plutonium and other radioactive elements. The waste remains radioactive for thousands of years. There is no safe and proper method for enduring disposal of nuclear wastes (Jain, 2019).

1.1.3 What is Zero Waste?

Zero waste means the conservation of all resources of accountable production, consumption, reuse and recovery of all goods, packaging and materials, and not burning them or discharging them to land, water or air that would threaten the surroundings or human health. Zero waste refers to waste management and designing approaches that emphasize waste interference as opposed to end-of-pipe waste management. It is a full systems approach that aims for a massive change in the way materials flow through society, resulting in no waste (Snow and Dickinson, 2001). Zero waste encompasses eliminating waste through recycling and reuse and focuses on restructuring production and distribution systems to reduce waste (Davidson, 2011). Zero waste is additional of a goal or ideal instead of a tough target. Zero waste provides guiding principles for frequently operating towards eliminating wastes (Spiegelman, 2006).

1.1.4 Why Zero Waste?

Day by day, the increase in population leads to an increase in food demand and other necessities, so it increases waste. Waste can cause serious problems: for example plastic waste produces toxic substances and it harm animals as well as aquatic cultures and environments. If waste is incinerated, gases from incineration may cause air pollution while ash from incinerators may contain heavy metals and other toxins.

Domestic waste tends to produce microbial pathogens which leads to infectious and chronic diseases. Hazardous waste contains hazardous chemicals, so it leads to chemical poisoning. Due to these serious problems there is a need for zero waste.

Zero waste promotes prevention and product design that consider the entire product life cycle. It strongly supports sustainability by protecting the environment, reducing costs and producing additional jobs in handling of waste. It saves money by reducing waste. It improves production processes. A zero-waste strategy would use far fewer new raw materials and send no waste materials to landfill. Any material waste would either return as reusable or recyclable materials or would be suitable for use as compost. Zero waste helps to reduce the need to create landfills. Reduction of landfills leads to reduction of incidences of diseases which are associated with toxins. It also can help to preserve the environment and prevent pollutants from entering the ecosystem.

1.2 ZERO WASTE MANAGEMENT PRACTICES

There are several methods to adopt zero waste management in the industrial unit like 5R's.

> The *Five R's* is a part of the waste management hierarchy. The Five R's are techniques for waste minimization. The essence of these approaches is characterized by a need to avoid, eliminate, prevent or significantly reduce the causes of environmental problems, as opposed to managing the

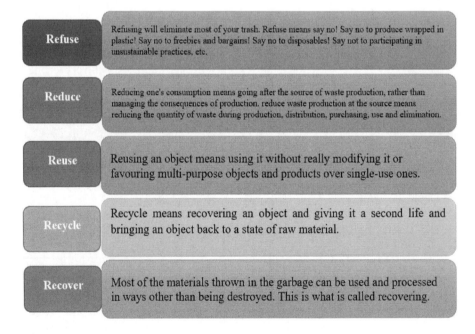

FIGURE 1.1 Waste Minimization Techniques of Five Rs

Introduction to Zero Waste

impacts, wastes and emissions arising more down the goods or service life cycle. This suggests a fundamental change in the nature of environmental interventions in terms of rationale, timing and specific approach. The Five R's, which are a main part of waste minimization, are listed in Figure 1.1 (Gertsakis and Lewis, 2003).

1.2.1 CLEANER PRODUCTION (CP)

Cleaner production means the reduction of pollution by means of pollution preventive measures applied to products and production processes. CP generates options for improvement in five categories: change of input materials, technology change, good operation practices product modification and on-site reuse and recycling. CP options can reduce the material, energy and water consumption per product and increase savings made on the cost of these natural resources. The cost for processing waste streams (including solid waste, wastewater and air emissions) will increase in the next future. Minimizing waste streams and a proactive compliance with laws and regulations can save money. And last but not least, most often with environmental measures, the efficiency of production processes will increase as well, resulting in higher levels of production output, or improvement of the product quality. CP can be approached in four phases: planning and organization, assessment, feasibility analysis and implementation and sustainability (GCPC, 2008).

1.2.2 MINIMUM USE OF NATURAL RESOURCES

This is the best option because the most effective way to limit the health effects and environmental impacts of a waste is not to create that waste in first place. Due to increase in urbanization and industrialization, it creates a load on natural resources. Likewise, due to population increase and increased use of fuel, vehicles are increasing day by day, which uses gasoline and gasoline is limited as it takes too much time for generation. All the natural resources including water, air, fuel and so forth are limited and depleting day by day due to over-exploitation. In the near future, it is possible that there will be shortage of fuel for daily life, no fresh water to drink, no fresh air to breathe and more. Therefore, we recommend to minimize the uses of natural resources at every place to sustain the life. For example, if you take a short, two-minute shower, it will save 30% water without compromising on the comfort of the user, because a showerhead uses as much as 16 litres per minute.

1.2.3 RAINWATER HARVESTING (RWH)

Rainwater harvesting is an easy process of collecting rainfall for future usage. The collected rainwater will be stored, utilized in several ways or directly used for recharge functions. With depleting groundwater levels and unsteady climatic conditions, RWH will go a long way to help mitigate these effects. If we capture the rainwater, it will facilitate recharging our native aquifers and reduce urban flooding, and most importantly it will ensure water availability in water-scarce zones. This water conservation methodology will be simply practiced in individual homes, apartments, parks, offices and temples too across the globe. From RWH, farmers can recharge their dry borewells, they can create water banks in drought areas, green their farms and increase sustainability of their water resources. RWH is an effective methodology in water-scarce times; it is also an easy practice.

1.2.4 Organic Waste Composting (OWC)

For centuries, gardeners and farmers have utilized organic waste, which is biodegradable, for producing stable and nutrient-rich compost for use in pots or directly for improving the soil. This application of the natural, exoergic process of aerobic decomposition, is acquainted and time-honoured. Currently, however, composting has been the recipient of increased attention as a potential means of treating biowaste on a municipal basis. Though the scale of such operations imposes certain restrictions of its own, generally, putrescible matter decomposes more efficiently and completely when oxygen is readily available. This results in proteins being degraded to nitrogen or ammonia gas and ultimately mineralized to nitrates, while fats and carbohydrates are broken down to carbon dioxide and water via organic acids. This is, of course, strictly a mass flow overview of the process, since a proportion of the material becomes incorporated into microorganism cells as the decomposers themselves multiply and grow. Even underneath optimized environmental conditions, there are a variety of rate-limiting factors in the process, which includes extracellular hydrolytic enzyme production, the speed of hydrolysis itself and the efficiency of oxygen transfer. These may, in turn, be influenced by other aspects such as particle size and nature of the biowaste material to be treated. In a practical application of organic waste composting, this can be a major consideration as the kind of biowaste to be composted can vary greatly, particularly when derived from municipal solid waste, since seasonal variation, local conditions and climate may produce a highly heterogeneous material. On the opposite hand, biowastes from food process or farming will be remarkably consistent and homogeneous (Gareth and Judith, 2003).

1.2.5 Zero Liquid Discharge (ZLD)

As the population continues to grow, people are putting ever-increasing pressure on water resources. So waters are being squeezed by human activities and urbanization and water quality is reduced. Poorer water quality means water pollution. Water pollution means one or more substances have built up in water to such an extent that they cause problems for animals or people or other natural resources/habitats. Water pollution is all about quantities: how much of polluting substance is released and how big volume of water is released. A small quantity of toxic chemical may have little impact if it is spilled into an ocean, but same chemical can have a much bigger impact if it is disposed into river or lake, because there is less clean water to disperse it. So, there is a need for ZLD.

Introduction to Zero Waste 9

ZLD is an ideal, closed loop cycle where no liquid effluent is released after the end-of-pipe treatment cycle, but effluent is reused or recycled and wastewater will be purified. There are many technologies to achieve ZLD. The processes are anaerobic digestion, reverse osmosis, multiple effect evaporator, mechanical vapour recompress, spray dryer, solvent strippers and so forth. Cost-wise, achieving ZLD is a costly proposition but it is now becoming a necessity because rivers need to be rejuvenated after recognizing that many industrial sectors are not able to achieve standards and this ultimately necessitates to work towards ZLD (CPCB, 2015).

1.2.6 Green Belt Development (GBD)

A green belt area refers to an area that is kept in reserve for an open area, most often around larger cities. The main purpose of the green belt is to protect the land around larger urban centres from urban slump, to maintain the designated area for forestry and agriculture and to provide habitat for wildlife. Trees facilitate removing carbon dioxide and alternative pollutants from the air and by introducing oxygen into the air, thereby rising air quality. A green belt development may facilitate in removing particulate matter from the air by trapping such particulate matter. A green belt reduces noise pollution. Working as a barrier, trees will either deflect, refract or absorb sound to reduce its intensity. The intensity reduction depends on the distance sound needs to travel from the source. Trees may reduce the moisture content and modify the climate which affects sound intensity. Trees also help reduce soil erosion through improvement of soil quality and bind soil particles, and help to contain water runoff (Telang, 2013).

Techniques for development of green belt are as follows:

1. *Social forestry* means tree plantation on unused and fallow land. Social forestry is for management and protection of forest and afforestation of barren and deforested lands. It is a democratic approach of forest conservation and usage.
2. *Community plantation* means plantation on underutilized locally owned land which is refurbished and maintained by any community.
3. *Mangrove plantation* means plantation of mangroves like shrubs or small trees near coastal saline or brackish water. Mangrove plantation has become hugely popular. The majority of planting efforts are failing, however. A more practical approach is to make the proper conditions for mangroves to grow back naturally. In this way, mangroves generally survive and perform better.
4. *Avenue plantation* means plantation on a straight path or road with a line of trees or large shrubs on both sides of the paths.
5. *Groove plantation* is plantation near the surroundings of a house.
6. *Anti-erosion bunds* are plantation of vegetation to reduce soil erosion. It is an excellent biological method to safeguard the landscape and the shape of the land. Cover crops such as vetch, rye and clover are excellent plants for erosion control.

7. *Barricading* means tree plantation for defence, like plantation on the sides near a gate or wall. We can say that in plantation for making borders, barricading small plants or shrubs will be used.
8. *Terrace gardening* is a method in which a terrace is used for gardening purpose. Terrace gardening provides clean air and makes the atmosphere cool for nearby houses. A raised terrace keeps a house dry and provides a transition between the arduous materials of the design and softer ones of the garden.
9. *Vertical gardening* is a garden that grows upward (vertically) using a trellis or other support system rather than on the ground (horizontally) – anything grown on a trellis or maybe a fence which is technically a part of a vertical garden. This technique will be used to create living screens between different areas, providing privacy for your yard or home. More recently, vertical gardens have also been used to grow flowers and even vegetables.
10. *Hydroponic system* is a method of growing plants without soil by using mineral nutrient solutions in a water solvent. Terrestrial plants are also grown with solely their roots exposed to the mineral solution, or the roots may be supported by an inert medium, such as perlite or gravel.
11. *Bioremediation* means use of living green plants for treatment of contaminated soil, sludges or groundwater by the removal, degradation or containment of the pollutants present. A large style of species from completely different plant teams, ranging from pteridophyte ferns, to angiosperms like sunflowers, and poplar trees, which employ a number of mechanisms to remove pollutants. Amongst these, some hyperaccumulate contaminants within the plant biomass itself, which can subsequently be harvested; others act as pumps or siphons, removing contaminants from the soil before venting them into the atmosphere; whereas others alter the biodegradation of comparatively massive organic molecules, like hydrocarbons derived from crude oil. However, the technology is comparatively new and so still in development (Gareth and Judith, 2003).
12. *Wildlife conservation* aims to conserve the wildlife in situ. The community forestry, social forestry, avenue plantation, groove plantation, thick green belt development and so forth are essential part of ecology to conserve the wildlife. National parks, sanctuaries, nature parks, biosphere reserves and transboundary protected areas as well as non-protected areas along or across borders work as inhabitation and it supports regional development based on nature conservation. Forests are natural habitats of wildlife. Due to urbanization, deforestation increases, so wildlife is gradually crowding up the endangered list due to habitat loss.

1.3 CONCLUSION

Waste management can be challenging for industrial, commercial and institutional sectors because waste management is closely linked to any organization's performance. However, a strategic waste management planning will help to define solid solutions. In many cases, the most efficient and cost-effective way is waste diversion

Introduction to Zero Waste

and waste minimization. New and improved technologies are emerging that can help manage waste in a more efficient way, which is more beneficial in the long run as well. Thus with these techniques, nowadays the focus has moved upstream, addressing the problem from the beginning; this starts at the point designing of waste, preventing it, reducing both the quantities and the uses of hazardous substances, minimizing and reusing resources and, where residuals still occur, keeping them concentrated and separated to preserve their potential value for recycling and recovery and prevent them from contaminating anything else with economic value after recovery. The main idea is to move away from waste disposal to waste management and from waste to zero waste.

REFERENCES

Abercrombie, T., 2019. What is liquid waste? e-Waste Disposal Inc. Available online at www.ewastedisposal.net/liquid-waste/

Baker, E., Bournay, E., Harayama, A. and Rekacewicz, P., 2004. Vital Waste Graphics. *UNEP*. Available online at www.unido.org/sites/default/files/2015-10/NCPC_20_years_0.pdf

CPCB, 2015. Guidelines on Techno – Economic Feasibility of Implementation of Zero Liquid Discharge (ZLD) for Water Polluting Industries. Available online at www.indiaenvironmentportal.org.in/files/file/FinalZLD%20water%20polluting%20industries.pdf

Davidson, G., 2011. Waste Management Practices: Literature Review. Office of Sustainability – Dalhousie University, Canada.

DEC, 2014. *What is Solid Waste*, Department of Environmental Conservation, New York. Available online at www.dec.ny.gov/chemical/8732.html

Gareth, M. E. and Judith, C. F., 2003. Environmental Biotechnology: Theory and Application, John Wiley & Sons, Ltd. UK.

GCPC, 2008. Cleaner production manual by Gujarat Cleaner production centre. Available online at http://www.gcpcenvis.nic.in/Manuals_Guideline/CP_Manual_Improving_Living_and_Working_Condition_of_People_around_Industries.pdf

Gertsakis, J. and Lewis, H., March 2003. Sustainability and the Waste Management Hierarchy. EcoRecycle Victoria. Available online at http://www.helenlewisresearch.com.au/wp-content/uploads/2014/05/TZW_-_Sustainability_and_the_Waste_Hierarchy_2003.pdf

Jain, R., 2019. Six most important sources of solid waste in India. Available online at www.shareyouressays.com/knowledge/6-most-important-sources-of-waste-in-india/110869

Prokerala, 2012. What are solid wastes? Available online at www.prokerala.com/going-green/solid-wastes.htm

Snow, W. and Dickinson, J., 2001. The end of waste: Zero waste by 2020. Available online at http://www.unep.or.jp/ietc/Focus/An%20End%20to%20Waste.pdf

Spiegelman, H., 2006. Transitioning to zero waste—What can local governments do NOW? Available online at www.rcbc.ca/files/u3/PPI_Zero_Waste_and_Local_Govt.pdf

Telang, S., 2013. Regulatory provisions for Green Belt Development in India. Available online at http://greencleanguide.com/regulatory-provisions-for-green-belt-development-in-india/

2 Zero Liquid Discharge
Industrial Effluent Management

Ashok K. Rathoure, Tinkal Patel and Devyani Bagrecha

CONTENTS

2.1 Introduction 13
2.2 Technologies of Zero Liquid Discharge 14
    2.2.1 Anaerobic Digestion 14
    2.2.2 Reverse Osmosis 14
    2.2.3 Multiple Effect Evaporator (MEE) 15
    2.2.4 Alternate Evaporative Process: Mechanical Vapour
           Recompression (MVR) 16
    2.2.5 Solvent Strippers 16
    2.2.6 Wastewater Spray Dryer 17
    2.2.7 Incineration 17
2.3 Case Study: ZLD System for Water Jet Looms (Weaving Industry) 17
    2.3.1 Grey Cloth 18
    2.3.2 Water Jet Weaving Machine 18
    2.3.3 Detail of Effluent Treatment Plant 19
           2.3.3.1 Collection Tank 19
           2.3.3.2 Bar Screen 19
           2.3.3.3 Oil Skimmer 19
           2.3.3.4 Flocculation and Chemical Dosing 19
           2.3.3.5 Dissolved Air Flotation (DAF) 19
           2.3.3.6 Reverse Osmosis (RO) 19
           2.3.3.7 Evaporator 20
    2.3.4 Water Requirement 20
    2.3.5 Indian Scenario of ZLD 22
2.4 Conclusion 22
References 22

2.1 INTRODUCTION

Zero liquid discharge (ZLD) is a process of water treatment in which all wastewater is purified and recycled, leaving zero discharge at the end of the treatment cycle.

Fresh water scarcity and concerns for environmental impact resulting from industrial wastewater discharges place a high degree of importance on recycling and reuse of water, an increasingly valuable resource. There are many technologies to achieve ZLD. The processes are anaerobic digestion, reverse osmosis, multiple effect evaporator, mechanical vapour recompress, spray dryer, solvent strippers and so forth. The process of achieving ZLD is costly (CPCB, 2015).

2.2 TECHNOLOGIES OF ZERO LIQUID DISCHARGE

2.2.1 Anaerobic Digestion

Anaerobic digestion (AD) is the natural process that breaks down organic matter in the absence of oxygen to release a gas known as biogas, leaving an organic residue called digestate. This is a well-established technology. Almost all distilleries have anaerobic digesters. Figure 2.1 shows an anaerobic digester. CSTR- and UASB-based digesters are more suitable for molasses-based distilleries. In the first step, the incoming flow of sludge is combined and heated for biological conversion. Residence time here ranges from 10 to 20 days. In the second step, the mixture is allowed to further digest and in the third step, settled sludge is dewatered and thickened. BOD removal efficiency is 85%–90%, COD removal efficiency is 55%–65%, specific biogas generation (NM^3/kg of COD consumed) is 0.45–0.55, methane content of biogas is 55%–65%, and H$_2$S content of biogas is 2%–4% (CPCB, 2015).

2.2.2 Reverse Osmosis

Reverse osmosis is the process by which an applied pressure, greater than the osmotic pressure, is exerted on the compartment that once contained the high-concentration

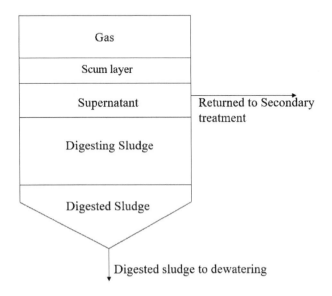

FIGURE 2.1 Anaerobic Digestion

Zero Liquid Discharge

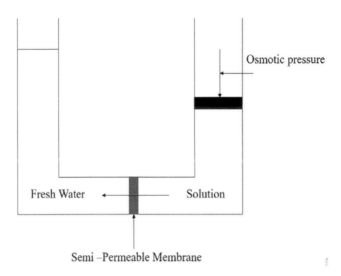

FIGURE 2.2 Reverse Osmosis

solution. This pressure forces water to pass through the membrane in the direction reverse to that of osmosis. Water now moves from the compartment with the high-concentration solution to that with the low concentration solution. Due to this, relatively pure water passes through membrane into the one compartment while dissolved solids are retained in the other compartment. Hence, the water in the compartment is concentrated or dewatered. Due to the resistance of the membrane, the applied pressures required to achieve reverse osmosis are significantly higher than the osmotic pressure (Ansa et al., 2014). Figure 2.2 shows process of reverse osmosis.

2.2.3 Multiple Effect Evaporator (MEE)

Evaporators can minimize the production of waste and increase the potential for valuable materials from those wastes. MEEs are common to industries that concentrate different products, regenerate solvents or separate solid-liquid mixtures. Process integration can help to choose the best configuration of MEE in order to achieve a more efficient process in the sense of energy use (Ghosna, 2012). Industries have installed a multiple effect evaporator for the treatment of industrial effluent with adequate capacity. The condensate water generated from the MEE shall be used in processing neutral effluent from a primary treatment plant passed through three stages. Figure 2.3 shows the three-stage process of MEE. The evaporator system and the evaporated water is collected in an evaporated water collection tank and then recycled to the plant after filtering through a sand filter and a carbon filter. In simple terms, evaporation is the process of concentrating a solution to remove the excess solvent so as to obtain a final product rich in solute concentration. And an evaporator is used to carry out this process by using steam as a heating medium in most of the cases (Gautami, 2011).

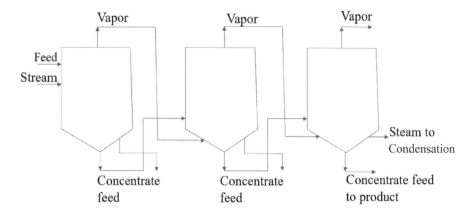

FIGURE 2.3 Multiple Effect Evaporator

2.2.4 Alternate Evaporative Process: Mechanical Vapour Recompression (MVR)

The heat pump of mechanical vapour recompression (MVR) could be widely applied to evaporation of solution. The MVR heat pump does not need a boiler to provide some drive vapour, so the pollutant released from coal-burning boilers reduces. In the places of evaporation, the MVR may be considered for adoption if there is enough electric power. It could reuse the heat of produced vapour evaporating from the solution, which shows the high efficiency of energy conservation. With the drop trend of electric comparing to vapour, the MVR heat pump will be used more than before. At present, in overseas areas without plenty of water, the MVR heat pump is applied to desalination of seawater by researchers, and some perfect fruit has been gained. In domestic salt industry, the MVR is forming its market and a few big factories have installed the equipment. The MVR systems of desalination or salt manufacturing are very complicated and their production capacity is quite large. Based on the need of medium and small evaporation in industry, this study is done to design a unit with compact structure and simple manipulation. All parts of the unit are homegrown. The compression ratio and temperature difference of heat transfer are not big. It is suitable to concentrate those solutions which are sensitive to temperature. It provides references for optimization of the latter systems by measuring and analyzing the performance parameters of the unit (Weike et al., 2013).

2.2.5 Solvent Strippers

Stripping is a process of physical separation where one or more parts are removed from a fluid stream from a vapour course. Start the cooling water pump, treated effluent transfer pump and feed pump and keep the feed rate to the stripping column constant. After reaching the 70% level in the kettle, stop the feed pump. Open the steam value (steam pressure should be 1.0–1.5 kg/cm^2) to the reboiler and drain the condensate water by opening the bypass valve. After removal of the condensate water, close the condensate bypass valve. Collect the top condensate in the collection

Zero Liquid Discharge

receiver. Start the feed pump and keep the required feed rate by adjusting the flow meter. Start the stripper bottom pump and transfer the bottom with required flow rate. Collect the condensate in the collection receiver up to 50% of the level. Start the reflux and collection pump (Popuri and Guttikonda, 2016).

2.2.6 WASTEWATER SPRAY DRYER

Spray drying is a process in which a liquid or slurry solution is sprayed into a hot gas stream in the form of a mist of fine droplets. In power generation, spray dryers are most commonly used in spray dryer absorption (SDA) applications in which an alkaline slurry is used to remove acid gases from a flue gas stream. Spray drying applications extend well beyond power generation and include the production of laundry detergents, pharmaceuticals, plastics, pigments, instant coffee, powdered milk and many more. Another spray drying application is salt drying. In salt drying applications, a liquid or slurry containing a significant concentration of dissolved salts is dried in a hot gas stream. During the drying process, as water is evaporated, the dissolved salts concentrate in solution (Klidas, 2016).

2.2.7 INCINERATION

The ZLD process generates thousands of tons of sludge as solid waste. Incineration is generally used to burn them with modern machinery fitted with air pollution control equipment. Waste material is converted into incinerator bottom ash, flue gases, particulates and heat, which can in turn be used to generate electric power. The flue gases should be cleaned of pollutants before they are dispersed in the atmosphere, while the bottom ash can be used in cement mix or disposed in landfill (Vyas, 2016).

2.3 CASE STUDY: ZLD SYSTEM FOR WATER JET LOOMS (WEAVING INDUSTRY)

The case study of a textile unit (weaving of grey cloth) at Om Textile Park, Kamrej Surat (Gujrat) India has been considered. They are manufacturing the grey cloth from yarn which is available from the local market. Knitting is a technique to turn thread or yarn into a piece of cloth. Knitted fabric consists of horizontal parallel courses of yarn, which is different from woven cloth. The courses of threads or yarn are joined to each other by interlocking loops in which a short loop of one course of yarn or thread is wrapped over another course. There are two types of knitting:

1. Weft knitting
2. Warp knitting.

Weft knitting is a method of forming a fabric in which the loops are made in a horizontal way from a single yarn and intermeshing of loops take place in a circular or flat form on a cross-wise basis. Warp knitting is a method of forming a fabric in which the loops are made in a vertical way along the length of the fabric from each warp yarns and intermeshing of loops take place in a flat form of length-wise basis. Weaving is a method of textile production in which two distinct sets of yarns or threads are interlaced at right angles to form a fabric or cloth. Other methods are knitting, crocheting, felting, and braiding and plaiting. The longitudinal threads are called the warp and the lateral threads are the weft or filling.

2.3.1 Grey Cloth

Grey cloth is made from weaving and knitting of yarn. Weaving is a method of textile production in which two distinct sets of yarns or threads are interlaced at right angles to form a fabric or cloth to produce the laminated fabric in a grey state, and thereafter dyeing and finishing the grey laminated fabric. The proposed production process is described in Figure 2.4.

2.3.2 Water Jet Weaving Machine

A water jet is the machine for weaving cloth (loom), which uses a jet of water to insert the weft (crosswise threads) into the warp (lengthwise threads). The force of air water carries the yarn from one side to the other. They are characterized in particular by high insertion performance and low energy consumption. In this technique a water jet is shot under force and, with it, a weft yarn. The force of the water as it is propelled across the shed carries the yarn to the opposite side. Figure 2.5 shows a water jet weaving machine.

FIGURE 2.4 Production Process

FIGURE 2.5 Water Jet Weaving Machine

Zero Liquid Discharge

2.3.3 Detail of Effluent Treatment Plant

2.3.3.1 Collection Tank

A collection tank, sometimes called an equalization tank, in any treatment plant serves the purpose of maintaining desired flow rate and for making combination homogeneous. Its main function is to act as a buffer: to collect the incoming raw sewage that comes at widely changing rates and pass it on to the rest of the ETP at a steady (average) flow rate. During the peak hours, sewage comes at a high rate. The equalization tank collects and stores this sewage and lets it out during the non-peak time when there is no/little incoming sewage.

2.3.3.2 Bar Screen

A bar screen is used to remove large solids like pieces of rags, fabric, yarn, lint, sticks and so forth that may cause damage to plant equipment.

2.3.3.3 Oil Skimmer

An oil skimmer is a device that separates oil or particles moving on a liquid surface. A common application is eliminating oil floating on water. Oil skimmers are not oil-water separator devices. They are used for oil spill remediation, as a part of oily water treatment systems, removing oil from machines, tools, coolant and removing oil from aqueous part gaskets.

2.3.3.4 Flocculation and Chemical Dosing

This is a chemical process that involves a neutralizing charge on the particles. This occurs when a coagulant is additional to water to destabilize colloidal suspensions. The coagulation and flocculation in wastewater treatment processes can be used as an initial step. Flocculation, a calm mixing stage, increases the particle size from sub microscopic micro floc to visible suspended particles. Micro floc particles collide, causing them to bond to produce larger, visible flocs called pin flocs. Floc size continues to build with additional collisions and communication with added inorganic polymers (coagulant) or organic polymers. Macro flocs are formed and high molecular weight polymers, called coagulant aids, may be added to help bridge, bind and strengthen the floc, add weight, and increase settling rate. Once floc has reached it optimum size and strength, water is ready for sedimentation.

2.3.3.5 Dissolved Air Flotation (DAF)

A dissolved air flotation (DAF) system is a water treatment equipment used to disperse water and contaminants in the wastewater. DAF System works with a two-step process, the first being the pre-treatment process and the second being a flotation process. The system works by the following method: during the pre-treatment process, chemicals are combined with the contaminant in the water to create a light, floatable floc in the mixing tank. After pre-treatment process, air (in micro bubble form) are introduced into the pre-treated water in the DAF unit. The air then will float the floc so it can be divided from the water with a skimmer.

2.3.3.6 Reverse Osmosis (RO)

Reverse osmosis is the process by which an applied pressure, greater than the osmotic pressure, is exerted on the compartment that once contained the high-concentration solution. This pressure forces water to pass through the membrane in the direction reverse to that of osmosis. Water now moves from the compartment with the high-concentration solution to that with the low concentration solution. Due to this, relatively pure water passes through membrane into the one compartment while dissolved solids are retained in the other compartment. Hence, the water in the compartment are concentrated or dewatered. Due to the resistance of the membrane, the applied pressures required to achieve reverse osmosis are significantly higher than the osmotic pressure. Figure 2.6 shows the flow diagram of ETP and Figure 2.7 shows structure of water jet ETP.

2.3.3.7 Evaporator
Evaporators after RO, which evaporate reject water.

2.3.4 Water Requirement

The total water demand for the proposed project will be about 256 KLD. The water demand for the park includes all forms of water use such as water required for production units, water required for workers, commercial use and in green belts. The water demand from groundwater will be 88.5 KLD. Water used in manufacturing process will be 192 KLD. For flushing, bathing and washing, 8 KLD water will be used. For drinking purposes, 25 L/hr RO will be installed. In the manufacturing process, 192 KLD water will be used. In green belt and dust suppression, 8 KLD water will be used. Figure 2.8 shows a water balance diagram of the weaving industry.

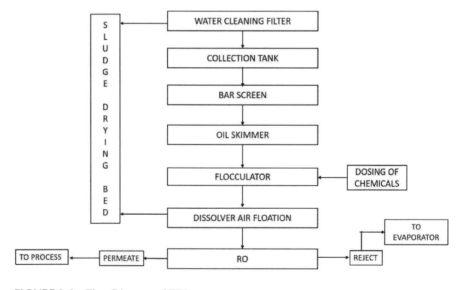

FIGURE 2.6 Flow Diagram of ETP

Zero Liquid Discharge

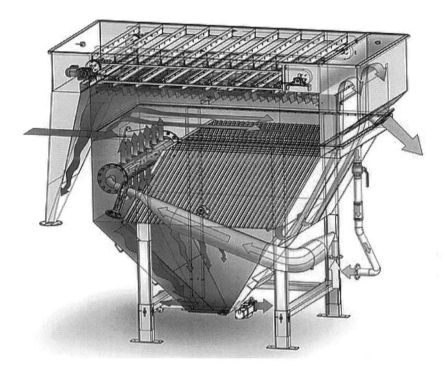

FIGURE 2.7 Sample of Water Jet ETP

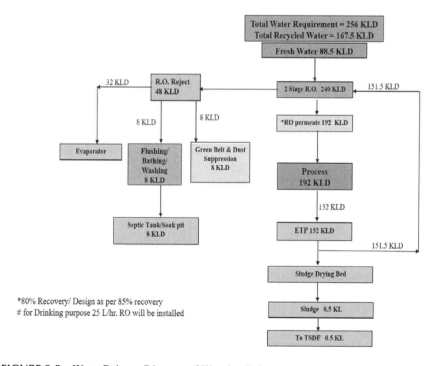

*80% Recovery/ Design as per 85% recovery
for Drinking purpose 25 L/hr. RO will be installed

FIGURE 2.8 Water Balance Diagram of Weaving Industry

2.3.5 INDIAN SCENARIO OF ZLD

India is taking aggressive action to curb severe water pollution, including the holy river Ganga. The recent three-year target set by the Indian government, known as the Clean Ganga project, imposes stricter regulations on wastewater discharge and compelling high-polluting industries to move towards ZLD. In 2015, the Indian government issued a draft policy that requires all textile plants generating more than 25 m^3 wastewater per day to install ZLD facilities. Dyeing plants in the city of Tirupur had already implemented ZLD in 2008, which recovered not only water but also valuable salts from textile wastewater for direct reuse in the dyeing process. According to a recent technical report, the ZLD market in India is valued at \$39 million in 2012 and is expected to grow continuously at a rate of 7% from 2012 to 2017. In India, textile, brewery, power, and petrochemical industries are the major application areas for ZLD installations (Chavan, 2017).

2.4 CONCLUSION

ZLD application is growing globally as an important wastewater management strategy to decrease water pollution and augment water supply. High cost and intensive energy consumption will remain the main barriers to ZLD adoption. Future growth of the ZLD market will heavily rely on regulatory incentives that outweigh its economic disadvantages. As the severe consequences of water pollution are increasingly recognized and attract more public attention, stricter environmental regulations on wastewater discharge are expected, which will push more high-polluting industries toward ZLD. The environmental impacts of ZLD need to be better understood. A life cycle assessment analysis of the energy demand and greenhouse gas emissions will provide additional insights into the cost-benefit balancing of ZLD. Along with advances in improving the energy and cost efficiencies of ZLD technologies, particularly by incorporating membrane-based processes, ZLD may become more feasible and sustainable in the future.

REFERENCES

Ansa, A., Sidra, Y. and Muhammad, U., 2014. *Report on Reverse Osmosis-Comsats Institute of Information and Technology*, Lahore Campus. 23–24. Available online at www.researchgate.net/publication/283734409_REPORT_On_Reverse_Osmosis_Desalination

Chavan, R. B., 2017. Advance research in textile engineering. *Towards Zero Liquid Discharge.* 2(1):1–2. Available online at file:///C:/Users/a/Downloads/fulltext_arte-v2-id1015.pdf

CPCB, 2015. Guidelines on techno—economic feasibility of implementation of ZLD (1–22). Available online at www.indiaenvironmentportal.org.in/files/file/FinalZLD%20water%20polluting%20industries.pdf

Gautami, G., 2011. Modeling and simulation of multiple effect evaporator system. 18–19. Available online at http://ethesis.nitrkl.ac.in/4417/1/main_copy_to_CD(PDF).pdf

Ghosna, J., 2012. Design of heat integrated multiple effect evaporator system. 1–63. Available online at http://ethesis.nitrkl.ac.in/3890/1/GHOSHNA_JYOTI-_THESIS.pdf

Klidas, M., 2016. Salt drying technology for zero liquid discharge. 1–12. Available online at www.babcock.com/products/-/media/07c4fca0079a4dcfb1c2d6a4976592a0.ashx

Popuri, A. K. and Guttikonda, P., 2016. *International Journal of ChemTech Research.* 9(11):80–86. Available online at www.sphinxsai.com/2016/ch_vol9_no11/1/(80-86)V9N11CT.pdf

Tong, T. and Menachem, E., 2016. The global rise of zero liquid discharge for wastewater management. 6846–6855. Available online at www.researchgate.net/publication/303872744_The_Global_Rise_of_Zero_Liquid_Discharge_for_Wastewater_Management_Drivers_Technologies_and_Future_Directions

Vyas, J., 2016. Industrial pollution prevention group centre for environment education, Ahmedabad. 1–24. Available online at www.ceeindia.org/ZLD%20Concept%20note.pdf

Weike, P., Luwei, Y. and Zhentao, Z., 2013. Operation characteristic of a mechanical vapor recompression heat pump driven by a centrifugal fan. 732–733, 165–171. Available online at file:///C:/Users/a/Downloads/10.1.1.903.3559.pdf

3 Zero Noise Pollution
Green Belt Development

Ashok K. Rathoure and Jahanvi Modi

CONTENTS

3.1 Introduction 26
3.2 Classes of Noise Related to Noise Pollution 27
    3.2.1 Atmospheric Noise 27
    3.2.2 Environmental Noise 27
    3.2.3 Occupational Noise 28
3.3 Effects of Noise 28
    3.3.1 Hearing Problems 28
    3.3.2 Health Issues 28
    3.3.3 Sleeping Disorders 29
    3.3.4 Cardiovascular Issues 29
    3.3.5 Trouble Communicating 29
    3.3.6 Effect on Wildlife 29
3.4 Measurement of Noise 29
    3.4.1 Measurement of Sound Levels 29
    3.4.2 Instrumentation 30
        3.4.2.1 Sound Level Meter (SLM) 30
    3.4.3 Noise Dosimeter 31
        3.4.3.1 Integrating Sound Level Meter (ISLM) 31
    3.4.4 Noise Monitoring System 31
3.5 Noise Standards in India 32
3.6 Mitigation Measures for Noise Pollution 33
3.7 Green Belt Development 35
    3.7.1 Mechanism 36
    3.7.2 Design Considerations 37
    3.7.3 Vegetated Solid Barriers 38
3.8 Case Study 40
    3.8.1 Green Belt Development (Action Plan) 40
    3.8.2 Selection of Plants for Green Belts 41
    3.8.3 Roadside Plantation 43
    3.8.4 Guidelines for Plantation 43
3.9 Recommendations 47
3.10 Conclusion 47
References 47

3.1 INTRODUCTION

It is known that 70% of the world's urban population lives in developing countries. Day by day, population increase and improvements in technology have brought about changes in the economic and social structure of societies. Much of these urban populations are vulnerable to the ill health effects of noise. Despite being a less frequently considered type of environmental pollution, noise has a major negative impact on the quality of life in cities. Especially dense transportation systems, including roads, railways and air traffic, characterize the modern urban environment. These systems have caused environmental noise pollution (McMichael, 2000; Cohen, 2006; Moudon and Wee, 2009).

Noise barriers are an effective method for reducing the noise (Kumar et al., 2006). Noise reduction occurs because of normal attenuation and excess attenuation. Normal attenuation is because of spherical deviation and friction between atmospheric molecules when noise progresses. This has been called the distance effect: in this, distance increases noise attenuation. Additionally, the effects of reflection, refraction, scattering and absorption because of an obstruction between a noise source and a receiver leads to extra attenuation. It is possible to reduce a significant amount of noise by creating sufficient areas with appropriate plant species (Guilford and Fruchter, 1973). Previous studies have shown the important role of nature on restorative experiences by contributing to the positive psychological well-being for humans (Herzog et al., 1997; Hartig et al., 2003; Islam et al., 2012). Furthermore, green plants operate as a buffer of undesirable situations. So, environmentalists rightly emphasize the existence of a perennial green enclosure in and around metropolitan areas and along roadsides (Rao et al., 2004). Green belts beautify the city and significantly attenuate the noise with their leaves and provide an effective trapping device for pollutants (Yang et al., 2008; Pathak et al., 2011). It is also considered to be one of the cheapest methods of pollution control (Pal et al., 2000; Kumar et al., 2006).

The main factor for the development of green belts is that various vegetation species have different levels of sensitivity toward different stressors, such as noise and air. According to the reaction of vegetation toward a specific stressor, they can be classified as sensitive and tolerant. To reduce the impact of the stressor, such as the pollution of air and noise from road traffic, tolerant species can be applied for green belt development. Tolerant vegetation species act as a pollution sink and the planting of these species in affected areas provides various environmental benefits (Rao et al., 2004).

In general, plants are able to attenuate sound levels in three ways:

1. Sound is reflected and diffracted by different parts of the plant, such as trunks, branches, and leaves. Sound energy will leave the line of sight between the source and the receiver when interacting with plants and lead to attenuated sound pressure levels (SPLs).
2. The mechanical vibration of plant parts by sound waves causes absorption by vegetation, through which sound energy converts to heat and noise reduction by a thermo-viscous boundary-layer effect on the plant surface.
3. The damaging interference of sound waves can reduce sound levels. For instance, the existence of the soil can cause destructive interference

between the direct contribution from the source to the receiver and a ground reflected contribution. This effect is identified as the acoustical ground effect or ground dip (Renterghem et al., 2012). Noise attenuation by plants with a sufficient area in the urban ecosystem is of great importance (Krag, 1979; Fang and Ling 2003; Rao et al., 2004). This highlights the fact that noise pollution is an important environmental issue and requires urgent investigation.

3.2 CLASSES OF NOISE RELATED TO NOISE POLLUTION

Noise is unwanted sound and a needless form of energy which is emitted by a vibrating body, and on reaching the human ear it causes the feeling of hearing through nerves. Not all sounds produced by vibrating bodies are audible. The limits of audibility are from 20 Hz to 20 kHz. Sounds of frequencies less than 20 Hz are called infrasonic, and sounds greater than 20 kHz are called ultrasonic. Noise may be continuous or irregular, and may be of high frequency or of low frequency, which is undesired for a normal human hearing. The discrimination between sound and noise depends upon the tendency and interest of the person receiving it, the ambient conditions and impact of the sound generated during that particular duration of time. Its pressure is measured in the logarithmic unit of decibels (dB), as the logarithmic scale permits a range of pressures to be described without using large numbers and it also represents nonlinear behaviour of the ear more convincingly. It is observed that noise can be perceived either physiologically or psychologically. When noise is sensed physiologically, humans subconsciously sense the vibrations of the sound waves in our physical body, whereas psychological perception of noise refers to when conscious awareness of a person shifts attention to that noise rather than letting it filter through instinctively where it goes ignored (Barthes, 1985).

3.2.1 ATMOSPHERIC NOISE

Atmospheric noise is a kind of radio noise caused by natural atmospheric processes, primarily lightning discharges in thunderstorms. It is primarily due to cloud-to-ground flashes, as the current is much stronger than that of cloud-to-cloud flashes. On a worldwide scale, 3.5 million lightning flashes occur daily. This is around 40 lightning flashes per second (NOAA, 2014). The addition of all these lightning flashes constitutes atmospheric noise. At very low frequency (VLF) and low frequency (LF), atmospheric noise often dominates, while at high frequency (HF), man-made noise dominates mainly in urban areas. From 1960s to 1980s, a universal effort was made to measure atmospheric noise and its variations (ITU, 1983; Lawrence, 1995; ITU, 2016).

3.2.2 ENVIRONMENTAL NOISE

Environmental noise is the abstract of noise pollution from outside sources, caused mainly by transport systems which includes a wheeled passenger vehicle that carries its own motor like buses, trains, trucks, cars, two- and three-wheelers, helicopters, watercraft, spacecraft and aircraft and various recreational activities like sports and music performances (USEPA, 1972; Hogan and Latshaw, 1973; European Commission, 2013). This class of noise is generally present in some form in all areas of human activity. The effects in humans of exposure to environmental noise may vary from emotional to physiological and psychological (Kinsler et al., 2000). Low-level noise is not necessarily harmful. However, the undesirable effects of noise exposure could include annoyance, sleep disturbance, nervousness, hearing loss and stress-related problems (WHO, 2013). Noise from transportation is generated by the engine or exhaust and aerodynamic noise compression and friction in the air around the body during motion. Recreational noise could be generated by a large number of different sources and processes. The background noise like alarms, people talking and bioacoustic noise from animals or birds also constitutes the environmental noise.

3.2.3 Occupational Noise

Occupational noise affects workers in the course of their jobs and is due to the work environment and/or to the machinery which they must operate. Industrial noise varies in loudness, frequency components and uniformity. It may be roughly uniform in frequency response and constant in level. Many machines in simultaneous operation are often like this. Other industrial or working place noise shows continuous background noise at relatively low levels with intermittently occurring periods of higher noise levels.

3.3 EFFECTS OF NOISE

3.3.1 Hearing Problems

Any unwanted sound that our ears haven't been designed to filter will cause issues inside the body. Our ears will absorb a particular variety of sounds while not obtaining broken. Long or recurrent exposure to sounds at or higher than 85 decibels will cause deafness. Man-made noises like jackhammers, horns, machinery, airplanes and even vehicles may be too loud for our hearing. Constant exposure to loud levels of noise can easily result in the damage of our eardrums and loss of hearing. It conjointly reduces our sensitivity to sounds that our ears obtain unconsciously to control our body's rhythm. The louder the sound, the shorter the amount of time it takes for noise-induced hearing loss to happen.

3.3.2 Health Issues

Excessive pollution in operating areas like offices, construction sites, bars and even in our homes will influence psychological health. Studies show that the incidence of aggressive behaviour, disturbance of sleep, constant stress, fatigue and high blood pressure are often connected to excessive noise levels. These successively will cause a lot of severe and chronic health problems later in life.

Zero Noise Pollution

3.3.3 SLEEPING DISORDERS

Loud noise can certainly hamper your sleeping and may lead to irritation and uncomfortable situations. Lacking an good night's sleep may lead to problems related to fatigue, and your performance may go down in the office as well as at home. A sound sleep is thus suggested to offer your body correct rest.

3.3.4 CARDIOVASCULAR ISSUES

High-intensity noise causes high blood pressure and increases heart beat rate, blood pressure levels, cardiovascular disease and stress-related heart problems, as it disrupts the normal blood flow.

3.3.5 TROUBLE COMMUNICATING

High sound will bother and not permit two individuals to speak freely. This may result in misunderstanding and you will have trouble understanding the opposite person. Constant sharp noise will provide you with a severe headache and disturb your emotional balance.

3.3.6 EFFECT ON WILDLIFE

Wildlife face a lot more issues than humans due to pollution. Animals develop higher hearing acuity since their survival depends on it. The unwell effects of excessive noise begin with reception. Pets react sharply in households wherever there's constant noise. They become frequently disoriented and face several behavioural issues. In nature, animals might suffer from deafness, which makes them simple prey and ends up decreasing their populations. Others become inefficient at looking, troubling the balance of the ecosystem. Species that rely on conjugation calls to breed are usually unable to listen to these calls thanks to excessive artificial noise. As a result, they're unable to breed which results in declining populations. Others need sound waves to echolocate and realize their method once migrating. Troubling their sound signals means that they stray simply and don't migrate when they ought to. To cope up with the increasing sound around them, animals have become louder, which can additionally augment the pollution levels (Nicks, 2018).

3.4 MEASUREMENT OF NOISE

3.4.1 MEASUREMENT OF SOUND LEVELS

Sound produced from any source is stimuli and it can be measured as sound pressure. Sound pressure range varies from 20 µPa to 200 Pa and it can be expressed on a scale based on the log of the ratio of measured sound pressure and a reference standard pressure sound level:

$$L = Log_{10} P \ / \ Po \left(bels \right) \tag{3.1}$$

30   Zero Waste

where

P = Measured quantity of sound pressure or sound power, or sound intensity.

P_o = Reference standard quantity of sound pressure, or sound power, or sound intensity (20×10^{-6} Pa)

L = Sound level in bels (B).

However, the unit bels (B) turns out to be a rather large unit, so a smaller unit of decibels (dB) is generally used.

$$L = 10 * \mathbf{Log_{10}} \, P / Po \, (dB) \tag{3.2}$$

$$1dB = 1 / 10 \, B$$

Sound pressure level:

$$Lp = 20 * \mathbf{Log_{10}} (Pr.m.s / 20\mu \, Pa) \tag{3.3}$$

(Source: CPCB, 2015.)

3.4.2 Instrumentation

The most common instruments used for measuring noise are the sound level meter (SLM), the integrating sound level meter (ISLM) and the noise dosimeter.

3.4.2.1 Sound Level Meter (SLM)

SLM consists of a microphone, electronic circuits and a readout display. The electro-acoustic transducer detects the little atmospheric pressure variations related to sound and changes them into electrical signals. These signals are then processed by the electronic circuitry of the instrument. The readout displays the sound level in decibels. The SLM takes the instantaneous sound pressure level at one instant in a very specific location. To take measurements, the SLM is held at arm's length at the ear height for those exposed to the noise. With most SLMs, it doesn't matter precisely however the electro-acoustic transducer is pointed at the noise supply. The instrument's guide explains a way to hold the electro-acoustic transducer. The SLM should be calibrated before and at every use. The manual also gives the calibration procedure. With most SLMs, the readings will be taken on either SLOW or FAST response. The response rate is that the fundamental measure over that the instrument averages the sound level before displaying it on the readout. Workplace noise level measurements should be taken on slow response. A pair of SLM is sufficient for industrial field evaluations. The more accurate and much more expensive Type 1 SLMs are primarily used in engineering, laboratory and research work. Any SLM that's less correct than a sort a pair of shouldn't be used for work noise measure. An A-weighting filter is generally built into all SLMs and can be switched ON or OFF. Some Type 2 SLMs provide measurements only in dB(A), meaning that the A-weighting filter is ON permanently. A standard SLM takes only instantaneous noise measurements. This is spare in workplaces with continuous noise levels. But in workplaces with

Zero Noise Pollution

impulse, intermittent or variable noise levels, the SLM makes it difficult to determine a person's average exposure to noise over a work shift. One solution in such workplaces is a noise dosimeter.

3.4.3 Noise Dosimeter

A noise dosimeter is a small, light device that clips to a person's belt with a small microphone that fastens to the person's collar, close to an ear. The measuring system stores the amplitude information associated and carries out an averaging method. It is useful in industries where noise usually varies in duration and intensity and where the person changes locations. A noise dosimeter requires the following settings:

1. Criterion Level: Exposure limit for eight hours per day five days per week. Criterion level is 85 dB(A) for several jurisdictions, 90 dB(A) for Quebec and 87 dB(A) for Canadian federal jurisdictions.
2. Exchange rate: 3 dB or 5 dB as specified in the noise regulation.
3. Threshold: Noise level limit below which the dosimeter does not accumulate noise dose data.

Wearing the dosimeter over a complete work shift gives the average noise exposure or noise dose for that person. This is sometimes expressed as a noise exposure level, $L_{ex,T}$. This is a logarithm that takes into consideration the exposure and therefore, the actual time worked. In the past, it was often expressed as a percentage of the maximum permitted exposure. If someone has received a noise dose of 100% over a work shift, this means that the average noise exposure is at the maximum permitted. For example, with a criterion level of 90 dB(A) and an exchange rate of 3 dB(A), an eight-hour exposure to 90 dB(A) gives a 100% dose. A four-hour exposure to 93 dB(A) is a 100% dose, whereas an eight-hour exposure to 93 dB(A) is a noise dose of 200%.

3.4.3.1 Integrating Sound Level Meter (ISLM)

The ISLM is similar to the dosimeter. It determines equivalent sound levels over a measurement. The main difference is that an ISLM does not provide personal exposures because it is handheld like the SLM and not worn. The ISLM determines equivalent sound levels at a specific location. It yields a one reading of a given noise, even if the actual sound level of the noise changes continually. It uses a pre-programmed exchange rate with a time constant that is equivalent to the SLOW setting on the SLM (CCOHS, 2014).

3.4.4 Noise Monitoring System

The Noise Monitoring System (NMS) is used for measuring real time noise since a large number of stations can be managed easily using this technology. NMSs are optimized for outdoor use with a small, custom designed enclosure and also designed for use in all climatic environments. The NMS consists of a weatherproof cabinet containing a noise level analyser and a battery, a communication device for transmitting data to nine receiving stations, a backplate and an outdoor microphone, all of which can be mounted on a mast. Some of the features and particulars of NMS as per CPCB (2015) are:

- NMSs are modular both in hardware and software.
- The NMS has been specifically designed to operate unattended in inhospitable environments, protecting the contents from weather, tampering, vandalism and so forth. The robust, durable, weatherproof cabinet includes a kit for fastening the cabinet to a wall or pole.
- Protection is also provided for the cabling, to reduce the risk of tampering or accidental damage.
- The NMS includes one battery, but two batteries can be used so that the NMS can function when there is no usable local power source or mains power has been disrupted. The batteries are charged whenever external AC or DC is applied to the NMS.
- The NMS can be powered from a variety of sources, such as solar panels, connected through the DC supply input.
- Data retrieval with automatic and manual operation and data storage in a SQL database, in order to allow the users to carry out their data analysis and data processing.
- The NMS supports GPS, so that with a standard commercial GPS receiver and antenna unit, longitude, latitude and height can be monitored and stored in the NMS with the noise measurements.
- Data from NMS is directly transferred to the main server (Central Receiving Station) via GPRS.

3.5 NOISE STANDARDS IN INDIA

The central government notified the Noise Pollution (Regulation and Control) Rules, 2000 (Table 3.1) as it is published in the Gazette of India, Extraordinary, Part-II—section 3(ii), vide S.O 123(E) dated 14.2.2000.

In reference to the aforementioned rules, the following responsibilities are vested with state governments, the district magistrate, police commissioner, or any other officer not below the rank of deputy superintendent of police:

1. Enforcement of noise pollution control measures and the due compliance of ambient air quality standards in respect of noise.
2. Restriction on the use of loudspeakers/public address system.
3. Restriction on the use of horns, sound-emitting construction equipment and bursting of firecrackers.
4. Prohibition of continuance of music sound or noise.
5. Authority shall act on the complaint and take action against the violator in accordance with the provisions of rules.
6. Disallowing sound producing instrument after 10 p.m. to 6 a.m. except in closed premises.

TABLE 3.1

Ambient Air Quality Standards in Respect of Noise Is Notified under Noise Pollution (Regulation and Control) Rules, MoEF (2000)

Area Code	Category of Area/Zone	Limit in dB(A) Leq*	
		Day time	Night time
A	Industrial area	75	70
B	Commercial area	65	55
C	Residential area	55	45
D	Silence zone	50	40

Notes:

1. Daytime shall mean from 6 a.m. to 10 p.m.
2. Night-time shall mean from 10 p.m. to 6 a.m.
3. Silence zone is defined as areas up to 100 meters around such premises as hospitals, educational institutes and courts. The silence zones are to be declared by competent authority.
4. Mixed categories of areas may be declared as one of the four aforementioned categories by the competent authority.

* dB(A) Leq denotes the time-weighted average of the level of sound in decibels on scale A which is relatable to human hearing.

A 'decibel' is a unit in which noise is measured.

'A', in dB(A) Leq, denotes the frequency weighting in the measurement of noise and corresponds to frequency response characteristics of the human ear.

Leq is the energy mean of the noise level over a specific period.

Source: Central Government notified the Noise Pollution (Regulation and Control) Rules, 2000.

7. The state government may permit loudspeakers or public address system in night hours (between 10:00 p.m. and 12:00 midnight not exceeding 15 days in the year).

3.6 MITIGATION MEASURES FOR NOISE POLLUTION

Numerous studies have already been conducted to investigate about the impact of noise pollution on human beings and other living creatures along with possible methods of mitigation of adverse impact on the biological community (Arana and Garcia, 1998; Morillas et al., 2002).

Here are some techniques which are used to mitigate noise pollution:

- *PPE*: The surest ways of preventing noise-induced hearing loss (NIHL) is to eliminate the source, or to reduce noise at the source by engineering methods. In such workplaces, workers may need to wear hearing protectors to reduce the amount of noise reaching the ears.

People ought to wear a hearing protector if the noise or sound at the workplace exceeds 85 decibels (A-weighted) or dBA. Hearing protectors reduce the noise exposure level and so the risk of hearing loss.

Types of hearing protectors are:

1. *Ear plugs* are inserted in the ear canal. They may be pre-moulded or mouldable (foam ear plugs). Disposable, reusable or custom molded ear plugs are available.
2. *Semi-insert ear plugs*, which consist of two ear plugs held over the ends of the ear canal by a rigid headband.
3. *Ear muffs* consist of sound-attenuating material and soft ear cushions that fit around the ear and hard outer cups. They are held together by a headband (CCOHS, 2017).

- *Acoustic Enclosures*: A noise barrier or acoustic shield reduces noise by interrupting the propagation of sound waves. With correct design and choice of material for the acoustic shield, noise reaching a noise sensitive receiver would be primarily through diffraction over the top of the barrier and around its ends. The acoustical shadow zone created behind the barrier is where noise levels are substantially lowered. To function well, the barrier must prevent the line of sight between the noise source and the receiver. Effective noise barriers will cut back noise levels by as much as 20 dB(A).

To perform well, a noise barrier should stop the line of sight between the noise supply and therefore the receiver. This is not forever possible, particularly with high-rise noise-sensitive uses. In this circumstance, noise enclosures are required to provide appropriate protection against environmental noise for the noise sensitive uses. In general, associate enclosure can reduce noise by over than 20 dB(A). Similar to noise barriers, a noise enclosure should be designed to serve both acoustic and aesthetic purposes. Table 3.2 provides some common types of noise barriers used to reduce noise pollution.

- *Vegetation (Green Belt)*: Vegetation has been proposed as a natural material to reduce noise outdoors. Belts of trees and bushes situated between the noise source and the receiver can reduce the noise level perceived by the

TABLE 3.2
Types and Characteristics of Enclosures (EPD-Hong Kong, 2016)

Type	Characteristics
Semi-enclosure	Effective in protecting high-rise sensitive receiver at one side of the carriageway.
Full Enclosure	Effective in protecting high-rise sensitive receivers located on both sides of carriageway.

receiver (Aylor, 1972). A green belt reduces the intensity of sound, worked as a barrier. Trees will either deflect, refract or absorb sound to reduce its intensity. The intensity reduction depends on the distance sound has to travel from source. Trees can modify the suitability of humidity and climate, which affect sound intensity.

3.7 GREEN BELT DEVELOPMENT

The development of a green belt around the noise polluting area is a useful measure in order to reduce noise pollution level (Aylor, 1972; Kumar et al., 2006). Different researchers have already studied the efficiency of green belts in the mitigation of noise pollution (Cook and Haverbeke, 1977; Krag, 1979; Rao et al., 2004). Published results on the effectiveness of tree and shrub barriers vary enormously, however; a review by Huddart (1990) shows that in some instances noise can be reduced by 6 dB over a distance of 30 m where planting is particularly dense. Leonard and Parr (1970) and Reethof (1973) found that a dense belt of trees and shrubs between 15 m and 30 m wide could reduce sound levels by as much as 6–10 dB. Another way in which noise may be made less intrusive is through the masking effect created by leaves, needles and branches in the wind. It is difficult to generalise, but a thick belt of densely planted trees and shrubs should provide a useful reduction in noise of several decibels. Longer life span is also an essential requirement to ensure the longevity of the developed green belt (Shannigrahi et al., 2004). As per the study made by some researchers for highway noise management through green belt development, an effective highway noise absorption and deflection can be done if the border plantings are lower towards the noise and higher towards the hearer (Rao et al., 2004; Pathak et al., 2011). The same principle can also be employed in the case of industrial noise pollution control. In the Bodhjungnagar Industrial Growth Centre, a green belt can be developed with such types of vegetations which have a higher elevation than the noise generating points and the adjoining locality, thus forming a shadow zone behind the barrier (Pathak et al., 2008). The arrangement of trees in the green belt is another important factor required to be considered during the development of the green belt. The arrangement of trees in periodic lattice is proved to be an effective means of sound attenuation by green belt (Martinez-Sala et al., 2006). Thus, it is clear from the literature studied that depending upon the agro-climatic condition of Bodhjungnagar Industrial Growth Centre, some plant species like *Caesalpinia pulcherrima, Bambusa sp., Azadirachta indica, Ficus benghalensis, F. religiosa* and *Acacia auriculiformis* can be used for development green belt in the growth centre area. These plants are also effective in prevention and control of air pollutants from any industrial sources.

Trees and shrubs can reduce noise levels, particularly at high frequencies, whereas a reduction in low-frequency noise levels can be attributed more to effect of the ground. Although planting trees will initially be more cost-effective than erecting a solid barrier, it would incur more ongoing management costs than a solid barrier. Tree and shrub belts offer many additional benefits over conventional techniques of controlling noise. Tree belts may develop into more effective windbreaks and provide more protection from the glare of the sun than mounds or fences. In addition,

FIGURE 3.1 A Visual Barrier between the Noise Source and the Hearer May Help Reduce the Perception of Noise

trees can also help purify the air, stabilize embankments with their roots, provide habitats for wildlife and improve the appearance of roads.

In order to achieve a significant noise reduction, a barrier consisting of trees and shrubs needs to be relatively wide (between 20 m and 30 m) (Figure 3.1). Such barriers are therefore best suited to areas where land is freely available for planting. However, the cost of land may be extremely high and in many instances is the main argument against the use of vegetation as a noise barrier. Nevertheless, a narrow strip of densely planted trees and shrubs of about 10 m wide could still give significant reductions in traffic noise level/of the order of 5 dB (Huddart, 1990). For comparison, a 3 m high solid barrier (e.g. a wall or a fence) erected on flat ground might be expected to give an attenuation of 15 dB immediately.

3.7.1 Mechanism

The decrease in sound by vegetation is attributed to the processes of reflection, scattering and absorption. Reflection and scattering from the surfaces of leaves, branches, trunks and the ground can alter the phase of sound, which can cause interference in the sound waves and a reduction in noise level (Figure 3.2). Thus, the more surfaces (leaves, needles and branches) there are within a tree belt, the better the reduction of noise will be, provided they are evenly distributed in the space between ground level and the tops of trees.

Leaves are the most efficient part of a tree for scattering sound, and it seems that large leaves are most effective than small leaves. Broad-leaved trees with large leaves will reduce noise more than conifers that have needle-like leaves (Tanaka et al., 1979). Since most broad-leaved trees lose their leaves in winter, conifers may give better year-round noise reduction, although the most effective trees are broad-leaved evergreens. Low shrubs and/or hedges along the edge of group of trees will reduce noise.

Zero Noise Pollution

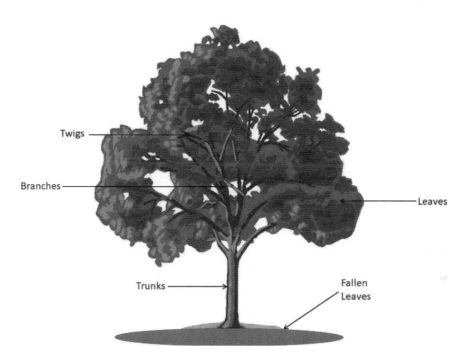

FIGURE 3.2 Illustration of How Plants Can Attenuate Sound

Tree bark is the most efficient part of a tree in noise absorption. The ground within a group of trees has a relatively large noise-absorbing capacity. The developmental stage of the trees is important in relation to their effectiveness in noise control. Noise reduction tends to increase with tree height up to 10 m to 12 m, after which attenuation decreases. This implies that a noise barrier comprising both trees and shrubs would be managed to ensure that the density of branches and foliage from ground level to 10 m remains high.

3.7.2 Design Considerations

Some important design parameters considered in the development of green belt to reduce noise level includes tree height, width of green belt, distance of green belt from the source, visibility and so forth. The vegetation species selected for such function are required to have a fast growth rate for quick development of a canopy, large size leaf for retention of pollutants and absorption of noise. Noise reduction is correlated with the width of a belt of trees (i.e. the wider it is, the greater the noise reduction). However, the amount of additional noise reduction declines with increasing distance. For example, from studies of traffic noise, Huddart (1990) found that a 10 m wide strip of trees planted close to a road gave an attenuation of about 5 dB more than the same width of grass, whilst a strip of trees 20 m wide only gave an attenuation of 6 dB more than grass (Figure 3.3). This appears to be because the interior of a wide group of trees is relatively free of foliage and small branches, especially at lower levels, and therefore somewhat hollow, whereas narrow strips of trees,

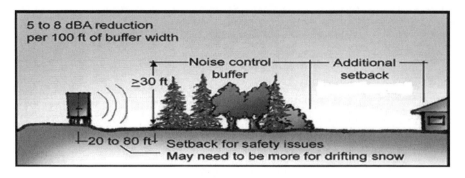

FIGURE 3.3 Design Consideration for Green Belt Development (for 100 ft. Buffer Width We Can Reduce Noise Level of 5 to 8 dB(A))

Source: USDA (2008).

especially young conifers, have foliage and small branches throughout, from top to bottom. These compensating factors probably account for the smaller than expected differences in sound level attenuation between wide and narrow belts. The length a tree and shrub belt extend may also influence its effectiveness in noise attenuation. Actual prescriptions are difficult however, as they will depend on the dimensions of the noise source (i.e. point or line source). Of more importance in the decrement of noise levels is the actual siting of the barrier: a screen placed relatively close to a noise source is more effective than one placed close to an area to be protected. At midway between the source and receiver, noise reduction is least. Also, a barrier is most effective when trees and shrubs are combined with soft rather than hard ground surfaces (i.e. grass instead of tarmac or gravel). Hard surfaces tend to reflect noise with little or no attenuation.

To maximise noise attenuation:

- A vegetation barrier should ideally form an irregular structure comprising:
  - Trees
  - Shrubs
  - Herbs
  - Litter layers of ornamental grasses.
- Particular attention should be paid to:
  - Density
  - Height
  - Amount of leaves in the shrub layer.
- Large-leaved plants will be more effective at reducing noise during spring and summer but evergreens will provide better year-round attenuation.

3.7.3 Vegetated Solid Barriers

In living walls, which generally consist of two parallel sets of posts which form the outer faces of the wall, between which willow branches are woven, in a similar way to a wicker basket, and as the weaving progresses the core is filled with soil. At each metre in height, internal irrigation pipes are installed and lateral rods for structural support. The woven willow then produces new shoots on the outside and roots within the internal core, providing a total covering of foliage within the first year after construction. A typical wall may have a width of about 2.5 m and a height of 4.0 m. Overall costs may be high; the willow requires cutting back annually but living walls may be a suitable option where space is limited, and where there needs to be a combination of greenery and noise reduction. The level of noise reduction provided by willow walls is similar to the reduced level of a solid noise barrier of similar height, because the soil core prevents sound leakage. Unlike a tree belt which takes time to become established, the benefits of such vegetated barriers are immediately available.

Lists of trees which are used for green belt as noise barriers are listed in Table 3.3.

TABLE 3.3

List of Trees for Noise Control (Garg et al., 2012)

S. No.	Common Name	Scientific Name
A	**Trees**	
1.	Bakul	*Mimusops elengi*
2.	Banyan	*Ficus bengalensis*
3.	Bher	*Ziziphus mauritiana*
4.	Ashoka	*Polyalthia longifolia*
5.	Anjur	*Ficus carica*
6.	Gul-Mohr	*Delonix regia*
7.	Tulip	*Thespesia populnea*
8.	Kadamp	*Anthocephalus cadamba*
9.	Arizona cypress	*Cupressus arizonica*
10.	Neem	*Azadirachta indica*
11.	Common Chinafir	*Cunninghamia lanceolate*
12.	Pipal	*Ficus religiosa*
13.	White Pine	*Pinus strobus*
14.	Tamarind	*Tamarindus indica*
15.	Virginia pine	*Pinus virginiana*
16.	Siris	*Albizzia lebbek*
17.	Leyland cypress	*Cupressocyparis leylandii*
18.	Atlas cedar	*Cerdrus atlantica*
19.	Pinus taeda	*Loblolly pine*
20.	Deodar Cedar	*Cedrus deodara*
21.	Sweet bay Magnolia	*Magnolia virginiana*
22.	Dawn Redwood	*Metasequoia glyptostroboides*

(Continued)

TABLE 3.3 (Continued)
List of Trees for Noise Control (Garg et al., 2012)

S. No.	Common Name	Scientific Name
B.	**Shrubs**	
23.	Bougainvillea	*Bougainvillea spectabillis*
24.	Basak	*Adhatoda vasica*
25.	Croton	*Croton devaricata*
26.	Mussaenda	*Mussaenda erythrophylla*
27.	Rangon	*Ixora coccinea*
28.	Tagar	*Tabernaemontana coronria*
29.	Lantana	*Lantana camara*
30.	Carrisa Holly	*Ilex cornuta 'carissa'*
31.	Dwarf Chinese	*Ilex cornuta 'rotunda'*
32.	Fortunes Osmanthus	*Osmanthus x fortunei*
33.	Dwarf Burford	*Ilex cornuta 'dwarf burford'*
34.	Dwarf Yaupon	*Ilex vomitoria 'nana'*
35.	Chinese Loropetalum	*Loropetalum chinensis*
36.	Dwarf Waxmyrtle	*Myrica cerifera 'pumilla'*
37.	Madhabilata	*Quisqualis indica*
38.	Pfitzer Juniper	*Juniper chinensis 'pfitzeriana'*
39.	Tecoma	*Tecoma stans*
C.	**Ornamental grasses**	
40.	Upland Sea Oats	*Chasmanthium latifolium*
41.	Zebra Grass	*Sinensis zebrinus*
42.	Pampas Grass	*Hibiscus rosasinensis*
43.	Switch Grass	*Cortaderia selloana*
44.	Dwarf Pampas Grass	*Panicum virgatum*
45.	Porcupine Grass	*Murraya peniculata*
46.	-	*Cortaderia selloana 'nana'*
47.	-	*Putranjeva roxburghi*
48.	-	*Ficus benjamina*
49.	-	*Miscanthus sinensis 'strictus'*
50.	-	*Cestrum nocturnum*

3.8 CASE STUDY

3.8.1 GREEN BELT DEVELOPMENT (ACTION PLAN)

Green belts are an effective mode of control of air pollution, where green plants form a surface capable of absorbing air pollutants and forming a sink of pollutants. Leaves with their vast area in a tree crown, sorbs pollutants on their surface, thus effectively

Zero Noise Pollution   41

reduce pollutant concentration in the ambient air. Often the adsorbed pollutants are incorporated in the metabolic pathway and the air is purified. Plants grown to function as pollution sink are collectively referred as green belts. An important aspect of a green belt is that the plants are living organism with their varied tolerance limit towards the air pollutants. A green belt is effective as a pollutant sink only within the tolerance limit of constituent plants. Planting pollutant-sensitive species along with the tolerant species within a green belt, however, do carry out an important function of indicator species. Apart from functioning as pollution sinks, green belts would provide other benefits, like aesthetic improvement of the area and providing suitable habitats for birds and animals.

3.8.2 SELECTION OF PLANTS FOR GREEN BELTS

The main limitation for plants to function as scavenger of pollutants are plants' interaction to air pollutants, sensitivity to pollutants, climatic conditions and soil characteristics. While making choice of plant species for cultivation in green belts, due consideration has to be given to the natural factor of bio-climate. Xerophyte plants are not necessarily good for green belts; their sunken stomata can withstand pollution by avoidance but are poor absorber of pollutants. Character of plants mainly considered for affecting absorption of pollutant gases and removal of dust particle are as follows.

- For absorption of gases

Tolerance towards pollutants in question, at concentration, that is not too high to be instantaneously lethal:

1. Longer duration of foliage
2. Freely exposed foliage
3. Adequate height of crown
4. Openness of foliage in canopy
5. Big leaves (long and broad laminar surface)
6. Large number of stomatal apertures.

- For removal of suspended particular matter
  1. Height and spread of crown
  2. Leaves supported on firm petiole
  3. Abundance of surface on bark and foliage
  4. Roughness of bark
  5. Abundance of axillary hairs
  6. Hairs or scales on laminar surface
  7. Protected stomata.

Table 3.4 provides a list of plants which are used in this case study for green belt development along the boundaries as wind and noise barriers.

TABLE 3.4
List of Plant Species for Green Belt Development along the Boundary as a Wind Barrier as Well as to Prevent Noise Pollution (CPCB, 2000)

Plant Species	Habit	Tolerance Limit	Stomatal Index	Mode of Regeneration
Acacia auriculiformis	Tree	Tolerant	10.9	Seeds
Acacia leucophloea	Shrub	T	12.01	Seeds
Ailanthus excelsa	Tree	T	13.01	Seeds, root cuttings
Alstona scholaris	Tree	T	15.23	seeds
Azadirachta indica	Tree	T	29.2	Seeds
Bauninia recemosa	Tree	T	25.68	Seeds
Bauhinia acuminata	Tree	T	22.31	Seeds
Bauhinia purpurea	Tree	T	23.58	Seeds
Bambusa vulgaris	Tree/Shrub	T	-	Cutting
Bougainvillea spectabilis	Shrub	T	32.53	Cutting
Caesalpinia pulcherrima	Tree	T	29.09	Seeds and cuttings
Callistemon citrinus	Small tree	T	127.49	Seeds
Cassia javanica	Tree	T	-	seeds
Cassia siamea	Tree	T	21.2	Seeds
Clerodendrum inerme	Shrub	T	18.02	Seeds/cuttings
Delonix regia (Gulmohur)	Tree	Sensitive	14.38	Seeds/stem cuttings
Dendrocalamus strictus	Shrub/ tall grass	T	-	Seeds/stem cuttings
Hibiscus rosa-sinensis	Small tree	T	23.32	Stem cuttings
Ixora arborea	Small tree	T	17.3	Stem cuttings
Ixora rosea	Small tree	T	20.30	Stem cuttings
Kegelia Africana	Small tree	T	12.90	Seeds
Lantana camara	shrub	T	12.13	Seeds/cuttings
Lowsonia intermis	Shrub	T	17.0	Seeds/cuttings
Mangifera indica	Tree	S	30.77	Seeds/budding/ grafting
Melia azadirachta	Tree	T	-	Seeds/stem cuttings
Nerium indicum	Shrub	T	15.7	Cuttings
Peltophorum pterocarpum	Tree	T	16.78	Seeds
Polyathia longifolia	Tree	Sensitive	22.27	seeds
Prosopis cineraria	Tree	T	18.1	Seeds/root suckers

(*Continued*)

Zero Noise Pollution

TABLE 3.4 (Continued)

List of Plant Species for Green Belt Development along the Boundary as a Wind Barrier as Well as to Prevent Noise Pollution (CPCB, 2000)

Plant Species	Habit	Tolerance Limit	Stomatal Index	Mode of Regeneration
Syzygium cumini	tree	T	20.60	Seeds
Tecoma satans	Shrub	T	23.08	Seeds/cuttings
Terminalia catapppa	Tree	T	20.9	Seeds
Thespesia populneoides	Tree	T	29.81	Seeds/cuttings
Thevetia peruviana	Shrub	T	27.8	Seeds

T – Tolerant, S – sensitive, (-) – Not available

3.8.3 ROADSIDE PLANTATION

Roadside plantation plays a very important role for greening the area, increasing the shady area, increasing aesthetic value and for eco-development of the area. The approach roads to the project site, colony and so forth can be planted with flowering trees. Trees can be planted to increase aesthetic value as well as shady area along the roads. The selected plant species list is given for roadside plantation (Tables 3.5 and 3.6).

3.8.4 GUIDELINES FOR PLANTATION

The plant species identified for green belt development can be planted using the pitting technique. The width of the green belt in the available land area may prove difficult for many industries to attain for one or more reasons. Hence it can be decided to have a green belt in places available around the industry (source-oriented plantation) as well as around the nearby habituated area (receptors-oriented plantation). The choice of plats for the green belt should include shrubs and trees. The intermixing of trees and shrubs should be such that the foliage area density in vertical is almost uniform. The pit size has to be either 45 cm × 45 cm × 45 cm or 60 cm × 60 cm × 60 cm. A bigger pit size will be considered at marginal and poor quality soil. Soil used for filling the pit should be mixed with well decomposed farm yard manure or sewage sludge at the rate of 2.5 kg (on dry weight basis) and 3.6 kg (on dry weight basis) for 45 cm × 45 cm × 45 cm and 60 cm × 60 cm × 60 cm size pits, respectively. The filling of soil has to be completed at least 5–10 days before actual plantation. Healthy saplings of identified species should be planted in each pit with the commencement of the monsoon. Provision for regular and liberal watering during the summer period during the commissioning stage of the plant will be arranged from the local available resources. After the proposed plant became operational, the authorities responsible for plantation will also make adequate measures for the protection of the saplings. The trees and shrubs selected from the aforementioned list based on its availability shall be planted as green belt of 10 m to 20 m width around the plant boundary. The plantation will be in this recommended pattern as listed in Tables 3.7 and 3.8 and Figure 3.4, which is a sample map for the green belt development plan.

TABLE 3.5

Species for Plantation along the Roadside

S. No.	Based on Colour	S. No.	Based on Colour
	Yellow Flowered Trees		
1.	*Acacia auriculaeformis*	10.	*Erythrina parcelli*
2.	*Acacia baileyana*	11.	*Laburnum anagyroides*
3.	*Acacia dealbata*	12.	*Michelia champaca*
4.	*Acacia decurrens*	13.	*Parkinsonia aculeata*
5.	*Acacia implexa*	14.	*Peltophorum pterocarpum*
6.	*Anthocephalus chinensis*	15.	*Pterocarpus dalbergioides*
7.	*Bauhinia tomentosa*	16.	*Schizolobium excelsum*
8.	*Cassia calliantha*	17.	*Tabebuia spectabillis*
9.	*Cassia fistula*	18.	*Thespesia populnea*
	Red Flowered Trees		
1.	*Brownea grandiceps*	5.	*Saraca asoca*
2.	*Erythrina blakei*	6.	*Spathodea campanulata*
3.	*Erythrina laurifolia*	7.	*Wrightia coccinea*
4.	*Erythrina variegate*		
	Scarlet Flowered Trees		
1.	*Barringtonia acutangula*	5.	*Callistemon lanceolatus*
2.	*Brassia actinophylla*	6.	*Delonix regia*
3.	*Brownea coccinea*	7.	*Stenocarpus sinuatus*
4.	*Butea monosperma*		
	Pink Flowered Trees		
1.	*Bauhinia purpurea*	5.	*Hibiscus collinus*
2.	*Cassia javanica*	6.	*Kleinhovia hospital*
3.	*Cassia nodosa (red)*	7.	*Lagerstroemia speciosa*
4.	*Cassia renigera*		
	Blue Flowered Trees		
1.	*Bolusanthus speciosus*		
2.	*Jacaranda acutifolia*		
	White Flowered Trees		
1.	*Albizia lebbeck*	8.	*Millingtonia hortensis*
2.	*Bauhinia acuminate*	9.	*Mimusops elengi*
3.	*Calophyllum inophyllum*	10.	*Plumeria alba*
4.	*Kydia calycina*		
5.	*Magnolia grandiflora*		
6.	*Magnolia pterocarpa*		
7.	*Mesua ferrea*		

Zero Noise Pollution

TABLE 3.6

List of Suitable Ornamental Climbers/Shrubs as Plantation inside Garden and Open Spaces between Different Units of Construction Project

Family	Scientific Name	Common English Name	Flowering Season
Bignoniaceae	*Bignonia ventusa*	Golden shower	January–February
	Bignonia capreolata	Trumpet flower	March–April
	Bignonia unguis-cati	Cat's claw	April
	Bignonia speciosa	Handsome flower	March–April
	Tecoma satans	Yellow bell	Throughout the year
	Tecoma radicans	Trumpet vine	Throughout the year
Caesalpiniaceae	*Caesalpinia pulcherrima*	Peacock flower	April–June
	Ixora coccinea	Scarlet ixora	Throughout the year
Rubiaceae	*Ixora rosea*	Pink ixora	August–September
	Ixora parviflora	Small flowered ixora	March–April
	Ixora barbata	Bearded ixora	April–May
	Ixora lutea	Yellow ixora	Throughout the year
Euphorbiaceae	*Euphorbia pulcherrima*	Christmas flower	December–January
Apocynaceae	*Thevetia peruviana*	Trumpet flower	Throughout the year
	Alemanda nerifolia	-	April–June
	Nerium Indicum	Oleander	Throughout the year
	Catharanthus roseus	Periwinkle	-
Malvaceae	*Hibiscus mutabilis*	Changeable rose	September–October
	Hibiscus schizopetalus	Coral hibiscus	April–September
	Hibiscus rosa—sinensis	Chinese rose	Throughout the year
Nyctaginaceae	*Bougainvillea*		Throughout the year
	spectabilis and		With seasonal bloom
	different varieties		

TABLE 3.7

Three Tier Plantation Management

Tire	Habit	Height (m)	Rows
1st Tier (Towards boundary)	Trees	10–20	3
2nd Tier (Middle layer)	Small tress	5–10	2
3rd Tier (Towards Plant)	Shrubs	1–5	Thick pattern

TABLE 3.8

Sample Budgetary Provision of Green Belt Development

S. No.	Year	No. of Trees for Proposed Project	Proposed Budget (Rs)	Remarks
1.	1st Year			
2.	2nd Year			
3.	3rd Year			
Total				

* Budget includes soiling, pitting, planting, irrigation and maintenance cost.

46 Zero Waste

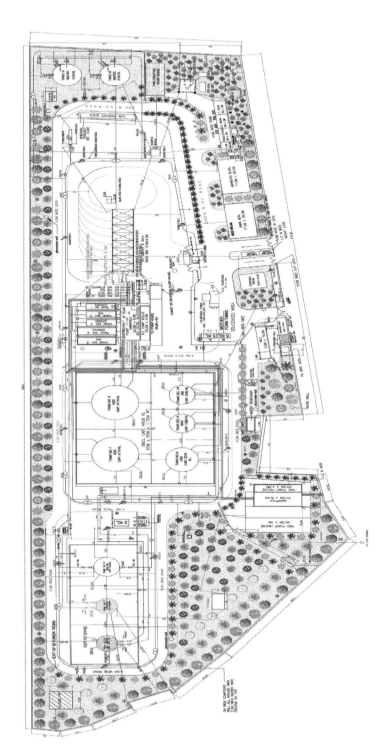

FIGURE 3.4 Sample Map of Green Belt Development

3.9 RECOMMENDATIONS

The recommendation for noise reduction of noise pollution in any area:

- Noise is more effectively reduced by complete screening of the source from view. Although gaps and partial views through a barrier may create an impression of greater noise reduction, they will allow noise to penetrate.
- A noise barrier should be planted as close to the noise source as possible.
- Widely spaced trees do not reduce noise effectively. Wide belts of high densities are required to achieve significant noise reductions.
- Effectiveness is closely related to the density of stems, branches and leaves. Use trees with dense foliage and branches that reach close to the ground. Alternatively plant an understory of dense shrubs or a surrounding hedge.
- Where year-round noise screening is desired use broad-leaved evergreens or a combination of conifer and broad-leaved evergreen species.
- Soft ground is an efficient noise absorber. Avoid hard surfaces—asphalt and concrete reflect virtually all incident sound at any angle. Cultivating ground before planting, and the addition of well-rotted organic matter to the soil surface may also help to reduce noise whilst vegetation becomes established.

3.10 CONCLUSION

We can conclude that vegetation or a green belt is the best solution to work as a noise barrier in the industries. Trees function as a green belt through the removal of noise in urban areas, which assists in restoring environmental quality and improving human health. The different plants and specified tree belt widths were significantly and positively correlated in noise reduction. The maximum reduction in noise levels was achieved by shrubs and trees of 100 m in width and the mixture of conifers and broad leaves of 100 m and 50 m in width. In each case, where possible, use trees that will develop dense foliage and relatively uniform vertical foliage distribution, or combinations of shrubs and taller trees to give this effect. Where the use of trees is restricted, use combinations of shrubs and tall grass or similar soft ground cover in preference to paved, tarmac or gravel surfaces to encourage absorption of noise rather than reflection.

REFERENCES

Arana, M. and Garcia, A., 1998. A social survey on the effects on environmental noise pollution in the city of Curitiba, Brazil. *Applied Acoustics*. 53:245–253.

Aylor, D. E., 1972. Noise reduction by vegetation and ground. *Journal of the Acoustical Society of America*. 51:197–205.

Barthes, R., 1985. *The Responsibilities of Farms: Critical Essays on Music, Art and Representation*. New York: Hill and Wang.

CCOHS, 2014. Noise-measurement of workplace noise document—on 5 March 2014 from Canadian centre for occupation, health and safety. Available online at www.ccohs.ca/oshanswers/phys_agents/noise_measurement.html

48 Zero Waste

CCOHS, 2017. Hearing protectors from Canadian centre for occupation, health and safety, 1 August 2017. Available online at www.ccohs.ca/oshanswers/prevention/ppe/ear_prot.html

Cohen, B., 2006. Urbanization in developing countries: Current trends, future projections, and key challenges for sustainability. *Technology in Society*. 28:63–80.

Cook, D.I. and Haverbeke, D.F., 1977. Suburban noise control with planting and solid barrier combinations. *Research Bulletin*. 100.

CPCB, 2000. Guidelines for developing green belts PROBES/75/1999–2000. Available online at http://cpcbenvis.nic.in/scanned%20reports/PROBES-75%20Guidelines%20For%20Developing%20Greenbelts.pdf

CPCB, 2015. National Ambient Noise Monitoring Network: NANMN/02/2015–16. Available online at http://cpcb.nic.in/openpdffile.php?id=UmVwb3J0RmlsZXMvTmV3SXRlbV8yMTlfU1RBVFVTX09GX0FNQklFTlRfTk9JU0VfTEVWRUxfSU5fSU5ESUEucGRm

EPD-Hong Kong, 2016. Noise mitigation—mitigation measures by environmental protection department—government of Hong Kong, October 2016. Available online at www.epd.gov.hk/epd/noise_education/web/ENG_EPD_HTML/m4/index.html

European Commission, 2013. The green paper on future noise policy by European commission. Available online at www.scribd.com/document/366415825/Noise-Pollution-a-Review (retrieved 7 September 2013).

Fang, C.-F. and Ling, D.-L., 2003. Investigation of the noise reduction provided by tree belts. *Landscape and Urban Planning*. 63(4):187–195.

Garg, N., Sharma, O., Mohanan, V. and Maji, S. Passive Noise Control Measures for Traffic Noise Abatement in Delhi, India. NISCAIR-CSIR, New Delhi, India. *Journal of Scientific & Industrial Research*. Vol. 71, pp. 226–234

Guilford, J.P. and Fruchter, B., 1973. *Fundamental Statistics in Psychology and Education*, McGraw-Hill, New York.

Hartig, T., Evans, G.W., Jamner, L.D., Davis, D.S. and Arling, T.G., 2003. Tracking restoration in natural and urban field settings. *Journal of Environmental Psychology*. 23(2):109–123.

Herzog, T.R., Black, A.M., Fountaine, K.A. and Knotts, D.J., 1997. Reflection and attentional recovery as distinctive benefits of restorative environments. *Journal of Environmental Psychology*. 17(2):165–170.

Hogan, C.M. and Latshaw, G.L., 1973. The relationship between highway planning and urban noise. In *Proceedings of the ASCE, Urban Transportation*, May 21–23, Chicago, IL.

Huddart, L., 1990. The use of vegetation for traffic noise screening, TRRL Report RR 238. Transport and Road Research Laboratory, Crowthorne.

Islam, M.N., Rahman, K.-S., Bahar, M.M., Habib, M.A., Ando, K. and Hattori, N., 2012. Pollution attenuation by roadside greenbelt in and around urban areas. *Urban Forestry and Urban Greening*. 11(4):460–464.

ITU, 1983. Characteristics and Applications of Atmospheric Radio Noise Data, Report 322-2, Geneva: International Telecommunications Union (ITU).1–72.

ITU, 2016. Recommendation *P. 372*–13 (09/2016): Radio noise. Available online at www.itu.int/rec/R-REC-P.372/en

Kinsler, L.E., Frey, A.R., Coppens, A.B. and Sanders, J.V., 2000. *Fundamentals of Acoustics*, John Wiley & Sons, New York. 359.

Krag, J., 1979. Pilot study on railway noise attenuation by belts of trees. *Journal of Sound and Vibration*. 66(3):407–415.

Kumar, K., Tyagi, V. and Jain, V.K., 2006. A study of the spectral characteristics of traffic noise attenuation by vegetation belts in Delhi. *Applied Acoustics*. 67(9):926–935.

Lawrence, D.C., 1995. *CCIR Report 322 Noise Variation Parameters*, Naval Command, Control and Ocean Surveillance Centre, RDT&E Division, NRaD Technical Document 2813; also, DTIC, San Diego, CA.

Zero Noise Pollution

Leonard, R. E. and Parr, S. B., 1970. Trees as a sound barrier. *Journal of Forestry.* XX:282–283.

Martinez-Sala, R., Rubio, C., Garcia-Raffi, L. M., Sanchez-Perez, J. V., Sanchez-Perez, E. A. S. and Llinares, J., 2006. Control of noise by trees arranged like sonic crystals. *Journal of Sound and Vibration.* 291:100–106.

McMichael, A. J., 2000. The urban environment and health in a world of increasing globalization: Issues for developing countries. *Bulletin of the World Health Organization.* 78(9):1117–1126.

MoEF, 2000. Noise Pollution (Regulation and Control) Rules, 2000 as it is published in the Gazette of India, Extraordinary, Part-II—section 3(ii), vide S.O 123 (E) dated 14.2.2000.

Morillas, J. M. B., Escobar, V. G., Sierra, J. A. M., Gómez, R. V. and Carmona, J. T., 2002. An environmental noise study in the city of Cáceres, Spain. *Applied Acoustics.* 63(10):1061–1070.

Moudon, A. V. and Wee, B. V., 2009. Environmental effects of urban traffic. In Garling, T. and Steg, L. (eds.), *Threats from Car Traffic to the Quality of Urban Life: Problems, Causes, and Solutions,* Elsevier, Amsterdam, Netherlands. 11–32.

Nicks, J., 2018. *The Adverse Effect of Noise Pollution on Human and Animal Health.* Published on 7 February 2018. Available online at https://helpsavenature.com/noise-pollution-effects

NOAA, 2014. Annual Lightning Flash Rate Map. Science on a Sphere, retrieved 15 May, 2014.

Pal, A. K., Kumar, V. and Saxena, N. C., 2000. Noise attenuation by green belts. *Journal of Sound and Vibration.* 234(1):149–165.

Pathak, V., Tripathi, B. D. and Mishra, V. K., 2008. Dynamics of traffic noise in a tropical city Varanasi and its abatement through vegetation. *Environmental Monitoring and Assessment.* 146:67–75.

Pathak, V., Tripathi, B. D. and Mishra, V. K., 2011. Evaluation of anticipated performance index of some tree species for green belt. *Urban Forestry and Urban Greening.* 10(1):61–66.

Rao, P. S., Gavane, A. G., Ankam, S. S., Ansari, M. F., Pandit, V. I. and Nema, P., 2004. Performance evaluation of a green belt in a petroleum refinery: A case study. *Ecological Engineering.* 23(2):77–84.

Reethof, G., 1973. Effect of plantings on radiation of highway noise. *Journal of the Air Pollution Control Association.* 23:185–189.

Renterghem, T. V., Botteldooren, D. and Verheyen, K., 2012. Road traffic noise shielding by vegetation belts of limited depth. *Journal of Sound and Vibration.* 331(10):2404–2425.

Shannigrahi, A. S., Fukushima, T. and Sharma, R. C., 2004. Anticipated air pollution tolerance of some plant species considered for green belt development in and around an industrial/urban area in India: an overview. *International Journal of Environmental Studies.* 61:125–137.

Tanaka, I., Kimura, R. and Simazawa, K., 1979. The function of forests in soundproofing. *Bulletin Tottori University Forests.* 11:77–102.

USDA, 2008. Conservation buffer, design guidelines for buffer, corridors and greenways. Available online at https://efotg.sc.egov.usda.gov/references/public/WA/2008_ConservationBuffers_DesignGuidelines.pdf

USEPA, 1972. Noise pollution and abatement act of 1972, senate public works committee, S. Rep. No. 1160, 92nd Cong. 2nd Session.

WHO, 2013. Guidelines for Community Noise. Retrieved 7 September.

Yang, J., Yu, Q. and Gong, P., 2008. Quantifying air pollution removal by green roofs in Chicago. *Atmospheric Environment.* 42(31):7266–7273.

4 Zero Defects in Woven Shirt Manufacturing
Application of Six Sigma Methodology Based on DMAIC Tools

R. Rathinamoorthy

CONTENTS

4.1 Introduction 51
4.2 Methodology 53
4.3 Results and Discussion 53
4.4 Six Sigma and DMAIC Application—A Case Study 55
    4.4.1 Define Phase 57
    4.4.2 Measure Phase 60
    4.4.3 Analyse Phase 61
        4.4.3.1 Process Analysis 61
        4.4.3.2 Analysis of the Defect—Down Stitch 63
        4.4.3.3 Analysis on the Defect—Skip Stitch 65
        4.4.3.4 Analysis of the Defect—Raw Edge 66
    4.4.4 Improve Phase 68
        4.4.4.1 Operator Quality Performance Card 68
        4.4.4.2 Operator Self-Inspection Report 68
        4.4.4.3 Signal System 68
        4.4.4.4 Corrective Actions 69
    4.4.5 Control Phase 69
4.5 Conclusion 71
References 72

4.1 INTRODUCTION

The successful running of any manufacturing company is based on their products' quality and customer satisfaction with their products. The success of all companies lies in their techniques: how they conquer their new customers and how well they keep their old customers satisfied. In manufacturing industries, the production of defective output results in higher consumption of raw material and wastes time and energy

(Dennis, 2002). Other researchers mentioned that defects increase the service, inspection/test, warranty, rework, and scrap costs (Slack et al., 2010). Six Sigma is one of the most commonly recognized tools for industrial problem solving and quality improvement process. It is also the highly reliable and accepted method for eliminating the defects out of manufacturing. Six Sigma is defined as a set of statistical tools adopted within quality management to construct a framework for process improvement (Goh and Xie, 2004; McAdam and Evans, 2004). The Six Sigma concept has enjoyed success throughout the business world over the last 20 years, contributing significantly to renowned corporations improving their net income (Manikandan et al., 2008). Six Sigma can reduce the variability in the products and so improve the profitability of the manufacturing firms (Sahoo et al., 2008; Yang et al., 2008).

The previous research works reports that the DMAIC (define, measure, analyze, improve and control) and design for Six Sigma (DFSS) are the two common methods for the implementation of Six Sigma, even though the fundamental objectives of the methods are different (Edgeman and Dugan, 2008). The other researchers also mentioned DMADV (define, measure, analyse, design and verify) methodology is the most common method used after the DMAIC approach (Banuelas and Antony, 2003). The DMAIC is a problem-solving method which aims at process improvement (Pande et al., 2000), DFSS and DMADV refer to new product development (Snee, 2004). The DMAIC approach was adapted in various manufacturing industries by different researchers for the different type of quality improvement purposes. The DMAIC approach was adopted to evaluate the influence of the process parameters on the sand-casting process (Kumaravadivel and Natarajan, 2013). Ploytip et al. (2014) reported the application of DMAIC techniques as a defect reduction tool in a rubber glove manufacturing process. Researchers also measured and optimized the cycle time of patient discharge process time in a hospital with the use of the DMAIC tool (Arun Vijay, 2014). Soković et al. (2006) evaluated the process improvement in the automotive part production industry using the Six Sigma tool.

As per the report of the Apparel Export Promotion Council (AEPC), India is the world's second largest textile and apparel exporter in the world. The industry contributes 2% to the GDP and 15% to the country's total exports earnings. During April–September 2018, total ready-made garments (RMG) exports from India stood at INR 52,810.51 crore (USD 7.53 billion) (Apparel and Garment Industry and Exports, 2018). The ready-made garment manufacturing industries are highly labour oriented, and so defect generation in the industry was unavoidable. Quality control in the garment industry is a big challenge and it maintains from the initial stage to the stage of the final finished garment (Cho and Kang, 2001).

A study on a branded knitted garment manufacturing unit of India indicates that the unit is experiencing about 11.18% defects in the process of manufacturing during 2013 (Textiles Committee, 2013). In order to reduce the defect generation in the apparel industry, various researches used the DMAIC approach in the apparel industry. Researchers analyzed a garment industry in Bangladesh using DMAIC approach and implemented Six Sigma concepts in the industry. They have reported that the defect percentage in the industry has been reduced from 11.67 to 9.672 and as a result, the sigma level has been upgraded from 2.69 to 2.8 (Zaman and Zerin, 2017). Hewan Taye Beyene (2016) conducted a study in the Ethiopian apparel industry using this approach

Zero Defects in Woven Shirt Manufacturing 53

and reported that they have identified skip stitches and broken stitches were the main causes for defective products. After implementing the DMAIC concepts the reduced the defect level from 3.51852 to 1.51852 and for skip stitches, they reduced it from 14.8125 to 3.8125. Other researchers reported that the implementation of the DMAIC approach helped them to reduce defect percentage from 12.61 to 7.7 and consequently the sigma level has been improved from 2.64 to 2.9255 (Uddin and Rahman, 2014).

Several other researchers also reported the application DMAIC concept in the apparel and textile industry and subsequent improvement in the quality and reduction in the defect levels for their respective operations (Patil et al., 2017; Tanvir Ahmed et al., 2013; Dewan et al., 2017; Muhammad et al., 2009; Aakanksha Upasham, 2006; Gijo et al., 2011). This work focused on the defect reduction in the ready-made garment manufacturing industry situated in Tamil Nadu, India, by the applications of DMAIC principles.

4.2   METHODOLOGY

The analysis was performed in an apparel industry which produces 100% cotton woven full sleeve casual shirts. The study consists of the following stages:

- Study of the operation sequence of the garment using the DMAIC tool.
- Implementing the define phase and analysis of the method of sewing for each operation.
- Implementing the measure phase of the study and identification of critical operations and defects, and recording the quality parameters for three months.
- In the analysis phase, study the trend of defects for the collected data and come up with potential causes for the defects.
- To improve the performance of the production floor, implement six sigma tools on the work floor and improve the quality of the production.
- During the control phase, maintain the implemented methods and measure the improvements in terms of quality improvement and defect reduction.
- Follow-up for further improvements.

4.3   RESULTS AND DISCUSSION

The study was conducted in an apparel manufacturing unit which manufactures woven full sleeve casual shirts made up of 100% cotton fabric. The study was mainly conducted in the major manufacturing line which manufactures for a leading international brand. The specification of the garments is as follows: the sewing thread used for the garment is of the fabrics' face side colour; the garment consists of 13 buttons attached to the garment, with 6 buttons in the front placket, 1 in the collar, 4 in the cuff and 2 extra buttons attached at the placket. It consists of 1 main label with the brand name, attached at the inside of the back yoke, and 1 care label with washing and ironing instructions attached at the side seam of the garment. A total of 34 operators are involved in the manufacturing of this shirt and the machines employed includes single-needle lockstitch machine, chain stitch machine, flatlock machine, the feed of the arm machine, button attaching machine and buttonhole sewing machine. The approximate standard minute value for completing this shirt is 26 minutes. The list of operations and their standard minute for each operation is listed in Table 4.1.

TABLE 4.1
List of Operations with SMV Details

Section	Operations	SMV	Target/HR
COLLAR	Collar run and trim	0.85	70
	Collar turn and top stitch	0.85	70
	Band hem	0.85	70
	Band set	0.85	70
	Band top, turn and ready	0.85	70
BACK	Yoke attach and top stitch	0.85	70
	Main label attach	0.85	70
SLEEVE	Sleeve under placket	1	60
	Sleeve top placket	1	60
CUFF	Cuff hem	0.92	65
	Cuff ready	0.92	65
	Cuff turn and top stitch	0.92	65
FRONT	Pocket iron	1	60
	Button placket sew and pocket hem	1	60
	Button hole placket lining attach	0.85	70
	Button hole placket iron	1	60
	Button hole placket top stitch	1	60
	Pocket mark	0.85	70
	Pocket attach	0.85	70
	Pairing	0.85	70
ASSEMBLY	Shoulder attach and top stitch	1	60
	Collar attach	1	60
	Collar finish	1.1	55
	Sleeve attach	1.1	55
	Sleeve gathering and wash care label attach	0.85	70
	Sleeve top stitch	1	60
	Side seam	1.1	55
	Bottom hem	1	60
	Cuff attach	1.1	55
	Button and button hole marking	0.85	70
	Button hole sew	1	60
	Button sew	1	60

4.4 SIX SIGMA AND DMAIC APPLICATION—A CASE STUDY

The Six Sigma and DMAIC approach is used to achieve zero defects in the process, as an initiative to the waste minimization in the woven shirt manufacturing unit. Based on the DMAIC methodology, the analysis was carried out in different stages that must be systematically analyzed according to the DMAIC model for the reduction in defects during the manufacturing process. For this study, a woven shirt manufacturing sewing line was considered. The manufacturing process of shirt consists of two different sections in the sewing process: sub-assembly line and assembly line. In the sub-assembly line, the components were attached or prepared as a separate process. The prepared or half-finished part/components will be assembled together in the assembly line by a sequence of operation to manufacture the final product. Understanding of the operation sequence of the product is a key feature in analyzing the manufacturing process. The sub-assembly and assembly sequence of the shirt is provided in Figures 4.1 and 4.2.

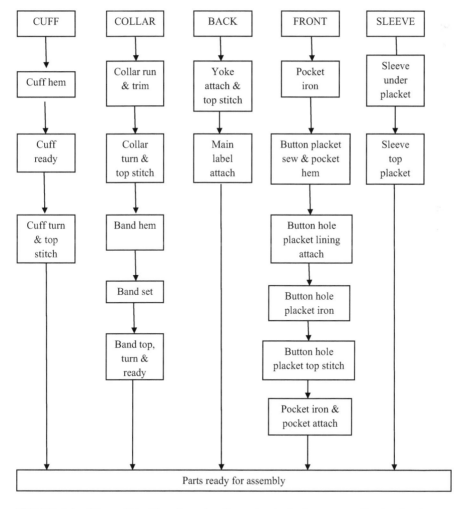

FIGURE 4.1 Woven Shirt Manufacturing Operation in the Preparatory Sewing Section

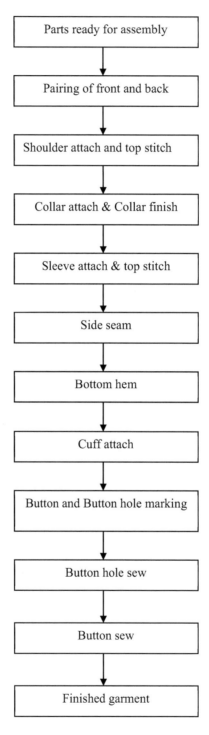

FIGURE 4.2 Woven Shirt Manufacturing Operation in Final Assembly Sewing Section

4.4.1 Define Phase

The first phase of the DMAIC process is the define phase. This particular part of the phase is identifying and defining the work objectives, scope and the boundaries of the project (Gijoet al., 2011). Previous researchers mentioned that the selected project should be focused on having a significant and positive impact on customers as well as obtaining monetary savings for the firm (Murugappan and Kenny, 2000). In this apparel manufacturing, the involvement of various factors like man, machine and materials create different sources for the generation of defects in the manufacturing process. This generation of defects considerably reduces the quality of the products manufactured and overall production numbers and also wastes time, material and money. So the cost of the production increases uncontrollably. Hence, the aim of the phase is to define the various types of defects that the company faces in the regular manufacturing process and to evaluate the potential causes behind the defects. The main objective is to analyze and identify the defects percentage in the current manufacturing facility and through the study to reduce it to a zero defect state by continued monitoring and adoption of process changes. The main challenge noted in this study was the numerous human factors involved in every step of the manufacturing process over the technical factors.

To conduct the study in a focused manner, the aim was narrowed down to only one product, that is a full sleeve woven formal shirt, for two reasons. The first reason is this is one of the larger order quantities and also this is the major manufactured product. The second reason is this particular product has the maximum number of operations in the manufacturing step as mentioned in the figure (out of different apparel that the company handled in its history). In the defining phase, the overall defect analysis was performed to identify the common defects that occur during the manufacturing process (for a period of six months) and the possible causes for the same by observation. The list of defects noted during the study and their causes are listed in Table 4.2.

TABLE 4.2
List of Critical Defects

Defect	Picture	Description	Common Occurrence
Skip Stitch		Stitch is not formed since there is no loop formation	Side seam, Buttonhole placket
Down Stitch		Stitch is not formed in the correct position such that it doesn't join the layers of fabric together	Collar attach, cuff attach, sleeve top placket

(Continued)

58   Zero Waste

TABLE 4.2 (Continued)
List of Critical Defects

Defect	Picture	Description	Common Occurrence
Raw Edge		Stitch doesn't bind the layers of fabrics such that the raw edge of the fabric is visible	Side seam, bottom hem, sleeve top stitch
Uneven Stitch		The stitch length is not uniform throughout the length of the stitch causing the difference in the SPI value	Anywhere in the garment
Roping		Folding of fabric along curved edges sometimes causes wrinkles in the fabric	Bottom hem
Puckering		Pulling of fabric due to improper tension causes wrinkled appearance	Anywhere in the garment
Piping		Extra width of stitching	Collar attach, Cuff attach
Centre out		The centre of two components doesn't match	Label attach

(Continued)

Zero Defects in Woven Shirt Manufacturing 59

TABLE 4.2 (Continued)
List of Critical Defects

Defect	Picture	Description	Common Occurrence
Shape out		The shape of the component is deformed	Sleeve placket, Pocket, Elbow patch
Nose		Protruding end of a component while attaching	Collar attach, cuff attach
Line Matchout		Line doesn't match in striped or checked garment	Button placket, sleeve placket
Stain		Dirt or stain present in the fabric	Anywhere in the garment
Fabric Defect		Weaving defect or damage in the garment	Anywhere in the garment
Missing Part		Missing of a component or a stitch	Pocket attach, Label attach, Button attach

4.4.2 Measure Phase

The measure phase of the current study is mainly focused on the reduction in defects in the manufactured product. The phase also used to define the metrics for the defect measurement. Based on the study conducted in the manufacturing firm for a period of three months, it is noted that all the mentioned defects have occurred randomly. The defect details are provided in Table 4.2. A sum of 385 defects noted in a production of 5,000 garments in a month noted by the company. This will give a defect per million opportunities (DPMO) value of 5,500 with a sigma level of 4.4. The Pareto chart analysis was performed as a next step of the process to identify the most troubling defects in the manufacturing activity. The number of occurrences per defect was plotted in the Pareto chart as in Figure 4.3.

From the chart, it can be clearly identified that down stitch and skip stitch are the two major defects noted at a higher percentage compared to the other defects. In particular these two defects together alone reason for the rejection of 71.9% of the garments. In this skip stitches have a share of 29.6% and down stitch occupies the rest of the percentage (42.3%). The down stitch defect rate was then converted into the quality level and sigma level. For skip stitch, it is noted as 21,150 DPMO and a sigma level of 3.53 (for order quantity of 5,000 garments). In the case of down stitch, the quality level is 30,241 DPMO and a sigma level of 3.38. As per the previous researchers, Lucas results, based on the sigma level noted the organizational department could be categorized as Industrial average with a 20%–30% of the cost of sales lost as poor quality (Lucas, 2002). These details of the measure phase emphasis that there is a lot of scope for improvement in the defect reduction in the manufacturing process.

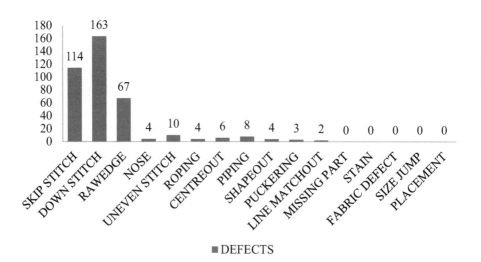

FIGURE 4.3 Top Defects for the Month of March 2018

4.4.3 ANALYSE PHASE

4.4.3.1 Process Analysis

The main purpose of the analysis phase in the DMAIC model is to evaluate and analyse the existing manufacturing system and to identify the root cause of the occurred problems and solving the same. The analysis can be performed in this phase in different manners by using the available tools like process mapping, brainstorming, cause and effect diagram, hypothesis testing, statistical process control and charts and simulation methods based on the adapted approach for the study (Pyzdek, 2003). In this analysis the brainstorming session was conducted in the analysis process among the team member and the cause and effect diagram was constructed for the results. Based on the study conducted in the manufacturing firm it is noted that there four different factors which have a major influence in the defect generation and they were listed in Figure 4.4.

Those factors like man, machine, material and method used to perform the sewing process. These identified factors were defined as measuring metrics for the defects. Hence, the identified defects were analyzed for their occurrence in the garment based on the identified factors. To evaluate the truthfulness of the defect occurrence, the analysis phase was conducted further two months and the results were given as in Figures 4.5.

From Figure 4.6 and Table 4.3, it can be understood that the majority of the defects that were caused during the consecutive months was down stitch with a count of 511 and contributing to about nearly 39.6% of the total defects that occurred. And when the causes for all defects were considered, it is seen that the maximum number of defects were caused due to human error followed by machine setting problems. Out of the total defects occurred, 45.34% of the defect happened due to manual errors and 31.83% defect dues to machine faults. When we consider the major defect, that

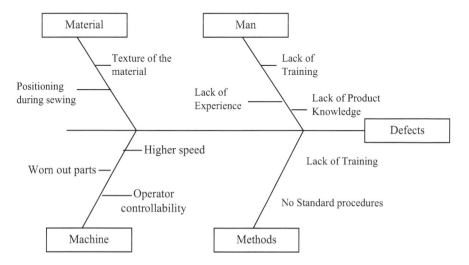

FIGURE 4.4 Cause and Effect Diagram for Defect Generation in the Sewing Department

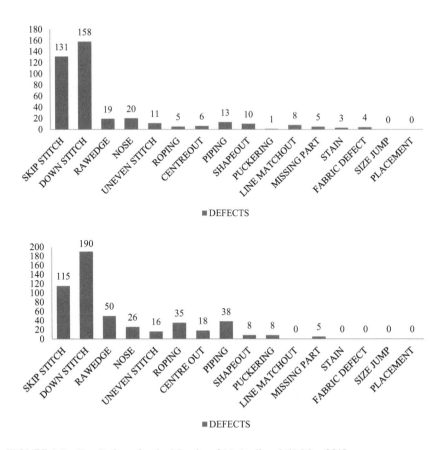

FIGURE 4.5 Top Defects for the Months of (a) April and (b) May 2018

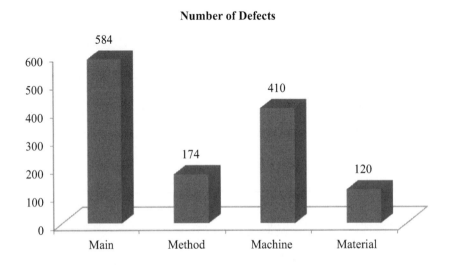

FIGURE 4.6 Defect Distribution Based on Causes

Zero Defects in Woven Shirt Manufacturing

TABLE 4.3

The Cumulative Defect Distribution Based on the Factors Identified

Defect	Causes				Sum
	Man	Method	Machine	Material	
Skip Stitch	166	43	122	29	360
Down Stitch	203	52	192	64	511
Raw Edge	47	48	28	13	136
Nose	35	15	0	0	50
Uneven Stitch	23	0	0	14	37
Roping	13	7	24	0	44
Centre Out	30	0	0	0	30
Piping	28	0	31	0	59
Shape Out	13	9	0	0	22
Puckering	4	0	8	0	12
Line Match Out	10	0	0	0	10
Missing Part	10	0	0	0	10
Stain	0	0	3	0	3
Fabric Defect	2	0	2	0	4
Size Jump	0	0	0	0	0
Placement	0	0	0	0	0
Sum	**584**	**174**	**410**	**120**	**1288**

is down stitch, out of 511 defects, 203 were caused due to the human error (39.72%). With respect to the major defect (511), 174 defects occurred due to the machine fault, it is 34.05% of the total major defects. The second major defect noted was skip stitch, which occurred 360 times out of 1,288 defects it is around 27.9% of the defect. The specific analysis in the sewing floor further revealed the findings that collar attachment and cuff attachment being the most critical operations, have been the major source of down stitches. On analyzing the causes for the defects, again the human error contributes the most; for both down stitches and skip stitches. By back tracking these two issues to the sewing floor, it is identified that the particular operator is semi-skilled and is not aware of the exact sewing procedure. Inadequate knowledge and training to the operator led to the high count of defects. The detailed analysis of the machine revealed that the machine was lack of proper machine maintenance due to the higher workload. Second, it is also noted that the improper fixation of particular sewing attachment also contributed to the defect. In total for sum of all defects in three months, it is noted as 17,173 DPMO and a sigma level of 3.6 (for order quantity of 15,000 garments). As per the previous researchers, Lucas results, based on the sigma level noted the organizational department could be categorized as Industrial average with a 20%–30% of the cost of sales lost as poor quality (Lucas, 2002).

4.4.3.2 Analysis of the Defect—Down Stitch

The study results revealed that the maximum number of defects caused is the down stitch. It accounts for nearly 39% of the overall defects occurred. The down stitch alone happened 511 times due to several factors in the past three months out of 1,288 defects occurred. The down stitch is a sewing defect which occurs when the stitch is not formed in the correct position such that it doesn't bind the two layers of fabric. The occurrence of the down stitch was analyzed with respect to the different operation using the Pareto chart and the results were provided in Figure 4.7.

From the results, it can be understood that the down stitch occurred at different operations in the shirt manufacturing. However, it is noted that the collar attachment and sleeve placket and cuff are the most common operations, which created the defect. The reason observed was the criticality of the process. In all the three above mentioned operation, there are three layers of material attached together in the wrong side and they were turned out and finished with the top stitch. The higher pressure related to the production and low operator skills are the main reason for the defects.

Defects by operators: From Table 4.3, it can be understood that the majority of the defects occurred due to man (39%). It is either the inadequate training or practice for the operators or the insufficient knowledge about the correct sewing procedure which causes the down stitch. Due to this, the placement of the different layers of fabric is not proper causing down stitch and opening at that place.

Defects by material: Down stitch can also be caused by the material. Due to the slippery and light weighted material used, handling by the operator during sewing was noted difficult, which sometimes ultimately lead to down stitch. According to the data, nearly 12% of the down stitches are caused by the material used, which is unavoidable in this case.

Defects by machine: The second major source for the down stitch is the machine, which causes around 37.5% down stitches. The major reason identified is improper placement of the folder during collar attachment, cuff attachment and the sleeve top

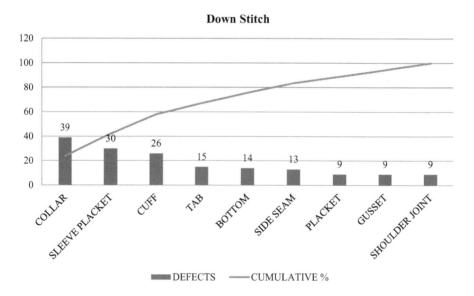

FIGURE 4.7 Down Stitch Pareto Analysis

Zero Defects in Woven Shirt Manufacturing

placket leads to 37.5% of the down stitch defects. When the pressure foot tension is not adequate to hold the layers of the fabric firmly in the correct position, slippage of the fabric may lead to down stitch. Sometimes the high speed of the machine, without the control of the fabric leads to down stitch. Poor quality of the pressure foot may also cause down stitch, which can be rectified by changing the presser foot.

Defects by method: The method or the procedure of stitching may affect the quality of the garment which is being sewn. In this case, about 10% of the defects caused due to the methods used for the operation. Even though the factory personal advices the effective methods for picking and aligning the material under the needle, few operators based on their body mobility restrictions and their hand movement easiness prefer their own way of the process. This might be the reason for the defects.

4.4.3.3 Analysis on the Defect—Skip Stitch

Skip stitch was noted as the second most commonly occurred fault with the production process. In three months, out of total 1,288 defects noted, 27.9% (360) defects due to the skip stitch issue. Skip stitch is defined as the defect which occurs when the machine misses the needle loop or the needle misses the looper thread, as a result of which the stitch doesn't form in the particular position.

From Figure 4.8, it can be seen that the skip stitch mostly occurred in either the side seam or in the placket attaching. Both operations operated with chain stitch machines. As the chain stitches are less secure than the lock stitches, these defects may lead to the unravelling of the entire row of the stitch. Hence, the importance of the defect increases multi-fold due to its nature. The results from Table 4.3 revealed that, as similar to down stitch, the major causes for the skip stitch are operator error and machine-related issues.

Defects by operators: Human error is one of the major contributing factors in the generation of skip stitch. Out of 360 occurred defects, 166 (46%) of the skip stitches occur due to operator inefficiency. This may be due to the lack of training to the

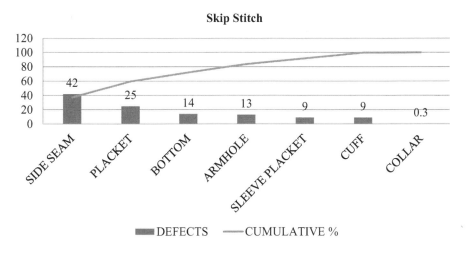

FIGURE 4.8 Skip Stitch Pareto Analysis

operators and their knowledge about the sewing procedure. Improper handling by the operator should be avoided to eliminate skip stitches in the garment.

Defects by method: With respect to the skip stitches, an incorrect method of sewing followed causes 11% of the skip stitches that have occurred, 43 defects out of 360 defects in three months. The important issue noted in this section is higher machine speed, which causes a miss of the stitch and the improper material feeding by the operators. Improper material handling leads to the small size of loop formation leading to missing of the stitch.

Defect by machine: In the case of skip stitch, next to operator error, the next most contributed factor is defects occurred due to the machine. The machine faults contribute to about 33% (122 defects out of 360) in three months. The major reasons noted in the production floor are, improper thread tension, incorrect threading in the machine, improper machine setting which includes improper pressure foot tension are all the causes for the skip stitch.

Defects by material: A minor part of the defects occurred due to the materials used in the production process. Materials like fabrics, needle and sewing threads account for nearly 8% of the total skip stitches count. Needle deflection or damaged needles or improper presser foot causes skip stitch. If the fabric is slippery and light weighted, there is a higher possibility of skip stitch occurrence. The selection of good quality yarn is very important to avoid skip stitch.

4.4.3.4 Analysis of the Defect—Raw Edge

The raw edge is another major defect which is commonly noted in the selected style of the garment. This is one of the most frequent defects noted among the skip and down stitch. The raw edge defect contributes to 10% of the total defects in the past three months. Raw edge is a defect which represents the stitch that doesn't cover the raw edges which may ultimately lead to fraying or unravelling of the edges of the fabric. This defect is most commonly seen in the side seam and armhole top stitch operations. The Pareto chart analysis for the defect raw edge is provided in Figure 4.9.

Defects by operator: Raw edge is a defect which is more predominant due to human carelessness or inadequate training to the operators. Proper handling of the material is very essential to avoid the raw edges. With respect to this defect occurrence, human error contributes to nearly about 34% of the total raw edges defects. The results of the observation revealed that the uneven alignment of the two panels of fabric while sewing creates the problem of raw edges.

Defects by machine: Raw edges due to machine defects are considerably less. It can be caused by damaged pieces of the machine. Damaged needles, damaged needle plates and improper machine maintenance will result in raw edges defects in the manufacturing. Twenty-one percent of the raw edges are caused due to machine problems. As the chains stitch machines used in this process, the speed of the machine, knife and needle setting positions also play a major role in causing raw edge defects. Improper folder setting leads to this defect because it doesn't fold the fabric properly resulting in showing off of the raw edges.

Defects by method: The major contributor for the raw edge defect is the method, about 34% of the raw edge defects were caused by the incorrect method of sewing followed by the operators. Uneven stitch margin will result in the inadequate amount of fabric for feeding inside the folder which will cause raw edges. Due to

Zero Defects in Woven Shirt Manufacturing

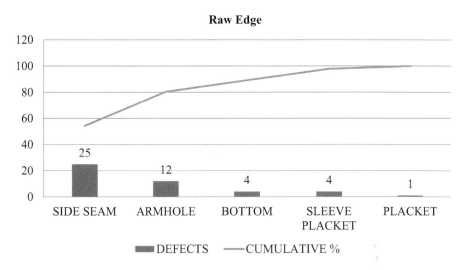

FIGURE 4.9 Raw Edge Pareto Analysis

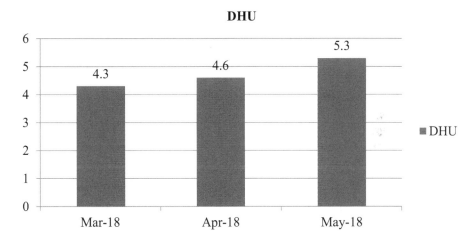

FIGURE 4.10 DHU Levels of the Analyzed Data

the improper positioning of the material in the chain stitch machine, the edges of the fabrics were not trimmed correctly. Results in the excess amount of fabric feeding in the folder, and created the raw edges.

Defects by material: It is one of the minor contributors to the defects. It contributes to only 9% of the total raw edge defects count in the past three months of the study. This is important because the improper cutting of the panels will lead to frayed edges while sewing which unravels and causes raw edges. Slippery fabrics are difficult to be fed into the folder and this may also lead to raw edges. Any damaged needle or machine part will be a cause for this defect. Based on the analysis of the result, the monthly defects per hundred unit (DHU), the report was developed for the selected style. The results are provided in Figure 4.10.

68　　　　　　　　　　　　　　　　　　　　　　　　　Zero Waste

The DHU values are calculated as per the following formula:

$$DHU = (\text{No. of defects found/Total no. of pieces checked}) \times 100\% \tag{4.1}$$

4.4.4 Improve Phase

After the root cause(s) has/have been determined, the DMAIC's improve stage aims at identifying solutions to reduce and tackle them. At this phase of the research, based on the root causes identified, to rectify the defects identified in the manufacturing activity and to improve the quality of the product and performance of the factory, different measures were performed in the shop floor. Various methods which come under the Six Sigma concepts were adapted to improve the quality. The systems include flag system, operator self-inspection card system, operator performance card system in the shirt manufacturing section.

4.4.4.1 Operator Quality Performance Card

The operator performance card (OPC) is implemented in the production floor. The card was hung in every sewing machine, which assigns it to each operator. It is asked to be filled by the inline quality checker. The inline quality checker must check eight pieces from each operator in the entire day and mark in the OPC card. If any defect is found in the checked pieces, it will be marked using a red coloured pen, and if defects not found it will be marked using a green coloured pen. When marked using a red pen, the corrective action which was taken for the operator by the inline quality checker should be mentioned in the card. The corrective action can be one among the six corrective actions:

- Operator retrained
- Garment accepted with deviation
- Operator changed
- Machine repaired/adjusted
- Garment accepted as second quality
- Panel/parts replaced.

4.4.4.2 Operator Self-Inspection Report

The operator self-inspection report (OSIR) is given to each operator in the line. This was introduced in the production line in order to have a record of the defects and reworks done by the operator. These defects, which are marked in this report, are not those which are passed to the end line and come back for rework, but those which are found by the operator during the self-inspection and which are altered then and there. It has details such as operator name, token number, the operation which is carried out by the operator, line number and month. It has a list of 34 different types of defects in which they have to be marked.

4.4.4.3 Signal System

Signal system is the one in which three colour flags (red, blue and yellow) are used. The red flag indicates criticality to quality; the blue flag indicates criticality

Zero Defects in Woven Shirt Manufacturing

to product; and the yellow flag indicates either operator change or a new operator. These flags are used on the machines to make the inline quality checkers aware of where they have to focus or concentrate more so as to ensure quality products. Apart from this, corrective actions were taken based on the root cause analysis done, to identify the causes of the problem.

4.4.4.4 Corrective Actions

Down stitch: From the root cause analysis done for down stitch, it is seen that majority of the down stitches are caused by human error and machine setting problems. Hence corrective actions are taken at those operations:

- Training of the operator
- Adjustment of the folder (up to 2 mm)
- Controlled feeding of fabric by the operator
- Correction of stitch margin.

Skip stitch:

- Training of the operator
- Controlled feeding of fabric by the operator
- Adjustment of thread tension
- Adjustment of machine setting
- Correction of stitch margin
- Change needle
- Control machine speed.

Raw edges:

- Correction of stitch margin by the previous operator
- Trimming of the frayed edges
- Training of operator for proper aligning
- Adjustment of the folder.

4.4.5 CONTROL PHASE

The aim of the control phase is to sustain the gains from processes which have been improved by institutionalizing process or product improvements and controlling ongoing operations (Omachonu and Ross, 2004; Stamatis, 2004). After implementation of improve phase, the method was followed for a month to adapt better. After a month of improve phase, the results of the same operations were evaluated and reported in Figure 4.11 after implementing the system of zero defects, it is noted that, similar to the previous months, again the three major contributors of defects were down stitch, skip stitch and raw edges with a reduced defect count of 44, 26 and 10, respectively. But when the causes for those defects were analyzed, the major contributor for the defects was machine setting problems. Figure 4.11 represents the defect analysis and comparison between the month of July and the average value of the previous three months.

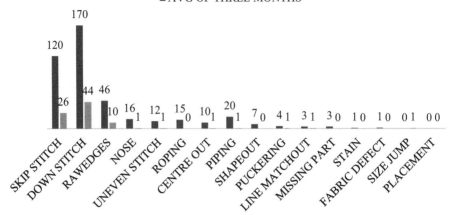

FIGURE 4.11 Defect Comparison of Three-Month Average and January

Comparing the results of the defect count, before and after the implementations of the systems, it is seen that the down stitch has reduced by 74.1% from 170 in the previous months to 44 in July. Similarly, skip stitch was reduced by 78.1% from 120 in the previous months to 26 in the month of July, while raw edges are reduced by 78% from 46 in previous months to 10 defects. This result is achieved by focusing on the critical operations which cause these major defects such as collar attachment, cuff attachment, side seam and bottom hem. The analysis of defects in the different heads like man, machine, material and methods are provided by comparing the average of three months with July in Figure 4.12.

From the results, it can be seen that the defects caused by human error are reduced by 84% from 195 in the previous months to 30 after the implementations. Similarly, the defects caused by machine setting problems were reduced by 74.4% from 137 defects during the previous months to 35 during the month of July. This result is achieved by training of the operator, adjustment of folders, checking for proper thread tension, proper machine maintenance and the replacement of damaged machine parts. After the implementation of the system initiated towards achieving zero defects, the monthly DHU level for the month of January has reduced to 1%. The system has been effective in reducing the defects in the line, thus reducing the time of reworks, improving productivity and also reducing the cost of poor-quality products. The results of the DHU were provided in Figure 4.13. For the improved performance, the DPMO and sigma levels were arrived. After the implementations total 87 defects were noted for the production of 5,000 garments, by assuming 5 possibilities of defects per products as an average, a 3,480 DPMO and a sigma level of 4.2 were achieved. As per the literature, the organizational department could be categorized as the industrial average with a 15%–20% of the cost of sales lost as poor quality (Lucas, 2002).

Zero Defects in Woven Shirt Manufacturing

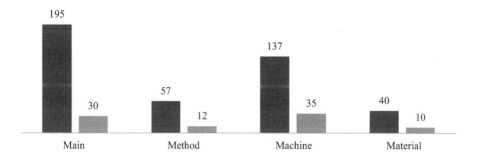

FIGURE 4.12 Cause Comparison of Three-Month Average and July Data

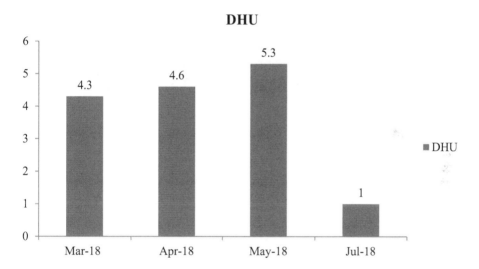

FIGURE 4.13 DHU Levels before and after Implementation

4.5 CONCLUSION

This case study was performed with Six Sigma–based framework using DMAIC methodology to improve the quality characteristic of an apparel product manufacturing industry. The study conducted for a period of six months as per the different phases represented in the DMAIC methodology. The operation sequences of the formal full sleeve shirt were thoroughly investigated and the major defects like down stitch, skip stitch and raw edges were identified. The root cause analysis revealed that the major causes for the occurrence of the defect are operator quality and machine related issues. Based on the analysis, in the improve phase, with the tools of the Six Sigma system, the quality improvement measures were implemented in the factory.

After the implementation process, the defects were measured again against the same parameter and it was noted that significant reduction in down stitch (74.1%), skip stitch (78.1%) and raw edges (78%) were observed. Further, the DHU value of the particular product manufacturing section is reduced from 5.3 (before implementation of DMAIC) to 1 after the DMAIC methodology implementation. Similarly, the sigma level of the sewing department increased from 3.6 to 4.2.

REFERENCES

Aakanksha Upasham, 2006. Minimizing defects in the sewing department leading to quality improvement. Masters diss., National Institute of Fashion Technology (Mumbai) India.

Ahmed, T., Acharjee, R. N., Rahim, A., Sikder, N., Akther, T., Khan, M. R., Rabb, F. and Saha, A., 2013. An application of pareto analysis and cause-effect diagram for minimizing defect percentage in sewing section of a garment factory in Bangladesh. *International Journal of Modern Engineering Research*. 3(6):3700–3715.

Apparel and Garment Industry and Exports, 2018. Available online at www.ibef.org/exports/apparel-industry-india.aspx

Arun Vijay, S., 2014. Reducing and optimizing the cycle time of patients discharge process in a hospital using six sigma DMAIC approach. *International Journal for Quality Research*. 8(2):169–182.

Banuelas, R. and Antony, F., 2003. Going from Six Sigma to design for Six Sigma: An exploratory study using analytic hierarchy process. *The TQM Magazine*. 15(5):334–344.

Cho, J. and Kang, J., 2001. Benefits and challenges of global sourcing: Perceptions of US apparel retail firms. *International Marketing Review*. 18:542–561. https://doi.org/10.1108/EUM0000000006045

Dennis, P., 2002. *Lean Production Simplified*, Productivity Press, New York.

Edgeman, R. L. and Dugan, J. P., 2008. Six Sigma from products to pollution to people. *Total Quality Management*. 19(2):1–9.

Gijo, E. V., Scaria, J. and Antony, J., 2011. Application of Six Sigma methodology to reduce defects of a grinding process. *Quality and Reliability Engineering International*. 27(8):1221–1234.

Goh, T. N. and Xie, M., 2004. Improving on the Six Sigma paradigm. *The TQM Magazine*. 16(4):235–240.

Hewan Taye Beyene, 2016. Minimization of defects in sewing section at Garment and Textile Factories through DMAIC methodology of Six Sigma (Case: MAA Garment and Textile Factory), Masters diss., Mekelle University. Available online at http://ieomsociety.org/ieom2017/papers/406.pdf

Kumaravadivel, A. and Natarajan, U., 2013. Application of Six-Sigma DMAIC methodology to sand-casting process with response surface methodology. *International Journal of Advanced Manufacturing Technology*. 69:1403–1420.

Lucas, J. M., 2002. The essential six sigma. *Quality Progress*. 35(1):27–31.

Manikandan, G., Kannan, S. M. and Jayabalan, V., 2008. Six Sigma inspired sampling plan design to minimise sample size for inspection. *International Journal of Productivity and Quality Management*. 3(4):472–495.

McAdam, R. and Evans, A., 2004. Challenges to Six Sigma in a high technology mass manufacturing environment. *Total Quality Management*. 15(6):699–706.

Muhammad, A., Baloch, A., Rehman, A. and Anees, M., 2009. How to minimize the defects rate of final product in textile plant by the implementation of DMAIC tool of Six Sigma, Masters diss., University of Borås, Sweden. Available online athttp://bada.hb.se/bitstream/2320/6914/1/Abid%20Rehman%20Anees.pdf

Murugappan, M. and Kenny, G., 2000. Quality improvement—the Six Sigma way, paper presented at APAQS. In *The First Asia-Pacific Conference on Quality Software*, October 30–31, Hong Kong.

Omachonu, V. K and Ross, J. E., 2004. *Principles of Total Quality*, 3rd ed., CRC Press LLC, Boca Raton, FL.

Pande, P. S., Neuman, R. P. and Cavanagh, R. R., 2000. *The Six Sigma Way: How GE, Motorola and Other Top Companies Are Honing Their Performance*. McGraw-Hill, New York.

Patil, N. S., Rajkumar, S. S., Chandurkar, P. W. and Kolte, P. P., 2017. Minimization of defects in garment during stitching. *International Journal on Textile Engineering and Processes*. 3(1):24–29.

Ploytip, J., Arturo Garza-Reyes, J., Kumar, V. and Lim, M. K., 2014. A Six Sigma and DMAIC application for the reduction of defects in a rubber gloves manufacturing process. *International Journal of Lean Six Sigma*. 5(1):2–21.

Pyzdek, T., 2003. *The Six Sigma Handbook: A Complete Guide for Green Belts, Black Belts, and Managers at All Levels*, McGraw-Hill, New York.

Sahoo, A. K., Tiwari, M. K. and Mileham, A. R., 2008. Six Sigma based approach to optimize radial forging operation variables. *Journal of Materials Processing Technology*. 202(1–3):125–136.

Slack, N., Chambers, S. and Johnston, R., 2010. *Operations Management*, 6th ed., Prentice-Hall, London.

Snee, R. D., 2004. Six Sigma: The evolution of 100 years of business improvement methodology. *International Journal of Six Sigma and Competitive Advantage*. 1(1):4–20.

Soković, M., Pavletić, D. and Krulčić, E., 2006. Six Sigma process improvements in automotive parts production. *Journal of Achievements in Materials and Manufacturing Engineering*. 19(1):96–102.

Stamatis, D. H., 2004. *Six Sigma Fundamentals: A Complete Guide to the System, Methods and Tools*, Productivity Press, New York.

Textiles Committee, 2013. Strategy for zero defect production of ready-made garments. Available online at http://ficci.in/events/22333/ISP/Nayak.pptx

Uddin, S. M. and Rahman, C. M. L., 2014. Minimization of defects in the sewing section of a garment factory through DMAIC Methodology of Six Sigma. *Research Journal of Engineering Science*. 3(9):21–25.

Yang, K. J., Yeh, T. M., Pai, F. Y. and Yang, C. C., 2008. The analysis of the implementation status of Six Sigma: An empirical study in Taiwan. *International Journal of Six Sigma and Competitive Advantage*. 4(1):60–80.

Zaman, D. M. and Zerin, N. H., 2017. Applying DMAIC methodology to reduce defects of sewing section in RMG: A case study. *American Journal of Industrial and Business Management*. 7:1320–1329. doi:10.4236/ajibm.2017.712093

Zaman, D. M., Zerin, N. H. and Rahman, M. M., 2017. Applying DMAIC methodology to reduce defects of Sewing section in RMG-A case study. In *Proceedings of the International Conference on Mechanical Engineering and Renewable Energy 2017 (ICMERE2017)*, Chittagong, Bangladesh. Available online at www.cuet.ac.bd/icmere/files2017f/ICMERE2017-PI-153.pdf

5 Biomedical Waste Disposal and Treatment

Rakesh K. Sindhu, Gagandeep Kaur and Arashmeet Kaur

CONTENTS

5.1 Introduction 75
5.2 Basic Terminologies 77
5.3 Biomedical Waste: Source 77
5.4 Biomedical Waste: Classification 78
    5.4.1 Based on the Type of Toxicity Levels 78
    5.4.2 Based on WHO Classification 79
5.5 Biomedical Waste: Points of Issue 80
5.6 Need for Biomedical Waste Management 82
5.7 Merits of the Management System 82
5.8 Biomedical Waste Management 82
    5.8.1 Biomedical Waste Generation 82
    5.8.2 Segregation of Biomedical Waste 83
    5.8.3 Storage of Biomedical Waste 83
    5.8.4 Transportation of Biomedical Waste 84
    5.8.5 Medical Waste Treatment and Disposal Treatment Processes 84
5.9 Various Methods Employed for the Disposal of Biomedical Waste 85
    5.9.1 Moist Heat Sterilization—Autoclave 85
    5.9.2 Method of Incineration 86
    5.9.3 Microwave Treatment 86
    5.9.4 Hydroclave Process Treatment 86
    5.9.5 Landfilling Process 86
5.10 Conclusion 87
References 87

5.1 INTRODUCTION

Science and technology have shown tremendous escalation since time immemorial for the betterment of mankind. Though health ecology of mankind is variegated with time, it has always remained a central concern for the people around the globe (Forget and Lebel, 2001). Technology has provided for the treatment of various health ailments or diseases which once were difficult to treat, and with advent of time certain diseases have been eradicated from the earth. While this advent has reaped benefits, it has also led to various harmful effects on the environment.

One of the increasing matters of contention is the increase in biomedical waste generated by various health centres (Mandal and Dutta, 2009). As a result of the unsustainable utilization of natural resources, the waste has been produced in large quantities unexpectedly. Health care waste has been reported to be one of the major negative contributing factors to the environment as well as human health due to their improper storage, transportation and disposal (Kumar, 2011). Biomedical waste is the waste which is generated in all health care institutions, such as clinics, hospitals, dental clinics, laboratories, blood banks and so forth, involved in the treatment of human or animal health and wellness. The waste generated during the diagnosis, prevention, cure and immunization techniques employed for the well-being of humans and animals or during the laboratory practices, testing conditions is referred to as biomedical waste (Dohare et al., 2013). There has been a rise in the generation of biomedical waste in recent years because of widespread utilization of disposal in medical care facilities. In addition, the improper handling of these wastes, including its storage, transportation and disposal, lead to potential health hazard risks to mankind as well as the environment (Rudraswamy et al., 2013). It has been reported that about majority of the countries around the world have distressing graphs of biomedical waste management system (Indupalli et al., 2015). Reports of WHO (World Health Organization) state that the waste generated by the health care institutions can be categorized as 5–10% infectious and the majority 58% constitute to be non-hazardous (WHO, 2015).

The waste generated by the hospitals is prone to infections which are easily transmitted from one person to another. These deadly infectious diseases include hepatitis B, hepatitis C, HIV, tetanus and so forth (Mathur et al., 2012). The onset of the spread of these infectious has shown an alarming rise in the recent decades due to improper handling procedures. Improper sterilization of needles, IV sets, glass bottles and syringes is one of the reasons for widespread infection. Moreover, insects, flies and rodents are contributing more to these diseases (Mishra et al., 2015). The major sites of the presence of this waste is around the residential areas and hospitals, unlike industrial waste which is generally found in the outskirts of the city. Direct contact with infected syringes often leads to the onset of diseases. Rag pickers and waste pickers are more prone to these infections during their searches for valuable waste. Sometimes they are at higher risk on their exposures to poisonous chemicals or objects causing serious damage to their health. Mainly developing countries are at a major risk due to lack of proper handling procedures and economic as well as technological developments. These countries are at a slower rate of development which creates such health hazards, compared to developed countries which are at a lower risk (Odumosu, 2016). Therefore, adequate management procedures must be followed in each country in order to reduce the environmental risk. Though the management of waste is not a legal issue, it requires a humanitarian concern in regard to the environment. Correct handling techniques would eventually lead to better ecosystem as whole. Proper education is required in order to curb the health-related hazard arisen due to improper disposal of waste (Diaz and Savage, 2003; Kalpana et al., 2016).

5.2 BASIC TERMINOLOGIES

1. *Waste*: Waste can be defined as any substance being disposed of or is intended for disposal according to the national regulations (Basel Convention, 1989).
2. *Clinical waste*: Waste generated during the investigation and treatment of a patient is referred to as clinical waste by Phillips 1989 (Patwary et al. 2009).
3. *Health care waste*: Health care waste can be defined as waste produced by research laboratories, analytical laboratories, faculties of research as well as health establishments. Moreover, it includes home care medical practices such as injections of insulin, dialysis, nebulizers and so forth (PAHO, 1994).
4. *Hospital waste*: Waste produced during the medical procedures of diagnosis, prevention, treatment and immunization for animals and humans or during the research purposes is referred to as hospital waste (Amin et al., 2013).
5. *Waste of hygiene*: It is also referred to as 'sanpro waste'. The waste which may be an offence to an individual who comes in contact with it, though they are not at the stage of causing an infection. These include waste generated from sanitary and human hygiene (Singh et al., 2017).

5.3 BIOMEDICAL WASTE: SOURCE

Biomedical waste is variable in origin. The type, quality and quantity of biomedical waste is variable according various health care factors such as size of hospitals, number of patients per day, location site and many more. It has been reported that 80% of the total biomedical waste is derived from health care, while pathological waste contributes to about 15%. Pharmacological waste constitutes to about 3% of the total biomedical waste. On the other hand, about 1% is the waste from sharp materials and about less than 1% for cytotoxic waste (Patwary et al., 2009). Figure 5.1 describe the source of biomedical waste.

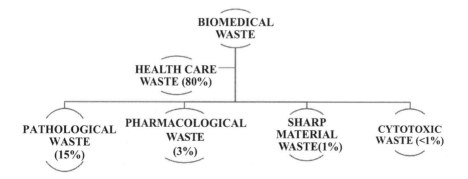

FIGURE 5.1 Sources of Biomedical Waste

Biomedical sources are categorized into the following major and minor sources.

Major Sources of Biomedical Waste

- Hospitals
    1. General hospitals
    2. Charitable hospitals
    3. Institutional hospitals
- Laboratories
    1. University or college laboratories
    2. Biotechnology laboratories
    3. Research centres
- Others
    1. Centres of dialysis
    2. Centres of transfusion
    3. Paediatrics clinics
    4. OPD clinics
    5. Emergency services
- Blood banks
- Mortuary centres
- Animal research laboratories
- Nursing homes.

Minor Sources of Biomedical Waste

- Small posts of first aid
- Small health care institutions
- Acupuncturists
- Dental clinics
- Small hospitals (Semwal, 2016).

5.4 BIOMEDICAL WASTE: CLASSIFICATION

5.4.1 BASED ON THE TYPE OF TOXICITY LEVELS

Biomedical waste can be categorized on the level of the toxicity of the waste produced. They may be:

1. Hazardous biomedical waste (Hegde et al., 2007; Hussain et al., 2018)

Hazardous biomedical waste can further be classified as:

a. Infected hazardous biomedical waste
   This category constitutes the hazardous biomedical waste which has the potential or the ability to produce infections. They are often referred to as potentially contaminated or infected waste. Infected hazardous biomedical waste constitutes contaminated dressing along with any

Biomedical Waste Disposal and Treatment   79

secretions from the body such as blood, fluids or pus. In addition, it includes culture stocks prepared in laboratories. These constitute about only 10% of the total waste biomedical waste generated. They are on the verge of contamination because of their removal from the body, organs, tissues or tumours during surgeries and so forth with the help of medical devices such as syringes, needles contaminated with blood or any material.

b. Toxic hazardous biomedical waste

Toxic hazardous biomedical waste includes the waste produced from invitro techniques of body imaging, body fluids and many more. They are wastes produced along with a radionuclide, thus increasing the potential toxic level. This waste consists of iodophors, glutaraldehyde and metal debris. Pharmaceutical waste includes anaesthesia, antibiotics, sedatives and so forth.

2. Non-hazardous biomedical waste

Non-hazardous biomedical waste is neither toxic nor potentially infected wastes. Rather they are simple waste generated from food leftovers, paper, fruits and so forth. It constitutes about 85% of total biomedical waste (Singh et al., 2014).

5.4.2 Based on WHO Classification

The World Health Organization categorizes biomedical waste into eight different categories. This classification is based upon a number of factors such as human exposure, ill effects on mankind, degree of transmission of infected waste and so forth. Moreover, they are categorized on the basis of their characteristics: physical, biological and chemical (Rudraswamy et al., 2013; Singh et al., 2017).

1. *Pathological waste*: WHO classifies pathological waste as the waste obtained from organs or body parts of human beings or animals. They constitute foetuses of humans, body fluids, blood, tissues and so forth. This waste is generated from equipment and wards which come in contact with an infected patient containing diseased pathogens (Chitnis et al., 2005).

2. *Chemical waste*: It is defined as the waste being obtained from noxious chemicals during research or diagnostic technique. In addition, they can be obtained from houses or as a means of disinfectant. Depending on the chemical property, they can be classified as hazardous or non-hazardous. Examples include disinfectants, formaldehyde, cadmium waste and so forth.

3. *General waste*: General waste according to WHO constitute any type of household waste produced being non-infectious in origin. These include kitchen waste, waste from laundries and so forth.

4. *Radioactive waste*: Radioactive waste is defined as the waste obtained from a nuclear reactor after the production of fuel. This is one of the most harmful wastes produced. They can be present in any form of solid, liquid or gas. These constitute all the waste produced during radiotherapy of a diseased individual, glassware, excreta, urine and so forth. Examples include alpha

TABLE 5.1

Types of Radioactive Waste

S. No.	Type of Radioactive Waste
1	Generators of radionuclides
2	Water or air tight sealed sources
3	Solid waste such as swabs, syringes, glassware, paper etc.
4	Radioactive residues
5	Radioimmunoassay utilized liquids
6	Liquid waste of low grade
7	Excreta of patients under radioactive treatment
8	Spills of radioactive substances
9	Gas releases from fuming cabinets

emitters and beta emitters (Khan et al., 2010). Radioactive waste is further classified according to World Health Organization (see Table 5.1).

5. *Infectious waste*: Infectious waste is defined as the waste produced during the diagnosis or treatment of human beings or animals or any of the procedures involved for the testing purposes. Examples include blood products, blood, infected bandages or dressings.

6. *Waste produced by pharmaceutical practices*: Pharmaceutical waste includes cartons of drugs, expired formulation packaging, leftover medications, sera, injections, vaccines and so forth.

7. *Waste produced by sharp materials*: These constitute the sharp-headed materials of equipment which is used for body or skin piercing for medical purposes. They are the major site for infection if proper sterilization procedures or disposal procedures are not followed. Examples include needles, syringes, blades, glass, reamers and so forth.

   Sharps can be further classified into (1) contaminated or non-contaminated sharps; (2) sharps of radioactive origin; (3) chemically infected sharps; and (4) laboratory glass sharps.

8. *Aerosol or pressurized containers*: The waste produced by these containers sometimes produces fatal effects. They explode upon incineration of accidental puncture. These include aerosol formulations, gas cylinders or cartridges and so forth.

5.5 BIOMEDICAL WASTE: POINTS OF ISSUE

The major significant issue which is related to biomedical waste is its improper disposal system by hospitals. Hospitals or other health care institutions dispose of the biomedical waste in an improper and regardless manner. This improper management of hospital or biomedical waste possess health hazards to humans and animals as well as the ecosystem, either directly or indirectly (Acharya et al., 2014). Figure 5.2 shows the improper waste management system. The harmful toxins produced by the

Biomedical Waste Disposal and Treatment

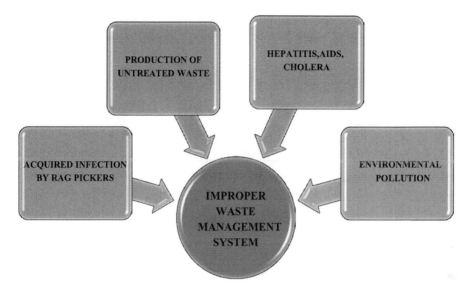

FIGURE 5.2 Improper Waste Management System

hospitals, such as xylene, mercury, formalin and so forth produce serious health hazards (Sharma, 1998). Inappropriate segregation methods employed to differentiate the types of biomedical waste creates havoc between the hospital and general waste generation. Moreover, this improper separation hampers the biomedical waste treatment process (www.medicinenet.com).

The environment as a whole is also adversely affected by the reckless disposal of biomedical waste. Environmental pollution is the key problem. In addition, the growth of microorganisms and obnoxious odours facilitate the disease transmission process of many diseases. These include hepatitis, cholera, AIDS, cancer and typhoid from contaminated syringes and blades. The major routes for contamination of these diseases are sweat, body fluids, water and so forth, which are carried from one person to another through vectors such as insects, flies, mosquitoes and so forth (CEET, 2008).

Negligent management of biomedical waste is a collective failure of the waste during various stages of treatment such as collection, separation, storage, transportation and so forth. The term 'improper handling' refers to the methods used for the handling of biomedical waste. For instance, lack of protective gear for the workers during the handling of toxic waste, high temperature from poor storage conditions of the biomedical waste, long-distance travelling problems and improper packaging conditions (Manyele and Tanzania, 2004). Most exposed individuals are as follows:

1. Medical staff (doctors, staff nurses, auxiliaries)
2. Stretch bearers, waste managers, carriers, pharmacists, lab technicians and so forth
3. Rag pickers, personnel processors, transport personnel (ICRC, 2011).

Emissions produced during the treatment of the biomedical waste in process such as incinerations or open cast burnings often lead to serious disorders of the respiratory system, cancer and so forth. Various toxins released are greenhouse gases, dioxin and various harmful pollutants (Gautam et al., 2010). Thus, biomedical waste needs proper handling techniques for reducing the ill effects to mankind as well as the ecosystem.

5.6 NEED FOR BIOMEDICAL WASTE MANAGEMENT

Biomedical waste management is a series of activities ranging from collection, storage, transportation and treatment of biomedical waste as a whole (Govt. of India, 1998). There is a need for proper and uniform biomedical waste management because of:

- Poor waste control methods leading to nosocomial infections
- Infections or injuries arise due to sharps
- Improper handling of disposables
- Drugs for disposals and repackings
- Defective process of incineration leading to environmental pollution
- Health hazards due to noxious chemicals and drugs and their improper handling (Chandra, 1999).

5.7 MERITS OF THE MANAGEMENT SYSTEM

1. Decline in infections and hospital-based infections
2. Decline in the mortality rate as well as disease rate
3. Decline in infection cost
4. Decline in the rate of occurrence of various occupational and hospital hazards
5. Healthier ecosystem
6. Better quality of life
7. Decline in the overall cost with appropriate waste disposal system (Mathur et al., 2012).

5.8 BIOMEDICAL WASTE MANAGEMENT

Biomedical waste management constitutes the following steps described in Figure 5.3.

5.8.1 BIOMEDICAL WASTE GENERATION

As already described, the biomedical waste produced by health care institutions can be categorized into various types based on the levels of toxicity. The waste generated is derived from various major and minor sources, owing to their consumption or the health care facilities they provide. In general, for an effective waste management system, proper management should be done starting from experts or doctors till

Biomedical Waste Disposal and Treatment

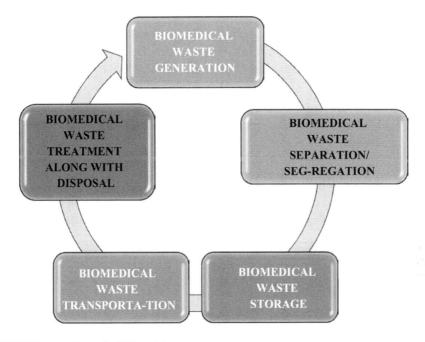

FIGURE 5.3 Biomedical Waste Management

the staff for waste cleaning. Thus, including chemists, nurses, personnel, inspectors, engineers, workers and so forth. The waste generated by each department should be collected and placed at specific site where it undergoes sorting. Liquid waste produced should first be arranged on the basis of reagents, chemicals and so forth, and then channelled into the drain (ISO, 2006; Mastorakis et al., 2011).

5.8.2 Segregation of Biomedical Waste

Waste segregation is the heart of biomedical waste management system in order to run an efficient system. Proper segregation of waste is directly dependent upon the type of treatment required for the waste. In addition, this will lead to proper storage, transport and disposal method (Defra.gov.uk, 2012). Waste segregation also help in reduction of waste volume for the waste management system (Mishra et al., 2016). 5.2 represents the waste subjected to different colours along with its treatment (Kharat, 2016).

5.8.3 Storage of Biomedical Waste

Storage and accumulating the generated biomedical waste are one of the essential key factors in the management system. The site for the storage of the biomedical waste is located somewhere between the site of waste generated and the site for the treatment or disposal of the biomedical waste. Therefore, in regard to this, any off-site close to the treatment is used. Storage of the biomedical waste is done in suitable

TABLE 5.2
The Waste Subjected to Different Colours along with Its Treatment (Kharat, 2016)

S. No.	Bag Colour	Utilization	Mode of Treatment
1.	Black	Solid chemical waste, ash produced by incinerators	Incineration
2.	White	Sharp waste: needles, blades, syringes, scalpels	Sterilization (dry heat), autoclave
3.	Red	Recycled waste: gloves, IV tubes, gloves, catheters	Microwave, autoclave
4.	Yellow	Human waste including body fluids, liquid chemical waste, blood contaminated dressings or beddings	Incinerators, pyrolysis
5.	Blue	Vials, ampoules	Autoclave/hydroclave

large variable containers with suitable labelling. The containers used must be non-reactive, and the waste as well as should be placed in such a way that they can be approached easily. In case of sharps, the type of biomedical waste the containers used must possess puncture-proof ability in regard to the spread of infections. It should be taken care that the biomedical waste should not be stored for more than 8–10 hours in case of large hospitals while more than 24 hours in case of nursing cares. Proper labelling and caution statement are essential on each storage bin used (Singh et al., 2014). Moreover, the storage floor should be timely followed with clean-up procedures in order to prevent spillage or infection caused by them. Disinfection at a regular interval is an essential factor (Silva et al., 2005).

5.8.4 Transportation of Biomedical Waste

Biomedical waste management has an objective of easy and safe transportation of the waste generated in order to avoid any health hazard to the environment as a whole. Only trained personnel with high experience should be employed for handling bio-medical waste during transportation. Transportation is a significant activity in the management system for the waste which is untreated at a specific site (Kautto and Melanen, 2004; Marinkovic et al., 2005). Transportation of the waste done by vehicle should be leakage proof and secured for the door opening. In addition, the interior should be combustion proof (Mastorakis, 2011). Also, complete information should be displayed for the transported and a hazard sign should be placed on the vehicle.

5.8.5 Medical Waste Treatment and Disposal Treatment Processes

The treatment process of biomedical waste involves Figure 5.4.

1. *Thermal process*: This treatment process of waste management involves the utilization of heat for the purpose of disinfection. They be categorized into:
   a. High heat systems
   b. Low heat systems—temperatures between 90–180°C.

Biomedical Waste Disposal and Treatment

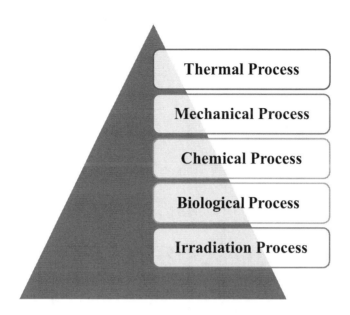

FIGURE 5.4 The Various Treatment Processes Involved in Waste Management

2. *Mechanical process*: In this process of waste treatment mechanical stress is used for the transformation of the waste into another for or to transform its characteristics of the biomedical waste. Two process involved are:
   a. *Shred*: Shredding refers to the destruction of plastic or paper in order to counter its reutilization.
   b. Compact: Compaction is done for the minimization of the waste or for the reduction of its volume.
3. *Chemical process*: Biomedical waste undergoes the series of the processes where chemicals are used for the disinfection. Various chemical used are hydrogen peroxide, ozone, chlorine dioxide and so forth.
4. *Biological process*: This process involves the utilization of enzymes for the biomedical waste treatment. It has been reported that the complete destruction of the constituents biologically will render the risk of contamination leaving behind the glass or plastic residues.
5. *Irradiation process*: The biomedical process of irradiation involves the use of ionizing radiation or UV rays in a vacuum chamber for the shredding of the waste (Chandra, 1999; Onursal, 2003; gov.uk).

5.9 VARIOUS METHODS EMPLOYED FOR THE DISPOSAL OF BIOMEDICAL WASTE

5.9.1 Moist Heat Sterilization—Autoclave

This method of sterilization is often referred to as steam sterilization carried out under low pressure. Autoclaving is defined as a process of low heat where the steam comes in direct contact with material for a specific time in order to disinfect. The waste which undergoes this type of treatment are sharps, biotechnology waste, microbiological waste and so forth. They have been categorized into three types:

1. *Pre-vacuum type*: In this type of sterilization vacuum pumps are utilized for the evacuation of air from the pre-vacuum system. The span reduces to about 30–60 minutes at 132°C.
2. *Gravity type*: In this type of sterilization gravity helps in the evacuation of air. It works at 121°C and about 15 psi pressure for the time period of 60–90 minutes.
3. *Retort type*: Retort autoclaves helps in conducting large volumes at higher pressure and temperature (OTA, 1990; Sah, 2007).

5.9.2 METHOD OF INCINERATION

The thermal process of incineration is carried out under high temperature for the conversion of materials into an inert form with the release of gases. Biomedical waste management involves three types of incinerators: rotary kiln, multiple hearth and controlled air. An optimum level of combustion is achieved by the primary as well as secondary chambers in these types of incinerators (Gravers, 1998). The major objective of incineration is the reduction of organic waste into inorganic waste with reduction of volume as well as weight. One of the demerits of incineration is that it requires large capital cost (Pathak, 1998; WHO, 2004).

5.9.3 MICROWAVE TREATMENT

This treatment of biomedical waste management is a wet thermal treatment required for the disinfection. Irrespective of the other thermal treatments. Microwave treatment concentrates the heat directly on the waste treatment thus leading to a higher degree of disinfection. They impart various benefits such as lack of harmful emissions and there is no requirement of chemicals (OTA, 1990; Research, 1988; Sah, 2007).

5.9.4 HYDROCLAVE PROCESS TREATMENT

Like the autoclave process, the hydroclave is a modified equipment of autoclave. In this process about 35–36 psi pressure is used and 132°C for 20 minutes is carried out. The waste in the hydroclave is subjected to various cycles such as start-up, heat up, sterilization, venting and dehydration for a time period of about 50 minutes. After this, the waste undergoes shredding before the disposal of the waste. This technology imparts various advantages such as lack of liquid waste, air emissions and chemicals (Sah, 2007; Dumitrescu, 2007).

5.9.5 LANDFILLING PROCESS

Landfilling is one of the process for the minimization of waste with least impact on the environment. The waste which is treated is disposed in the landfill. Along

Biomedical Waste Disposal and Treatment

with non-incineration methods, the landfill process is a low-cost method of disposal treatment process, though it increases the risk of environmental and human health hazard. Various types of leachates are solid leachates, liquid landfills and gas land-fills. Hence, landfills are not an optimum method for the management of biomedical waste (Butt et al., 2008; Özkan, 2013).

5.10 CONCLUSION

Biomedical waste management has been an elemental factor in the prevention and treatment of infections in an efficient manner. For a sound quality biomedical waste management proper process of waste generation, segregation, storage, disposal and transportation is required. Segregation is the heart of BMW management. Adequate training and health education must be provided for the proper handling of the bio-medical use. Development of the effective system required with the proper implementation of rules and regulations will eventually lead to a reduction in health hazards for the environment and human health. To ensure balance in the ecosystem and biological biodiversity, biomedical waste management is an essential feature (Chakraborty, 2013).

REFERENCES

Acharya, A., Gokhale, V. A. and Joshi, D., 2014. Impact of biomedical waste on city environment: Case study of Pune, India. *Journal of Applied Chemistry.* 6(6):21–27.

Amin, R., Gul, R. and Mehrab, A., 2013. Hospital waste management. *Professional Medical Journal.* 20(6):988–994.

Basel Convention, 1989. Available online at www.basel.int/SearchResults/tabid/835/Default. aspx?Search=waste (Accessed on 2017).

Butt, T. E., Lockley, E. and Oduyemi, K. O., 2008. Risk assessment of landfill disposal sites— State of the art. *Waste Management.* 28(6):952–964.

CEET, 2008. *Biomedical Waste Management Burgeoning Issue.* Centre of Excellence in Environmental Toxicology.

Chakraborty, S., Veeregowda, B., Gowda, L., Sannegowda, S. N., Tiwari, R., Dhama, K. and Singh, S. V., 2013. Biomedical waste management. *Interaction.* 12(2):592–604.

Chandra, H., 1999. Hospital waste: An environmental hazard and its management. *C. Newsletter of ISEB*, India. 5(3):80–85.

Chitnis, V., Vaidya, K. and Chitnis, D. S., 2005. Biomedical waste in laboratory medicine: Audit and management. *Indian Journal of Medical Microbiology.* 23(1):6–13.

Defra, 2012. Waste segregation and national colour coding approach. Available waste segregation and National colour coding approach. Available online at www.defra.gov.uk/environment/waste/special/index.htm

Diaz, L. F. and Savage, G. M., 2003. Risks and costs associated with the management of infectious wastes. Available online at www.wpro.who.int/environmental_health/documents/docs/LFDRiskassessmentDec03Final.pdf

Dohare, S., Garg, V. K. and Sarkar, B. K., 2013. A study of hospital waste management status in health facilities of an Urban area. *International Journal of Pharma and Bio Sciences.* 4(1(B)):1107–1112.

Dumitrescu, A., Vacarel, M. and Qaramah, A., 2007. Waste management resulting from active medical devices. *Journal of Environmental Protection Ecology,* 116–119.

Forget, G. and Lebel, J., 2001. Ecosystem approach to human health. *International Journal of Occupational and Environmental Health, Supplement.* 7(2).

Gautam, V., Thapar, R. and Sharma, M., 2010. Biomedical waste management: Incineration vs. environmental safety. *Indian Journal of Medical Microbiology.* 28(3):191.

Govt. of India, Ministry of Environment and Forests Gazette notification No. 460, New Delhi: 1998. 10–20. Available online at www.moef.nic.in/legis/hsm/biomed.html

Gravers, P. D., 1998. Management of hospital wastes-an overview. *Proceedings of National Workshop on Management of Hospital Waste.* 8(4):276–282.

Hegde, V., Kulkarni, R. D. and Ajantha, G. S., 2007. Biomedical waste management. *Journal of Oral and Maxillofacial Pathology.* 11(1):5–9. Available online at www.gov.uk/healthcare-waste

Hussain, A., Gupta, S. and Kohli, K. S., 2010. Biomedical waste management in India: A Review. *International Journal of Advance Research in Science and Engineering.* 7(1):79–94.

ICRC, 2011. Medical waste management, Geneva, Switzerland. Available online at https://shop.icrc.org/medical-waste-management-1256.html

Indupalli, A. S., Motakpalli, K., Giri, P. A. and Ahmed, B. N., 2015. Knowledge, attitude and practices regarding biomedical waste management amongst nursing staff of Khaja Banda Nawaz institute of medical sciences, Kalburgi, Karnataka. *National Journal Community Medicine.* 6(4):562–565.

ISO 10014, 2006. Quality management—guidelines for realizing financial and economic benefits. Available online at www.iso.org/obp/ui/#iso:std:iso:10014:ed-1:v1:en

Kalpana, V. N., Prabhu, D. S., Vinodhini, S. and Devirajeswari, V., 2016. Biomedical waste and its management. *Journal of Chemical and Pharmaceutical Research.* 8(4):670–676.

Kautto, P. and Melanen, M., 2004. How does industry respond to waste policy instruments—Finnish experiences. *Journal of Cleaner Production.* 12(1):1–11.

Khan, S., Syed, A. T., Ahmad, R., Rather, T. A., Ajaz, M. and Jan, F. A., 2010. Radioactive waste management in a hospital. *International Journal of Health Sciences.* 4(1):39.

Kharat, D. S., 2016. Biomedical waste management rules. A review. *International of Advanced Research and Development.* 1:48–51.

Kumar, S. M., 2011. Hospital waste management and environmental problems in India. *International Journal of Pharmaceutical & Biological Archive.* 2(6).

Mandal, S. K. and Dutta, J., 2009. Integrated bio-medical waste management plan for Patna city. Institute of Town Planners. *India Journal.* 6(2):1–25.

Manyele, S. V. and Tanzania, V., 2004. Effects of improper hospital-waste management on occupational health and safety. *African Newsletter on Occupational Health and Safety.* 14(2):30–33.

Marinkovic, N., Vitale, K., Afric, I. and Janev Holcer, N., 2005. Hazardous medical waste management as a public health issue. *Arhiv za higijenurada I toksikologiju.* 56(1):21–32.

Mastorakis, N. E., Bulucea, C. A., Oprea, T. A., Bulucea, C. A. and Dondon, P., 2011. Holistic approach of biomedical waste management system with regard to health and environmental risks. *Development, Energy, Environment, Economics.* 5(3):287–295.

Mathur, P., Patan, S. and Shobhawat, S., 2012. Need of biomedical waste management system in hospitals - An emerging issue - A review. *Current World Environment.* 7(1):117–124.

Mishra, K., Sharma, A. and Sarita, A. S., 2016. A study: Biomedical waste management in India, IOSR. *Journal of Environment Science, Technology and Food Technology.* 10(5):64–67.

Mishra, S., Sahu, C. and Mahananda, M. R., 2015. Study on Knowledge, Attitude and Practices (KAP) of bio-medical waste management at Veer Surendra Sai Institute of Medical Science and Research (VIMSAR), Burla. *International Journal of Emerging Research in Management &Technology.* 4(11):2278–9359.

Odumosu, B. T., 2016. Biomedical waste: Its effects and safe disposal. In *Environmental Waste Management*, CRC Press. 81–93.

Onursal, B., 2003. Health care waste management in India. *Lessons from Experience.* 1–2.

Biomedical Waste Disposal and Treatment

OTA, 1990. Special report on medical waste treatment methods, finding the Rx for managing medical wastes, NTIS order #PB91–106203, 1990. Available online at www.princeton.edu/~ota/disk2/1990/9018/9018.PDF

Özkan, A., 2013. Evaluation of healthcare waste treatment/disposal alternatives by using multi-criteria decision-making techniques. *Waste Management & Research.* 31(2):141–149.

PAHO, 1994. *Hazardous Waste and Health in Latin America and the Caribbean*, Pan American Health Organization, Washington, DC.

Pathak, S., 1998. Management of hospital waste: A Jaipur Scenario. *Proceedings of National Workshop on Management and Hospital Waste.* 1:31–33.

Patwary, M. A., O'Hare, W. T., Street, G., Elahi, K. M., Hossain, S. S. and Sarke, M. H., 2009. Country report: quantitative assessment of medical waste generation in the capital city of Bangladesh. *Waste Management.* 29:2392–2397.

Research Triangle Institute, 1988. Review and evaluation of existing literature on generation. Management and Potential Health Effects of Medical Waste, Health Assessment Section, Technical Assessment Branch, Office of Solid Waste, U.S. Environmental Protection Agency.

Rudraswamy, S., Sampath, N., Doggalli, N. and Rudraswam, D. S., 2013. Global scenario of hospital waste management. *International Journal of Environmental Biology.* 3:143–146.

Sah, R. C., 2007. Bio-medical waste management practice and POPs in Kathmandu, Nepal. Center for Public Health and Environmental Development of Kathmandu, Nepal. Available online at www.noharm.org/details.cfm

Semwal, R., 2016. Dharmendra. Environmental concern and threat investigation due to malpractices in biomedical waste management: A review. In *Proceedings of IRF International Conference*, 28 February.

Sharma, A. K., 1998. *Bio Medical Waste (Management and Handling) Rules*, Suvidha Law House, Bhopal.

Silva, C. E., Hoppe, A. E., Ravanello, M. M. and Mello, N., 2005. Medical wastes management in the south of Brazil. *Waste Management.* 25(6):600–605.

Singh, H., Rehman, R. and Bumb, S. S., 2014. Management of biomedical waste: A review. *International Journal of Dental and Medical Research.* 1(1):14–20.

Singh, S., Sahana, S., Anuradha, P., Narayan, M. and Agarwal, S., 2017. Decoding the coded, an overview of -bio medical waste management. *International Journal of Recent Scientific Research.* 8(7):18066–18073.

WHO, 2004. *Safe Health-Care Waste Management: A Policy Paper*, WHO, Geneva.

WHO, 2015. *Water, Sanitation and Hygiene in Health Care Facilities: Status in Low- and Middle-Income Countries*, World Health Organization, Geneva.

6 Best Practices in Construction and Demolition Waste Management

Ashok K. Rathoure and Hani Patel

CONTENTS

6.1 Introduction 92
6.2 Construction and Demolition (C&D) Waste 92
    6.2.1 Types of Construction and Demolition (C&D) Waste Generators 92
    6.2.2 Types of Construction and Demolition (C&D) Waste 93
    6.2.3 Contributors of Construction and Demolition (C&D) Waste 93
    6.2.4 Various Components of Construction and Demolition (C&D) Waste 93
    6.2.5 Typical Composition of Indian Construction and Demolition
        (C&D) Waste 94
6.3 Construction and Demolition (C&D) Waste Management 95
    6.3.1 Collection of Construction and Demolition (C&D) Waste 95
    6.3.2 Concept of 3R for Waste Management 96
    6.3.3 Storage of Construction and Demolition (C&D) Waste 96
    6.3.4 Segregation of Construction and Demolition (C&D) Waste 97
    6.3.5 Collection and Transportation of Construction and Demolition
        (C&D) Waste 97
    6.3.6 Reduce of Construction and Demolition (C&D) Waste 97
    6.3.7 Reuse of Construction and Demolition (C&D) Waste 98
    6.3.8 Recycling of Construction and Demolition (C&D) Waste 99
        6.3.8.1 Mobile Plant 99
        6.3.8.2 Semi-Mobile Plant 99
        6.3.8.3 Stationary Plants 99
    6.3.9 Applications of Construction and Demolition (C&D) Waste 100
    6.3.10 Disposal of Construction and Demolition (C&D) Waste 100
6.4 Construction and Demolition (C&D) Waste Handling 101
6.5 Estimation of Construction and Demolition (C&D) Waste Generation
    for India 101
6.6 Construction and Demolition (C&D) Waste—An Overview of
    Construction Industry in India 102

6.7 Uncertainty in Quantum of Generation of Construction and Demolition (C&D) Wastes from MoUD (Ministry of Urban Development).................. 103
6.8 Conclusion 103
References 104

6.1 INTRODUCTION

As per Construction and Demolition (C&D) Waste Management Rules 2016, construction means the process of erecting of building or built facility or other structure, or building of infrastructure including alteration. As per Construction and Demolition Waste Management Rules 2016, demolition means breaking down or tearing down buildings and other structures either manually or using mechanical force (by various equipment) or by implosion using explosives. As per Construction and Demolition Waste Management Rules 2016, construction and demolition waste means the waste comprising building materials, debris and rubble resulting from construction, remodelling, repair and demolition of any civil structure. Waste is generated at different stages of the construction process. Waste during construction activity relates to excessive cement mix or concrete left after work is over, rejection/demolition caused due to change in design or wrong workmanship and so forth. Construction waste is bulky, heavy and mostly unsuitable for disposal by incineration or composting (Patel et al., 2014).

In India, recent year construction and demolished concrete waste handling and management is a difficult drawback long faced by many areas of the country. It is very challenging and hectic problem that has to be tackled in an indigenous manner because of strict environmental laws and lack of dumping sites in urban areas, construction and demolished waste disposal is a great problem. It is necessary to completely recycle construction and demolished concrete waste in order to protect natural resources, environment and reduce environmental pollution (Singh and Kumar, 2014). The growing population within the country and requirement of land for alternative uses has reduced the availability of land for waste disposal. Re-utilization or recycling is an important strategy for management of such waste (Patel et al., 2014).

6.2 CONSTRUCTION AND DEMOLITION (C&D) WASTE

6.2.1 Types of Construction and Demolition (C&D) Waste Generators

1. Large Amount of C&D Waste Generators
   - Airports
   - Runway construction
   - Government infrastructure projects (e.g. roads, highways, bridges, flyovers, government buildings)
   - Flats
   - Residential projects
   - Parks
   - Malls
   - Industries
2. Small amount of C&D waste generators
   - Houses and small buildings.

Best Practices in C&D Waste Management

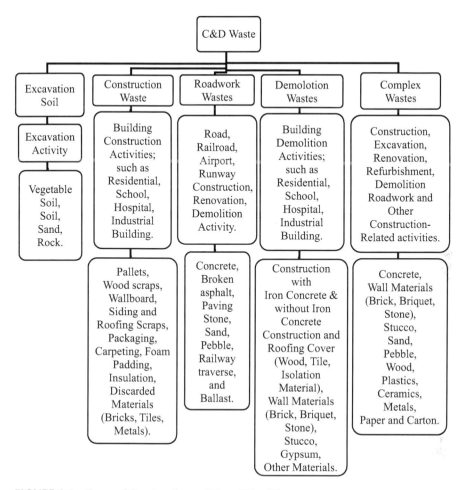

FIGURE 6.1 Types of Construction and Demolition Waste

6.2.2 Types of Construction and Demolition (C&D) Waste

The construction and demolition waste are generated from different activities such as excavation, construction of new projects and demolition/renovation of old buildings. Some types of C&D waste as per their source of generation is shown in Figure 6.1.

6.2.3 Contributors of Construction and Demolition (C&D) Waste

Contributors to C&D waste are shown in Figure 6.2.

6.2.4 Various Components of Construction and Demolition (C&D) Waste

Various compounds of C&D waste as per Patel et al. (2014).

1. Major components
   - Cement concrete
   - Bricks

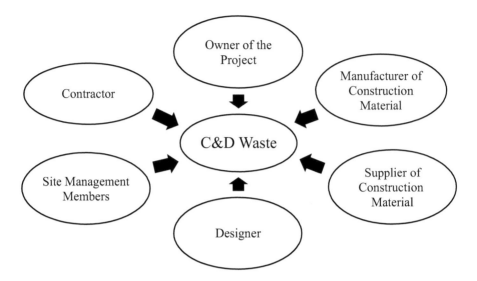

FIGURE 6.2 Contributors of Construction and Demolition Waste

- Cement plaster
- Steel (from RCC, door/window frames, roofing support, railings of staircase etc.)
- Rubble
- Stone/timber/wood (marble, granite, sandstone).
2. Minor components
    - Conduits (iron, plastic)
    - Pipes (galvanized iron, iron, plastic)
    - Electrical fixtures (copper/aluminium wiring, wooden baton, Bakelite/plastic switches, wire insulation)
    - Panels (wooden, laminated)
    - Others (glazed tiles, glass panes).

6.2.5 Typical Composition of Indian Construction and Demolition (C&D) Waste

As per CPCB (2017), the composition of C&D waste can vary depending on the age of building being demolished/renovated or the type of buildings being constructed. C&D waste generation figures for any region varies as it depends largely on the type and nature of construction/demolition project activities which may be regional/site/project specific. Under Rule 4(3) the segregation by bulk C&D waste generators shall be done into four streams such as:

1. Concrete
2. Soil
3. Steel, wood and plastics
4. Bricks and mortar.

Best Practices in C&D Waste Management

Demolition waste characteristics as per CPCB (2017): In India, when old buildings are demolished the major demolition waste is soil, sand and gravel accounting for bricks (26%) and masonry (32%), concretes (28%), metal (6%), wood (3%) others (5%). Bricks, tiles, woods and iron metal are sold for reuse/recycling.

Excavations, concrete, masonry and wood together constitute over 90% of all C&D waste.

The typical composition of Indian C&D waste as per CPCB (2017): The major constituents are concrete, soil, bricks, wood, asphalt and metal. Brick and masonry, soil, sand and gravel account for over 60% of total waste. The percentages of major constituents in C&D Waste in shown in Table 6.1.

6.3 CONSTRUCTION AND DEMOLITION (C&D) WASTE MANAGEMENT

6.3.1 COLLECTION OF CONSTRUCTION AND DEMOLITION (C&D) WASTE

As per Patel et al. (2014) collection of construction and demolition (C&D) waste.

- Collection of C&D waste can be done by trucks having containers of different sizes.
- The size of the container depends upon the demolition area/part.
- For handling very large volumes, front-end loaders in combination with sturdy tipper trucks may be used so that the time taken for loading and unloading is kept to a minimum.
- For small generators of construction debris (e.g. petty repair/maintenance job), there may be two options:
  1. Specific places for such dumping by the local body
  2. Removal on payment basis.
- In case of small towns where skips and tipping trailers are not available, manual loading and unloading should be permitted.
- In case of large towns where C&D waste generates in large amount, zoning of the towns is necessary. By multiple pickup points of C&D waste we can easily do collection of C&D waste in large cities.

TABLE 6.1
Typical Composition of Indian C&D Waste (CPCB, 2017)

Material	Composition
Soil, Sand and Gravel	36%
Brick and Masonry	31%
Concrete	23%
Metals	5%
Bitumen	2%
Wood	2%
Others	1%

- Close co-ordination between the sanitary department, municipal engineering department and town planning department is essential if there is no consolidated solid waste management department to take care of the construction and demolition waste in addition to other municipal garbage.

6.3.2 Concept of 3R for Waste Management

The conception of 3R refers to reduce, reuse and recycle, largely within the context of production and consumption which is well known today. It is something like using recyclable materials in actual practice, reusing of raw materials if possible and reducing use of resources and energy. These are typically applied to the hole life cycles of product and services—starting from design and extraction of raw materials to transports, manufacture, use, dismantling and disposal. The waste management hierarchy within the 3R concept is shown in Figure 6.3 (Gayakwad and Sasane, 2015).

6.3.3 Storage of Construction and Demolition (C&D) Waste

These wastes are best stored at the source of generation. If they're scattered around or thrown on the road, they not only cause obstacle to traffic but also add to the workload of the local body. All attempts should be made to stick to the following measures as per CPCB (2017):

- All construction/demolition waste should be stored within the site itself. A proper screen should be provided so that the waste doesn't get scattered and doesn't become an eyesore.
- Attempts should be made to stay the waste segregated into different heaps as far as possible so that their further gradation and recycle is facilitated.
- Material which can be reused at the same site for the purpose of construction, levelling, making road/pavement and so forth should also be kept in separate heaps from those which are to be sold or landfilled.
- For large projects involving construction of bridges, flyovers, subways and so forth, special provision should be made for storage of waste material.

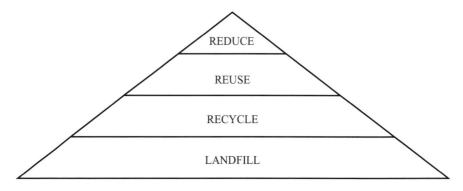

FIGURE 6.3 Waste Management Hierarchies

Best Practices in C&D Waste Management

6.3.4 SEGREGATION OF CONSTRUCTION AND DEMOLITION (C&D) WASTE

Segregation can be carried out at source during construction and demolition activities or can be achieved by processing the mixed material to remove the foreign materials. Segregation at source is most efficient in terms of energy utilization, economics and time. Gross segregation of construction and demolition wastes into road work materials, structural building materials, salvaged building parts and site clearance waste is necessary. Additional segregation is required to facilitate reuse/recycling of materials like wood, glass, cabling, plastic, plaster board and so on before demolition in order to produce recycled aggregate that will meet the specification (THRI, 2014).

6.3.5 COLLECTION AND TRANSPORTATION OF CONSTRUCTION AND DEMOLITION (C&D) WASTE

Construction and demolition debris is stored in skips. Then skip lifters fitted with hydraulic hoist system are used for efficient and prompt removal. In case, trailers are used, then tractors may remove these. For handling very large volumes, front-end loaders in combination with sturdy tipper trucks may be used so that the time taken for loading and unloading is kept to the minimum (THRI, 2014).

6.3.6 REDUCE OF CONSTRUCTION AND DEMOLITION (C&D) WASTE

As per Patel et al. (2017) during the C&D planning stage there are several activities that might avoid waste generation, which can include:

1. 3R concept should be taken for reducing the C&D waste.
2. The waste management plan should be applied in each and every stage of construction which can reduce the C&D waste generation.
3. Charges should be applied on C&D waste generators and for dispose of waste.
4. The municipal by-laws need to be harshly applied and modified.

In construction planning stage, several activities that might avoid waste generation as per Couto and Couto (2002):

- Co-ordination between designers and construction companies should be attended in the definition of materials and construction products.
- Promoting adequate communication among owners, project designers and contractors. Lack of communication is commonly the cause of partial demolition and removal of applied material, contributing towards needless output of debris.
- Arrival of materials and products should be planned, according to available place on site and to production flow, to avoid excessive stocks and deterioration of goods and packs.
- Stockpiles of sand, gravel, soil and other material should be situated so that they don't spill and can't be washed onto the adjacent street.
- Label packages of materials as it comes in, and record the date for reception of materials that deteriorate easily, so that the first to come in are used first.

In construction stage several activities that might avoid waste generation as per Couto and Couto (2002):

- Ordering pre-cut, prefabricated materials that are the correct size for the job.
- Reduce packaging by returning to the supplier, or requesting reusable packaging such as cardboard or metal instead of plastic.
- Bulk buy to avoid excess packaging (however ensuring site requirements are not exceeded, avoiding the environmental impact of transportation and excess storage).
- Orders to suppliers of materials should respect sizing needs so that size adjustments can be avoided during construction.
- Keep the site tidy to reduce material losses and waste.
- Promote good practice awareness as part of health and safety induction/ training for workers on-site.
- Reusable shuttering materials with eventual wreck value should be preferred even if investment costs are higher.

6.3.7 Reuse of Construction and Demolition (C&D) Waste

There is a need to understand the reuse and recycle potential of these C&D waste which on one hand will generate potential business opportunity, employment generation and above all environmental sustainability. Useful products like reinforcement, mild steel, doors and windows, structural steel, bricks and other metal items can be taken out easily and again put to reuse without much processing. It is well said that reuse is the most beneficial form of recycling waste products. In developing countries like India and China where there is poverty and massive requirement of low-cost housing, these products can be consumed easily and also reduce cost of construction of affordable houses (Bansal et al., 2016). There are several opportunities for waste reuse as per Couto and Couto (2002) are following:

- Careful demolition can maximize the reuse value of materials, particularly fittings, floorings and timber linings.
- Sort demolition materials and identify the materials that can be reused, and grade accordingly to quality and re-usability.
- Reuse rock, soil and vegetation on site for landscaping.
- Stockpile the materials for removal and reuse off site, ensuring adequate provision for sediment and erosion control (ensuring minimal impact to the aesthetic quality of the surrounding environment).
- Reuse materials from the demolition stage.
- Buy used materials from reclamation yards where possible re-usable shuttering materials with eventual wreck value should be preferred even if investment costs are higher.
- Re-usable shuttering materials with eventual wreck value should be preferred even if investment costs are higher.

Best Practices in C&D Waste Management

6.3.8 Recycling of Construction and Demolition (C&D) Waste

As per THRI (2014), construction and demolition waste are bulky, heavy and mostly unsuitable for disposal by incineration or composting. The growing population and demand of land for different uses has reduced the availability of land for waste disposal. Reutilization is an important strategy for management of such waste. Apart from mounting problems of waste management, other reasons that support adoption of recycling strategy are reduced extraction of raw materials, improved profits and reduced environmental impact. In the present context of increasing waste production and growing public awareness of environmental issues, recycled materials from demolished concrete or masonry can be profitably used in different ways within the building industry.

Concrete appears in two forms in waste. Structural components of building have reinforced concrete, whereas foundations have mass non-reinforced concrete. Excavations produce top soil, clay, sand and gravel. Excavation materials may be either reused as filler at the same site after completion of work, in road construction or in stone, gravel and sand mines, land fill construction, structural fill in low lying areas to help in future development, in garden and landscaping. Concrete and masonry constitute more than 50% of waste generated. It is reused in block/slab form. Recycling of this waste by converting it to aggregate offer dual benefit of saving landfill space and reduction in extraction of natural raw material for new construction industry. Basic methodology of recycling of concrete and masonry waste is to crush the debris to produce a granular product of given particle size. Plants for processing of demolition waste are differentiated based on mobility, type of crusher and process of separation.

As per THRI (2014), three types of recycling plants available are mobile, semi-mobile and stationary plant.

6.3.8.1 Mobile Plant

In the mobile plant, the material is crushed and screened. Ferrous impurities are separated through magnetic separation. The material is transported to the demolition site itself and is suited to process only non-contaminated concrete or masonry waste.

6.3.8.2 Semi-Mobile Plant

In the semi-mobile plant, removal of contaminants is carried out by hand and in the end product is also screened. Magnetic separation for removal of ferrous material is carried out. End product quality is better than that of a mobile plant. Above plants are not capable to process a source of mixed demolishing waste containing foreign matter like metal, wood, plastic and so forth.

6.3.8.3 Stationary Plants

Stationary plants are equipped for carrying out crushing, screening as well as purification to separate the contaminants. Issues necessary to be considered for erection of stationary plants are plant location, road infrastructure, availability of land space, provision of weigh bridge, provision for storage area and so forth. Different types of crushers are used in recycling plants, namely, jaw-crusher, impact crusher and impeller-impact crusher.

6.3.9 Applications of Construction and Demolition (C&D) Waste

Recycled aggregate is used as general bulk fill, sub-base material in road construction, canal lining, playground, fills in drainage projects and for making new concrete to a lesser extent. While using recycled aggregate for filler application, care should be taken that it is free from contaminants to avoid risk of groundwater pollution use of recycled aggregate are sub-base for road construction is widely accepted in most of nations. Bricks and masonry arise as waste during demolition. These are generally mixed with cement, mortar or lime. It is used in the construction of road base and drainage layer, and mechanical soil stabilizers due to its inertness after crushing and separation. Tile material recycling is almost identical to bricks. Tile is usually mixed with brick in the final recycled product. Metal waste is generated during demolition in the form of pipes, light sheet material used in ventilation system, wires and sanitary fittings and as reinforcement in the concrete. Metals are recovered and recycled by remelting. The metals involved on-site separation by manual sorting or magnetic sorting. Aluminium can be recovered without contamination; the material can be directly sold to a recycler. Wood recovered in good condition from beams, window frames, doors, partitions and other fittings is reused. However, wood used in construction is often treated with chemicals to prevent termite infestation and warrants special care during disposal other problems associated to wood waste are inclusion of jointing, nails, screws and fixings. In fact, wood wastes have a high market value for special reuses (furniture, cabinets and floorings). Lower quality waste wood can be recycled/burned for energy recovery. Scrap wood is shredded in-site/in a centralized plant. Shredded wood is magnetically sorted for scrap metal. Wood chips are stored so as to remain dry and can be used as fuel. Also, it's used in the production of various press boards and fibre boards and used for animal bedding. Bituminous material arises from road construction, breaking and digging of roads for services and utilities. Recycling of bituminous material can be carried out by hot or cold mixing techniques either at location or at a central asphalt mixing plant it offers benefits like saving in use of asphalt, saving of energy, reduction in aggregate requirement etc. Other miscellanies materials that arise as waste include glass, plastic, paper and so forth can be recovered and reused (THRI, 2014).

6.3.10 Disposal of Construction and Demolition (C&D) Waste

Being predominantly inert in nature, C&D waste doesn't create chemical or biochemical pollution. Hence maximum effort should be made to reuse and recycle them as indicated earlier. The material can be used for filling/levelling of low-lying areas. Within the industrialized countries, special landfills are sometimes created for inert waste, which are commonly located in abandoned mines and quarries. The same can be attempted in our country also for cities, which are located near open mining quarries or mines where normally sand is used as the filling material. However, proper sampling of the material for its physical and chemical characteristics should be done for evaluating its use under the given circumstances (CPCB, 2017).

Best Practices in C&D Waste Management

6.4 CONSTRUCTION AND DEMOLITION (C&D) WASTE HANDLING

In India, contractors play an important role in waste management. Contractual arrangement require that demolition wastes have to be disposed of by the contractor at his cost. Other than new construction, renovation or repair of buildings, demolition of an existing building/structure is the main cause of waste generation from the construction industry. In India, services of demolition contractors are taken when an old building is to be demolished due to deterioration of the building or to make way for construction of a new building. According to the Technology Information, Forecasting and Assessment Council (TIFAC) study (Shrivastava and Chini, 2010):

- Items recovered during demolition are sold in the market at a discount with respect to price of new material.
- Items that cannot be reused are disposed to landfill site.
- Some municipal corporations allow C&D waste in their landfills, while others want to minimize it to prolong useful life of landfill sites.
- Different constituents of waste are not segregated prior to disposal.
- Builders/ owners bear the cost of transportation, which at present, ranges between USD 6 and USD 13 per truckload depending on the distance of demolition site from landfill area.
- Municipal authorities incur cost of USD 1.50 to USD 2 per tonnes of waste, but presently no charge is levied by them on the owner or builder
- Though directives exist for disposal of waste to landfill areas, presently penal action against violators is practically not taken.

6.5 ESTIMATION OF CONSTRUCTION AND DEMOLITION (C&D) WASTE GENERATION FOR INDIA

Various methods have been employed to quantify the C&D waste generation at both regional and project levels. TIFAC has developed some estimations on C&D waste generation which recognizes that the generation is project specific as follows:

1. Range 40–60 kg per sq. metre of new construction
2. Range 40–50 kg per sq. metre of building repair
3. Range 300–500 kg per sq. metre for demolition of buildings.

From the above, it may be noted that the highest waste generation comes from demolition of buildings. C&D waste generation figures for any region fluctuate as it depends largely on the type and nature of construction/demolition activities of the project concerned. Various approaches for estimation of C&D waste generation in literature include the following:

1. The following five categories of existing C&D waste quantification methodologies are reported:
   a. Site visit method
   b. Waste generation rate method

c. Lifetime analysis method
d. Classification system accumulation method
e. Variables modelling method

Approach to estimate C&D waste is through materials flow analysis is embedded in above methods (CPCB, 2017).

2. Estimation of C&D waste generation based on per capita multipliers or waste generation rate model (CPCB, 2017).

6.6 CONSTRUCTION AND DEMOLITION (C&D) WASTE—AN OVERVIEW OF CONSTRUCTION INDUSTRY IN INDIA

Scientific processing and utilization of C&D waste has achieved isolated successes in India. Delhi was the first city to implement a C&D waste management plan through a pilot processing facility developed under a public-private partnership (PPP) in 2010. After the initial success of the pilot plant processing waste at 500 tonnes per day (TPD), the capacity of the plant has been increased to 2,000 TPD. To minimize transportation distances and associated costs, Delhi planned to have a distributed network of processing facilities in different zones of the city. Accordingly, two more smaller (500 TPD and 150 TPD, respectively) plants have recently come online (2017–2018), with planning for more under way. Ahmedabad was the second city in India to implement C&D waste processing, by adopting a similar PPP model as that in Delhi. A 300 TPD processing facility was launched in 2014, the capacity of which was increased to 600 TPD 2016 after successful operation and now to 1,000 TPD in 2018 (Ministry of Housing Affairs, 2018).

In both Delhi and Ahmedabad, the Design Build Operate Finance and Transfer (DBOFT) model is being followed. The municipal corporation contracts a private party and this authorized agency is responsible for both transportation and processing of the C&D waste and develops the necessary infrastructure with its own financing. The municipal corporation offers land to the contracted party for establishing the processing facility and also designates a series of intermediate collection points at favourable locations throughout the city. The authorized agency collects C&D waste from these designated collection points as well as from unauthorized dumps, as directed by the urban authority, and transports it to the processing facility. The municipal corporation pays the authorized agency an agreed fee per tonne of waste that is collected and transported. The authorized agency may also collect fees directly from large generators (such as Metro Rail) for waste collection; however, if generators bring waste to the processing facility at their own expense, the agency accept it without charge. Therefore, the private partner has two sources of revenue—the tipping fee from the ULB and the sale of recycled products made from C&D waste. This ensures the viability of the enterprise. However, in both Delhi and Ahmedabad, market uptake of recycled products made from C&D waste remains an ongoing challenge (Ministry of Housing & Urban Affairs, 2018).

6.7 UNCERTAINTY IN QUANTUM OF GENERATION OF CONSTRUCTION AND DEMOLITION (C&D) WASTES FROM MOUD (MINISTRY OF URBAN DEVELOPMENT)

Uncertainties in estimating the quantum of C&D waste generation can be attributed to several reasons like different methods adopted to estimate quantum of C&D waste generated, varying pace of developmental activities in cities, redevelopment of cities due to rapid urbanization wherein demolition activities become necessary. Some estimations of C&D wastes are provided below:

From MoUD:

1. 10 MT–15 MT (MT = million tonnes) per year by MoUD (2000).
2. Approximately 25–30 million tonnes of C&D wastes are generated annually in India, of which 5% is processed.
3. The amount of C&D wastes in India has been estimated to be 10–12 million tonnes annually and the proportion of concrete estimated as 23% to 35% of total waste. Considering 30% of C&D wastes of 12 million tonnes as concrete, and 50% of the concrete as coarse aggregate, the total available recycled concrete aggregate (RCA) in India is of the order of 1.8 million tonnes annually.

The quantum of generation of C&D waste estimates available from other sources are summarized below:

1. 12 MT–15 MT by TIFAC (2001)
2. 10 MT–12 MT by MoEF (2010)
3. 12 MT by CPCB
4. 165 MT–175 MT per annum between 2005–2013 (BMTC)

Forecast estimates: Presently, C&D waste generation in India accounts up to 23.75 million tonnes annually and these figures are likely to double up to 2016 (CPCB, 2017).

6.8 CONCLUSION

Construction waste management is required for a country to develop in a sustainable manner. It helps to address issues related to environment, social and economy. Once the root causes of waste generation are notified, it can either be avoided or minimized to benefit the world for better future. The exploitation of potential resources from construction and demolition (C&D) wastes is yet another opportunity and future profession in the construction industry in India. Waste minimization and waste management programs are in its infancy in India. It's possible to reduce the quantity of C&D waste generated by identifying the potential waste early in the design period. C&D waste is used as the coarse aggregate in new concrete. C&D waste concrete may be an alternative to the conventional concrete. C&D waste management system helps to Improve organization's public image. It also improves the market for recycled content products and helps the community meet local and state waste reduction goals.

REFERENCES

Bansal, A., Mishra, G. and Bishnoi, S., 2016. Recycling and reuse of construction and demolition waste: Sustainable approach. 1–7. Available online at www.researchgate.net/publication/314065807_Recycling_and_Reuse_of_Construction_and_Demolition_waste_sustainable_approach

Couto, A. and Couto, J. P., 2002. Process management, guidelines to improve construction and demolition waste management in Portugal. 285–308. Available online at http://cdn.intechweb.org/pdfs/9673.pdf

CPCB, 2017. Guidelines on environmental management of Construction & Demolition (C&D) wastes. 1–72. Available online at http://cpcb.nic.in/openpdffile.php?id=TGF0ZXN0RmlsZS8xNTlfMTQ5NTQ0NjM5N19tZWRpYXBob3RvMTkyLnBkZg==

Gayakwad, H. P. and Sasane, N. B., 2015. Construction & demolition waste management in India. 2(3):712–715. Available online at www.irjet.net/archives/V2/i3/Irjet-v2i392.pdf

Ministry of Housing & Urban Affairs, 2018. Strategy for promoting processing of Construction and Demolition (C&D) waste and utilisation of recycled products. 1–31. Available online at http://niti.gov.in/writereaddata/files/CDW_Strategy_Draft%20Final_011118.pdf

Patel, S., Pansuria, A., Shah, V. and Patel, S., 2014. Construction and demolition waste recycling. *International Journal for Innovative Research in Science & Technology*. 1(7):266–286. Available online at www.ijirst.org/articles/IJIRSTV1I7108.pdf

Patel, V.B., Pitroda, J. and Bhagat, S.S., 2017. Organizing and reducing construction and demolition waste: A review. 5–9. Available online at www.researchgate.net/publication/320145974_Organizing_and_Reducing_Construction_and_Demolition_waste_A_Review

Shrivastava, S. and Chini, A., 2010. Construction materials and C&D waste in India. 72–76. Available online at www.irbnet.de/daten/iconda/CIB14286.pdf

Singh, M. K. and Kumar, D., 2014. Physical properties of construction & demolished waste concrete. *International Journal for Innovative Research in Science & Technology*. 2(8):122–123. Available online at www.ijsrd.com/articles/IJSRDV2I8054.pdf

THRI, 2014. Construction & demolition. Waste to resources: A waste management handbook. 27–29. Available online at http://cbs.teriin.org/pdf/Waste_Management_Handbook.pdf

7 Plastic Waste Management Practices

Savita Sharma and Sharada Mallubhotla

CONTENTS

7.1 Introduction 105
7.2 Different Categories of Plastics 106
7.3 Plastic Waste and Its Effects on Environment 107
7.4 Plastic Waste Sources 107
7.5 Profile of Plastic Industry in India 108
7.6 Practices of Plastic Waste Management Approaches 108
    7.6.1 Plastic Waste Management by Conventional Practices 108
        7.6.1.1 Recycling 108
        7.6.1.2 Landfilling 110
        7.6.1.3 Incineration 110
7.7 Plastic Waste Management by New Technologies 110
    7.7.1 Plastic Coated Bitumen Road 110
    7.7.2 Co-Processing of Plastic in Cement Kilns 111
    7.7.3 Plastic Waste Conversion into Liquid Fuel 111
    7.7.4 Plasma Pyrolysis Technology 111
7.8 Biodegradable or Degradable Plastics 111
7.9 Alternative Ways to Reduce Impact of Plastic Waste 112
7.10 Conclusion 112
References 112

7.1 INTRODUCTION

Plastics are high molecular weight synthetic polymers mostly obtained from petro-chemicals. Due to their various features like opacity, plasticity and low cost, it is pref-erable to use in manufacturing of different items of day-to-day life. Globally around 150 million tons of plastic is produced per year and approximately eight million tons of plastic products are consumed by India every year because of their wide range of applications in industrial, health care, automotive, construction, textiles, electronics and household products (Kumar et al., 2017). The most important utilization of plastic is in packing and in packaging applications; about 40% plastic materials are used worldwide. Due to its versatility, it has been successfully used in effective packag-ing of food items like spices, bread, rice, wheat flour, milk, edible oils and different

106   Zero Waste

pharmaceutical products. Besides its various applications, plastics are also responsible for serious health and environmental issues (Siddiqui and Pandey, 2013).

7.2 DIFFERENT CATEGORIES OF PLASTICS

Plastics are categorized into two different types (Chavan, 2013):

1. *Recyclable plastic or thermoplastic* includes those plastics which became pliable or soft when heated and solidify when cooled. By applying pressure, they can be shaped or molded in any form and retain the respective state on cooling. In India they contribute to 80% of total plastic waste and are recycled. Some of them are described below:

   Polyethylene terephthalate (PET) is tough, strong, clear, heat, fire and chemical resistant general-purpose thermoplastic belonging to the polyester family. They are used in packing of food containers, mineral water, soft drink bottles, films and as fibres for clothing. They are designated as number one recyclable plastic.

   Polyvinyl chloride (PVC) is durable, versatile, economical, fire resistant, and energy saving plastics. Polymerization of vinyl chloride leads to its production and manufactured in flexible plasticized form. PVC is recognized as one of the world's third-largest thermoplastic by volume after polyethylene and polypropylene, used in industries such as building, packaging, agriculture, transport, automobile, medical and construction. They are also used in products ranging from domestic appliances, furniture, piping, films, footwear, wire and cables.

   High-density polyethylene (HDPE) is light weight, strong, corrosion resistant, durable, eco-friendly and long-term reliable plastic. Co-polymerization of ethylene and a small amount of another hydrocarbon leads to its production. It is used for making buckets, mugs, storage bins, films, carrier bags, food containers, pipes and so forth.

   Low-density polyethylene (LDPE) plastics have low density, low melting point, resistance to chemicals and moisture, are cost-effective and have excellent dielectric properties. They are used for making of nursery bags, carrier bags, milk pouches, cable and wire insulation.

   Polypropylene (PP) is chemical resistant, low density, low cost and good processability plastics. They are linear hydrocarbons produced from propene monomers. They are used both as plastic and fibre in consumer goods, furniture market, automotive industries and in industrial applications. PP is the cheapest plastic among all available types.

   Polystyrene (PS) is rigid, has high clarity and a hard, glassy surface. It is produced by polymerization of styrene and used for making circuit boards, sockets, plugs, switch boards, toys, cutlery, foam, ice cream cups, wall tiles, tumblers and dairy containers.

2. *Non-recyclable or thermoset* plastics cannot be softened or remolded when heated. They are resistant to high temperature and remain in a solid

Plastic Waste Management Practices

state once hardened. These types of plastics are cost-effective but are non-recyclable. Some commonly known thermoset plastics are silicones, phenolics, polyesters and epoxies.

7.3 PLASTIC WASTE AND ITS EFFECTS ON ENVIRONMENT

Due to increase in developmental activities, changing lifestyles and increasing population, the rate of solid waste is increasing. In total municipal solid waste (MSW), plastic waste contributes a large proportion and it is estimated that approximately 9% out of the total MSW is in the form of plastic waste. Inconvenient decomposition of plastics has resulted in various environmental issues. A few of them are as follows:

- On burning, plastic leads to emission of toxic gases like chlorine, dioxin, carbon monoxide, furans, hydrochloric acid, styrene, benzene, butadiene, acetaldehyde and nitrides. These harmful gases are responsible for air pollution, which in turn affects biodiversity.
- They have barrier properties, so their inappropriate dumping makes lands infertile and also spoils the beauty of cities.
- They choke drains and pose problems during the rainy season.
- When mixed with garbage, plastics interfere in waste processing facilities and also cause problems in landfill practices.
- They emit harmful gases during the manufacturing process, which leads to serious health issues.
- Plastic recycling industries are also a threat on environment as they are operating in non-confirming areas.

At the end, plastic waste enters into seas and oceans from land by drainage, sewage system or by rivers and resulting in its pollution. It was recorded that pollution in the water bodies near Kerala has effected more than 267 marine species (seabird, sea turtle and marine mammal species) worldwide, also resulting into their extinction (Laist, 1997).

7.4 PLASTIC WASTE SOURCES

Sources of plastic include both durable and lightweight plastic products. Examples of durable plastic products are parts of washing machines, televisions, blenders, computers and furniture. Before discarded as waste, these products must be recycled at least six times and some recycled plastic granules of certain proportion are mixed with virgin plastic in a molding shop to form finished products. Lightweight plastic waste includes single-use plastic products which are generally generated at community centres, apartments, canteens, hotels, household and event venues – carry bags, PET bottles, packing films, milk pouches, cups, plates, disposables and so forth. Some other sources of plastic waste include automobile service centres, hospitals and construction sites (Deshmukh and Borade, 2015).

7.5 PROFILE OF PLASTIC INDUSTRY IN INDIA

In India, the plastic industry is growing at a remarkably fast rate, and maximum growth is observed in the agriculture, sanitation, electronic, industrial and furniture sectors (Chenjen and Hulming, 2008). Plastic packaging is preferable over paper, glass and wooden packaging. For foodstuff packaging, mostly PET and LDPE plastics are used, while PVC and HDPE plastics are widely used in the agriculture, sanitation and furniture sectors. From the current scenario of India, it is suggested that plastic industry has a promising growth in the years to come in the packaging sector.

7.6 PRACTICES OF PLASTIC WASTE MANAGEMENT APPROACHES

Accumulation of plastics (non-biodegradable) in drainage leads to blocking of streams which ultimately leads to unhygienic environmental conditions and serious health problems. Moreover, animals ingest plastic along with their food and it sometimes leads to their death (Balan et al., 2010). Therefore, disposal of plastic waste is a serious threat, and proper waste management strategies must be planned in such a way that various sources of plastic waste generation are taken care of properly. Various practices which were utilized for plastic waste management are categorized broadly into two types—conventional or traditional and new technologies—whose details are summarized in Figure 7.1 and described below.

7.6.1 Plastic Waste Management by Conventional Practices

These practices of waste management include recycling, incineration and landfilling.

7.6.1.1 Recycling

Recycling of plastic waste must be carried out in such a way that it minimizes the level of pollution throughout the process and also help in conservation of energy and

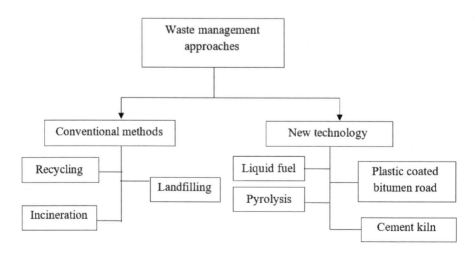

FIGURE 7.1 An Overview of Plastic Waste Management Practices

make the process more efficient. This is done by selecting the right type of motor having minimum capacity, assigning right temperature, optimizing emission of pollutants and converting the by-products from pollutants into reusable articles. India is ranked highest in terms of plastic recycling process (60%) in the world, out of 20% is the world average plastic waste recycling capacity (Oehlmann, 2009). Historically, recycling technologies of plastic waste have been classified into four types—primary, secondary, tertiary and quaternary.

7.6.1.1.1 Primary

The process involves processing of a waste plastic into a product by melting, molding and solidification, wherein the product shows similar characteristic features with those of the original product.

7.6.1.1.2 Secondary

It involves processing or recycling of waste plastic into a product by melting and extrusion. However, the products prepared thereby don't show similar characteristic features with that of the original product.

7.6.1.1.3 Tertiary

This type of recycling involves physical and chemical methods which include thermolysis (catalytic cracking, pyrolysis etc.) and depolymerization (hydrolysis, aminolysis, acidolysis etc.) for production of fuels and basic chemicals from waste plastic.

7.6.1.1.4 Quaternary

In this method, recycling energy is retrieved from plastic by its incineration or burning. Until now this process has not been utilized in India. Before recycling or reprocessing of plastics from municipal solid waste, recyclers must select the suitable waste by segregation method. Segregation is done as per codes stated in the guidelines of Bureau of Indian Standards (BIS) (IS: 14534:1998). Collection of waste plastic is done prior to segregation, either by informal sector (Kabariwala, scrap dealers or waste pickers) or by the formal sector (municipal corporations). The next step after collection and separation is processing. Processing of used plastic waste (post-consumer waste) involves washing, shredding, agglomeration and granulation before recycling, whereas factory waste (pre-consumer waste) should be directly recycled. Post-consumer plastics from furniture, landscaping, construction, soft toys and shipping sectors have a great potential on environment and are largely used for the production of fabrics from them. Fabrics obtained from PET bottles are used for making scarves, denim, pillows and clothing materials (T-shirts). Interestingly, the Indian cricket team jersey is also made out of such recycled plastic (Marar, 2017). In India, many issues are associated with plastic recycling like small and medium-sized recycling enterprises, financial limitation, inappropriate recycling and so forth. Moreover, too much recycling of plastic makes it unsuitable for further reprocessing and some factories also cannot afford pollution control facilities, therefore they must discontinue plastic production and their recycling process (Jose, 2008).

7.6.1.2 Landfilling

Landfill is a specialized area designed for deposition of waste material and it is a traditionally utilized method for this activity. As population size is increasing day by day, so in some countries space has become limited for construction of landfills and areas near such designated sites are becoming less valuable for marketing (Ready, 2005; Akinjare et al., 2011). An advantage of landfill method is that it will restrict the environmental hazard of plastic waste by their proper decomposition, but on the other hand it will also lead to pollution, thus also destroying the ecology of the site (Oehlmann, 2009). Carbon dioxide and methane are the main gases emitted from landfills and these gases are well known for their detrimental environmental issues like global warming, degradation of ecosystem quality and acidification of soils (Damgaard et al., 2011).

7.6.1.3 Incineration

It is clear that land is running out for building more landfills. So the need for plastic waste management through landfilling is reduced by introducing the process of incineration. Incineration involves the burning of waste up to a point at which nothing is left behind except ashes (Hickmann and Lanier, 2003). It is carried out in an incinerator which is built in such a way that the heat produced during combusting remains within the chamber and this heat will further help in complete burning of plastic waste (Innocent et al., 2013). But on the other hand, this process leads to release of many toxic gases like dioxins, polychlorinated biphenyls and furans into the atmosphere and no proper control on reduction of gases is so far available (Gilpin et al., 2003). As a result, this method is not preferably used for waste management and has faced rigorous criticism due to release of toxic components. Moreover, incineration is recognized as a costly method amongst all other conventional methods.

7.7 PLASTIC WASTE MANAGEMENT BY NEW TECHNOLOGIES

New technologies used for plastic waste management include plastic coated bitumen road, co-processing of plastic in cement kilns, plastic waste conversion into liquid fuel and plasma pyrolysis.

7.7.1 Plastic Coated Bitumen Road

Plastics are highly durable and degraded at a very slow rate, so shredded plastic waste is mixed with bituminous and used for flexible road construction. The plastic in the bituminous mixes not only strengthen the roads by correcting various defects like pot holes, ruts, corrugation and so forth; but also increase the durability of roads in an economical way (Manju et al., 2017). Apart from providing stability to roads, this process also helps in eco-friendly waste management by utilizing a higher amount of plastic waste. For constructing such type of roads, first plastic garbage in the form of carry bags, PET bottles and disposable cups are collected. They are then shredded into appropriate sizes and mixed into heated bitumen. Melted mixture of plastic waste and bitumen in the form of an oily coat when laid on surfaces of roads, looks

Plastic Waste Management Practices 111

like normal tar road (Siddiqui and Pandey, 2013). These roads reduce the requirement of bitumen by about 10%, and plastic-coated bitumen is a better source for road construction as compared to plain bitumen (Dhodapkar, 2008).

7.7.2 Co-Processing of Plastic in Cement Kilns

Co-processing involves the use of waste plastic in industrial processes as an alternative fuel or raw material (AFRs). As compared to incineration and land filling methods of waste disposal, co-processing is more eco-friendly and an efficient method. In this method, no residue is left after treatment and it also reduces emission of harmful toxic gases. In various countries, co-processing of waste plastics in cement kilns for management of both hazardous and non-hazardous wastes has been successfully studied. Nowadays, central pollution control boards (CPCB) and state pollution control boards (SPCBs) provide consent for the use of plastic wastes as AFRs in different cement plants (Siddiqui and Pandey, 2013). This technology has emerged as an advantage for cement industry and also for authorities involved in waste management.

7.7.3 Plastic Waste Conversion into Liquid Fuel

It is estimated that non-renewable fuel resources like coal, natural gas and petroleum shall last for around 1,000 years. Increase in their consumption has also increased their prices. So conversion of waste plastic into liquid fuel is a suitable alternative for reducing annual energy demands. For this purpose, a broad range of plastic materials like unsorted, unwashed or those which are difficult to recycle materials can be used (Joshi and Punia, 2013).

7.7.4 Plasma Pyrolysis Technology

Complete destruction of medical and hazardous waste is done safely by plasma pyrolysis. This technology integrates thermochemical properties of plasma with pyrolysis process and the procedure takes place either in negligible or even absence of oxygen. Pyrolysis is done in a shorter period of time and it is suitable for disposal of all types of waste plastics (Dave and Joshi, 2009).

7.8 BIODEGRADABLE OR DEGRADABLE PLASTICS

Plastics produced from plant or animal resources like starch, collagen, cellulose, triglycerides or soy protein polyesters are termed as biodegradable or degradable plastics. These bio-based plastics are very good alternatives for minimizing plastic waste. Their usage on a large scale in different sectors must be promoted as they are helpful in conservation of resources which are non-renewable like coal, petroleum and natural gas (Chenjen and Hulming, 2008). Moreover, these plastics do not require any proper waste management strategy because they get degraded when exposed to air or sun for longer hours. Demand for bio-based products has increased day by day and their acceptance in the commercial market depends upon the response of customers and on their biodegradability percentage. Research on these plastics is continuous in order to spread awareness among consumers regarding their benefits and environmentally friendly nature. In India, bioplastics are greatly utilized in surgical implants, milk sachets, pharmaceuticals, recreational activities, medical services, shopping bags, the food service sector and so forth. But as compared to their synthetic counterparts, the associated cost for their production is high. Various research funding is involved in the development of a strategy for their cost-effective adequate production and supply of biomass feedstock. Besides animal and plant sources, some microalgae species also showed potential for production of degradable plastics (Bhattacharya et al., 2018).

7.9 ALTERNATIVE WAYS TO REDUCE IMPACT OF PLASTIC WASTE

There are number of other ways to mitigate the bad impacts of plastic waste (Chaturvedi, 2000). These are:

- Source reduction by making product more durable so that it can be reused
- Implementation of taxation on plastic bags
- Energy recovery from incineration
- Recycling sector reorganization
- Increasing use of landfilling approaches
- Extended manufacturer duties
- Awareness through education

7.10 CONCLUSION

Owing to the increasing global demand of convenience of plastic products and urbanization, plastic consumption has increased at a much faster rate. Various unscientific practices utilized for waste management has resulted into adverse environmental issues. So proper planning and strategies must be designed for their sustainable utilization and for effective waste management. By using advanced recycling methods and instrumentation, emission of greenhouse gases from recycling units must be reduced and efforts made towards use of more recycled products in place of virgin plastic. Recycling is the best way for their waste management and it will also help in improving environmental conditions and the need of the hour: global warming.

REFERENCES

Akinjare, O. A., Oluwatobi, A. O. and Iroham, O. C., 2011. Impact of sanitary landfills on urban residential property value in Lagos state, Nigeria. *Journal of Sustainable Development.* 4:48–60.

Balan, S., Robert de, S., Mark, G., Stephan, M. W. and Sushmera, M., 2010. Modelling carbon footprints across the supply chain. *International Journal of Production Economics.* 128:43–50.

Bhattacharya, R. R. N. S., Chandrasekhar, K., Deepthi, M. V., Roy, P. and Khan, A., 2018. *Challenges and opportunities: Plastic waste management in India.* The Energy and Resources Institute (TERI), New Delhi. 1–18.

Plastic Waste Management Practices

Chaturvedi, B., 2000. *Polybags: The Enemy Within*, Oxford and IBH Publishing, New Delhi.

Chavan, M.A.J., 2013. Use of plastic waste in flexible pavements. *International Journal of Application or Innovation in Engineering and Management.* 2:540–552.

Chenjen, C. and Hulming, W., 2008. Green component life cycle value on design and reverse manufacturing in semi closed supply chain. *International Journal of Production Economics.* 113:528–545.

Damgaard, A., Manfredi, S., Merrild, H., Steen, S. and Thomas, H.C., 2011. LCA and economic evaluation of landfill leachate and gas technologies. *Waste Management.* 31:1532–1541.

Dave, P.N. and Joshi, A.K., 2009. Plasma pyrolysis and gasification of plastics waste - A review. *Journal of Scientific and Industrial Research.* 69:177–179.

Deshmukh, Y.P. and Borade, A.B., 2015. Plastic waste management-present practice and future possibilities. *International Journal on Recent and Innovation Trends in Computing and Communication.* 3:010–012.

Dhodapkar, A.N., 2008. Use of waste plastic in road construction. *Indian Highways, Technical Paper, Journal.* 31–32.

Gilpin, R., Wagel, D. and Solch, J., 2003. Production, distribution and fate of polychlorinated-dibenzo-*p*-dioxins, dibenzofurans and related organo-halogens in the environment. In *Dioxins and Health*, 2nd ed., John Wiley & Sons.

Hickmann, H. and Lanier, J.R., 2003. *American Alchemy: The History of Solid Waste Management in the United States*, Forester Press.

Innocent, A.J., Chamhuri, S. and Anowar, A.H.B., 2013. Incineration and its implications: The need for a sustainable waste management system in Malaysia. *International Journal of Environmental Sciences.* 4:367–374.

Jose, M.C., 2008. Dynamics of supply chain networks with corporate social responsibility through integrated environmental decision making. *European Journal of Operation Research.* 184:1005–1031.

Joshi, A.R. and Punia, R., 2013. Conversion of plastic wastes into liquid fuels – A review. *Recent Advances in Bioenergy Research.* 3:444–454.

Kumar, S., Smith, S.R., Fowler, G., Velis, C., Kumar, S.J., Shashi, A., Rena, K.R. and Christopher, C., 2017. Challenges and opportunities associated with waste management in India. *Royal Society Open Science.* 4:160–164.

Laist, D.W., 1997. Impacts of debris: Entanglement of marine life in marine debris including a comprehensive list of species with entanglement and ingestion records. In Coe, J.M. and Rogers, D.B. (eds.), *Marine Debris-Sources, Impacts and Solutions*, Springer Verlag, New York. 99–139.

Manju, R., Sathya, S. and Sheema, K., 2017. Use of plastic waste in bituminous pavement. *International Journal of ChemTech Research.* 10:804–811.

Marar, A., 2017. *Recycling Turns Plastic into Pillows, Denims and Team India Gear*, The Indian Express, Pune.

Oehlmann, J., Schulte, O.U., Kloas, W., Jagnytsch, O., Lutz, I., Kusk, K.O., Wollenberger, L. et al., 2009. A critical analysis of the biological impacts of plasticizers on wildlife. *Philosophical Transactions of the Royal Society B: Biological Sciences.* 364:2047–2062.

Ready, R.C., 2005. Do landfills always depress nearby property values? Working paper, The Northeast Regional Center for Rural Development.

Siddiqui, J. and Govind, P., 2013. A review of plastic waste management strategies. *International Research Journal of Environment Sciences.* 2:84–88.

8 Industrial Waste Management System

Rakesh K. Sindhu, Gagandeep Kaur and Arashmeet Kaur

CONTENTS

8.1 Introduction 115
8.2 Industrial Waste: Classification 117
    8.2.1 General Classification.. 117
    8.2.2 Based on the Toxicity of the Waste 117
        8.2.2.1 Characteristic of Hazardous Waste 118
        8.2.2.2 Listed Waste.. 119
        8.2.2.3 Mixed Waste 120
        8.2.2.4 Universal Waste 120
8.3 Characteristics of Industrial Waste... 120
    8.3.1 Physical Characteristics 120
    8.3.2 Chemical Characteristics... 121
8.4 Upshots of Industrial Wastes 122
8.5 Waste Management Approach 122
8.6 Main Objective for the Management of Industrial Waste.......................... 124
    8.6.1 Measures Employed for Controlling the Generation of Waste......... 124
8.7 Hierarchical Relationship of Waste Management 124
8.8 Assortment and Treatment of Industrial Waste.. 125
    8.8.1 Segregation of Waste 125
    8.8.2 Methods for the Waste Treatment... 125
        8.8.2.1 Liquid Industrial Waste.. 125
        8.8.2.2 Solid Industrial Waste.. 126
8.9 Conclusion 128
References... 129

8.1 INTRODUCTION

The onset of industrialization has proved to be a cardinal factor for the development of mankind. Industrial development is of fundamental importance for the development of a country economically, socially or as a whole. Industries have proved to be a primary source for the preeminent growth of a country in terms of productivity and income (national as well as per capita income), creating vast opportunities of employment, agricultural growth, productivity growth, which in turn helps in increasing the

GDP or the income of any country (Vaishnavi, 2016). The commencement of revolution of industries have resulted in accelerated development of economy as well as nimble up the urbanization around the globe. This in turn has resulted in a massive rise in generation of industrial waste, as its release is directly proportional to the rise of utilization of resources in the industry (Blanchard, 1992; Gerbens-Leenes et al., 2010). Subsequently, this industrial revolution has proved to be an accelerating graph for a number of industries such as iron, steel, electronic, agro-chemical and paper industries to name a few (Vaishnavi, 2016).

Industrial waste can be defined as the generation of waste by any industrial process or operation which can be solid, semi-solid, liquid, gas or any particulate matter which may impart contravening effects on the living as well as the mankind. The best suited example for this is acid rain, which is caused due to oxides of sulphur and nitrogen from exhaust pipes or chimneys. The most practical example for acid is the deterioration of one of the seven wonders of the world, the Taj Mahal, which resulted in yellowing of the heritage structure (Anand, 1999; Nebel and Wright, 1999). Improper disposal of this industrial waste is an emerging menace in today's world. The handling, collection, storage and disposal of waste products produces various environmental risk factors to the earth (Zhu et al., 2008). The act of disposal of wastes in open spaces or water resources by the small as well as large-scale industries often leads to generation of toxic or hazardous effects on the environment (Vaishnavi, 2016). Various environmental problems include the contamination of groundwater, ultimately leading to unfit water for drinking. In addition, various metals such as arsenic pollute the soil, leading to cultivation of unfit crops ultimately affecting the humans after its consumption. Improper planning of landfills often creates a havoc in the neighbourhood and also the lack of management of these landfills may lead to serious hazardous issues for the living (Hand, 2010).

Therefore, there is a sudden need which has arisen due to this improper disposal of industrial waste. In respect to this various effective measure are taken by the government of each country for the management and appropriate disposal of this waste so that the effect caused by these left out waste causing spread of diseases, pollution and many more are controlled. The major problem of improper disposal of the industrial waste arises due to the following reasons:

1. Lack of proper disposal site for the industrial site.
2. The rate of generation of solid industrial waste is much greater than its disposal.
3. Lack of competent facilities or measures for the collection and treatment of the waste produces.
4. Lack of strict measures by the regulatory bodies of the country.
5. The presence of industries in non-conforming area without of the consent of specific regulatory bodies such as the State Pollution Control Board in India.

There has been an estimation that nearly 12 billion tons of waste is produced by the industries yearly by the United States, which is nearly equal or more than 40 tons of waste produced as whole by each citizen in the country (David, 1992). In developing

Industrial Waste Management System 117

countries like India, irrespective have a specific regulatory framework for the waste management, universal 3 Rs—reuse, reduce and recycle—are rarely used. Hence there is an urgent and serious requirement for the effective stringent measure for the managing this curb. Though the present methods for the disposal of waste do not yield positive benefits to the environment. It is a Herculean task to take out measure at one stance for the large volume of industrial toxic waste (Babu and Ramakrishna, 2003).

The waste management measure should be in accordance to the objectives which were laid in the Rio conference, 1992, Agenda 21—Earth Summit as follows:

1. Maximization of the act of recycling.
2. Minimizing the production of waste.
3. Carrying out and promoting the waste of waste in an environment friendly way.

(Roghaia, 2003)

In addition to the local regulatory bodies it is the duty of industry themselves to adopt various measures for the disposal of waste upon the instructions of the regulatory authorities. Hence a two-tier approach should always be carried for the prevention and control of industrial waste (Ramachandra and Saira, 2004).

8.2 INDUSTRIAL WASTE: CLASSIFICATION

8.2.1 General Classification

1. *Industrial waste obtained from powder industry*: This industrial waste includes coal ash; lag from power plants containing coal are utilized as fuel.
2. *Industrial waste obtained from metallurgical industry*: This industrial waste include slag which is obtained from metallurgical processes.
3. *Industrial waste obtained from chemical industry*: This includes unreacted raw materials, disarmed catalysts, additives and various left out products during a chemical reaction. In addition, they also include washing and refining discharge waste.
4. *Industrial waste obtained from mining industry*: They include waste which is produced during the duration of tailing and mining.
5. *Industrial waste obtained from power (light) industry*: They include wastes originated from alkali waste, residues from animals and sludge during the operations of manufacturing.
6. *Industrial waste obtained from oil-chemical industry*: The waste obtained from oil chemical industries comprises of organic solvent waste, left out catalysts, mud-oil and slag during the operation of processing of oil (Zhang et al., 2015).

8.2.2 Based on the Toxicity of the Waste

Hazardous waste: Industrial hazardous waste can be defined as any reactive, toxic or corrosive waste produced during any industrial operation or process. Waste is often regarded as toxic if it is proved to cause detrimental effects on the living when exposed via skin, inhalation or by oral route. These include metals such as lead, cadmium, thermostats, mercury, DDT, power plants, petroleum refiner wastes and so forth. The series of regulations for the control of the hazardous industrial waste is referred to as the universal waste rule (Vaishnavi, 2016). Hazardous waste can be sub-classified into various categories as in Figure 8.1.

8.2.2.1 Characteristic of Hazardous Waste

An industrial hazardous waste is regarded as characteristic when it possesses a specific menace according to the regulations. An industrial waste is regarded as characteristic if it imparts the following characteristics in Figure 8.2 (Syamala et al., 2018).

1. *Corrosiveness*: A waste of industrial origin is regarded as corrosive when it has a pH of about 2 or less, or it has a pH more than or equal to about

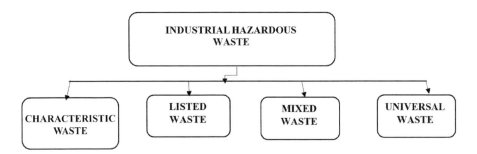

FIGURE 8.1 Types of Industrial Hazardous Waste

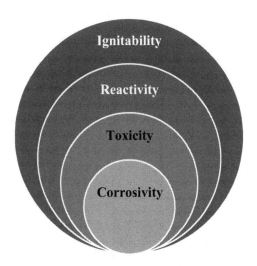

FIGURE 8.2 Characteristics of Hazardous Waste

12.5. For instance, HCl (hydrochloric acid) has application in cleaning of metal parts. Another instance follows sodium hydroxide (NaOH) which has a higher pH and is utilized in degreasing metal parts.

2. *Toxicity*: Those wastes which when ingested or absorbed in the body prove fatal. When these wastes are disposed, they leach the soil and result in groundwater pollution. They include waste containing lead, mercury and so forth.

3. *Reactivity*: A waste of industrial origin is considered reactive if it has the ability to react with water vigorously and has the ability for the generation of toxic wastes, if the waste generated explodes on exposure to flame or heat. Various instances include the metal sodium, gunpowder, sulphides containing waste and cyanide containing waste.

4. *Ignitability*: A waste is considered ignitable if it has a flashpoint less than 60°C. In addition, it has the ability to spontaneously combust. Oils and solvent waste are included in this.

8.2.2.2 Listed Waste

Listed waste (Babu and Gupta, 1997; Kansas, 2013; Syamala et al., 2018) has been set by the Environmental Protection Agency (USEPA) which constitutes the following types:

1. F-List (waste from non-specific sources)
2. K-List (specific industries waste)
3. P&U List (commercial waste).

8.2.2.2.1 F-List (waste from non-specific sources)

The waste obtained during various operation or processes of industry and during the manufacturing process are included in F-List. Various illustrations which are included in the F-List are:

1. F019—Treatment of wastewater from the chemical process of aluminium coating
2. F006—Treatment of wastewater from process of electroplating
3. Treatment of wastewater from chemicals constituting dioxin.

8.2.2.2.2 K-List (specific industries waste)

The K-List of the listed industrial waste includes waste which is produced by discrete industrial processes. Waste from production of chemicals, explosives, pesticides, pigment are included in the K-List.

8.2.2.2.3 P&U List (commercial waste)

The list of the hazardous waste includes disposed waste products, residues of the containers, spilled materials and many more. The P&U list differs in the terms of toxicity. The products on the U list are less toxic in regard to the products included on the P list. The P list is comparatively less common. Examples include acetone, toluene, warfarin and so forth.

8.2.2.3 Mixed Waste

Mixed waste is defined as those industrial hazardous waste containing components of hazardous system and radioactive substances.

8.2.2.4 Universal Waste

These wastes comprise mercury-containing products, pesticides, batteries lamps, thermostats and so forth (Energy Institute, 2014).

8.2.2.4.1 Non-Hazardous Waste

Industrial non-hazardous waste can be defined as the generated waste which is not toxic. Hence it possesses no threat to humans. In certain cases, industrial waste may cause harmful effects to the general population in addition to the workers in the industry. This include an example of cotton industry. Cotton dust causes respiratory ailments to workers which contains micro-organisms irritating the respiratory tract producing allergies. Non-industrial hazardous waste includes animal excreta, blood, pulps, carcasses and many more (Vaishnavi, 2016).

8.3 CHARACTERISTICS OF INDUSTRIAL WASTE

Interestingly, liquid industrial waste characteristics can be characterized on the basis of physical and chemical aspects as follows.

8.3.1 Physical Characteristics

Physical characteristics as given by Metcalf and Eddy (1991) and Munter (2003) are as follows:

Colour: The colour of the generated wastewater is a qualitative phenomenon for determination of the degree of industrial liquid waste. Colour standards are used for the comparison of the generated wastewater. Different colours determine the level of the industrial waste complexity in Table 8.1.

Content of solid present: This characteristic refers to the presence of solids (in minute quantity) or in dissolved form possessing risk to the environment. The amount solids present in the industrial wastewater can be determined by the weighing method. In case of a volatile element present, they are first ignited and then weighed. The solids which can be settled are measured.

Temperature: The estimation of temperature is an important characteristic for the determination of industrial wastewater characteristics. Generally, the temperature

TABLE 8.1
Colour of Industrial Waste (Munter, 2003)

S. No.	Colour	Comments
1	Black	Presence of sulphides (mainly ferrous sulphates)
2	Dark Grey	Bacterial decomposition of wastewater
3	Light brown	Wastewater less than 6 hours

Industrial Waste Management System 121

varies greatly according to the location and season. The temperature varies from 13 to 25 degrees Celsius in the warmer regions to about 7–19 degrees Celsius in the colder regions.

Odour: Determination of odour is an important aspect in the physical characteristics of industrial wastewater. Characteristic odour arises due to the presence of various components such as hydrogen sulphide, mercaptan, dimethyl sulphides and indole.

8.3.2 Chemical Characteristics

The chemical characteristic as given by Bond et al. (1974) of the industrial wastewater includes:

1. *Organic chemicals*: In order to determine the organic content present in the industrial liquid waste various laboratory methods are employed. These include the following methods:
   a. BOD—Biochemical oxygen demand
   b. Chemical oxygen demand
   c. Total organic carbon
   d. Mass spectroscopy

2. *Inorganic chemicals*: The determination of inorganic chemical consists of nitrites, phosphorus (organic and inorganic), ammonia, nitrates, sulphates, chlorides and so forth. In addition to these, the determination of heavy metals is also done including iron, zinc, cobalt, zinc and copper. The determination of these heavy metals is done as they produce toxic effects to the environment.

3. *VOC—Volatile organic carbons*: Certain compounds such as xylenes, benzene, dichloromethane, toluene and trichloroethane are referred to as volatile organic carbons. These compounds have the tendency to produce industrial pollution. The cause for this pollution arises due to improper landfills, leakage of storage tanks and so forth.

TABLE 8.2
Industrial Effluents and Their Sources

S. No.	Industrial Effluent	Source
1	Ammonia	Manufacturing of chemicals, coke
2	Copper	Plating of copper
3	Cyanides	Manufacturing process of gases and plating.
4	Chlorine	During bleaching of textiles, milling of paper
5.	Hydrocarbons	Petrochemical industries
6.	Compounds Containing Nitro Group	Manufacturing of explosives
7.	Phenols	Chemical manufacturing

They also include organic components such as formaldehyde,1,3 butadiene,1,2 dichloroethane, hexachlorobenzene and so forth. 8.2 is a list of industrial effluents and their sources.

8.4 UPSHOTS OF INDUSTRIAL WASTES

On humans: Industrial waste imparts serious outcomes on the health of human beings. In correlation to the effect of harmful released effluents on the soil and plants, the capacity of yielding crop has been reduced to about 10%. The basic mechanisms behind this is the decrease in the content of chlorophyll, thus altering the metabolic pathway. Another cause for the harmful effect on the human life is the release of industrial discharge in to the water bodies directly. This untreated industrial discharge lead to various health hazards such as respiratory disease, cancer and many more. The third reason for the effect on human health is accounted for the polluted groundwater. Thus, the improper generation and management of industrial waste is a serious issue to be taken in to charge (Anand and Sibyala, 2017).

On flora: The effect of industrial discharge on plants is a serious one which affects various plant processes. The waste produced including heavy as well as trace metals directly act on the land in the form of sludge, thus used as soil manure for the growth and development. However, the effects are very treacherous (Webber, 1981). This waste generation has tremendously affected the germination of seeds (moisture condition), Plant growth, flowering, fruiting etc. (Noggle and Fritz, 1991). In addition to this the wastewater from industries is sometimes utilized as a mode of irrigation for the fields. Thus, it creates a serious menace leading to the toxicity of crops or plants with significant amount of metals (Koc, 1976). For instance, the growth of rice (germination and growth) showed a decline when the discharge released from the industry was more than 25% (Sahai and Neelam, 1987).

On the marine ecosystem: Industrial effluents have affected the marine ecosystem the most. Biochemical flow has been altered by the metals which eventually affect the water system from the rivers to the oceans. Additionally, the sea water composition as well as the sediments around have been greatly affected. Second, the contamination of the marine ecosystem as whole including flora and fauna which in turn enters the human food chain. Thus, the marine ecosystem is directly or indirectly the most affected ecosystem (Anand and Sibyala, 2017).

8.5 WASTE MANAGEMENT APPROACH

Management of waste of the industry is an important aspect for a healthy environment. Though this cannot be achieved by integration and mutual coordination between the citizen of the country. One individual is not sufficient to change the world completely. The management of industrial waste has been viewed as a two-tier approach. This two-tier approach includes prevention along with the control of industrial waste generated. Minimizing waste is a better method for the management of industrial waste instead of the treatment of industrial waste or the by-products produced. Minimizing the waste generated can be achieved by reducing or recycling the waste generated at each step of the process of obtaining a finished product

(Vaishnavi, 2016). A typical strategy employed for the industrial management can be accounted as follows:

1. Audits of source of waste generated including quantity, quality, composition and any threats and procedures followed for waste management;
2. Storage and handling procedures required for risk assessment;
3. Waste recycling, reduction and reuse procedures;
4. BPEO—Best Practicable Environmental Option required for the management of left out waste;
5. Audit required for contractors for the management of waste (Hand, 2010).

In addition to this, installing equipment producing zero waste, equipment modification enhancing recycling and recovering the waste; maintaining the regulations for prevention of waste. In case for the prevention of hazardous waste they should be replaced by non-hazardous raw material, segregating the waste, eliminating the oil spills or leak source (Vaishnavi, 2016). Following the flow cycle, Figure 8.3 represents the schedule for the management of industrial waste (Woodard, 2001).

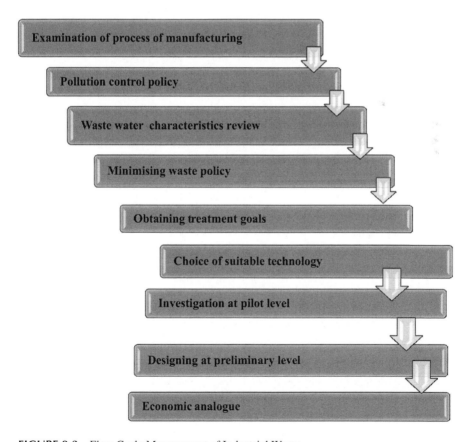

FIGURE 8.3 Flow Cycle Management of Industrial Waste

8.6 MAIN OBJECTIVE FOR THE MANAGEMENT OF INDUSTRIAL WASTE

Waste management at an industrial setup usually includes the following listed goals:

1. Reestablishment and recycling of products or substances again into the production process for the generation of secondary product;
2. Introduction of the waste generated into the environmental rotation;
3. Establishment of the two-tier approach for the reducing the total amount of waste generated by means of the 3 Rs—Reduce, Reuse and Recycle;
4. Approaching to an idea concerning the reasons for the oscillation in waste generation patterns;
5. Reducing the amount of waste for the ejection on the landfills supremely made for the purpose (AEVG, 2008).

8.6.1 MEASURES EMPLOYED FOR CONTROLLING THE GENERATION OF WASTE

Management of sources at preliminary level: Controlling the variations at a basic level of the industrial process which involves the possible choices of the raw materials as well as the operations involves for analyzing the released gases prior to their discharge from the industry. In addition to this proper mixing of the pollutants release the stock height is increased to a height of about 38 metres.

Industrial location selection process: The choice of an industrial site should be analyzed carefully in correlation to various factors such as climate, topography and so forth prior to the setting of the industry.

Process of treatment of generated industrial waste: Proper treatment processes should be employed before the disposal or introduction of waste into the ecosystem.

Elevation in rate of plantation: Increasing the rate of plantation around the region of an industry helps in the reduction of various industrial pollutants, smoke and dust.

Employing strict government control: Government regulatory policies should be employed rigorously in order to maintain a qualm in industries for reducing the amount the waste generated and to keep within the limits prescribed by the control board of each country.

Analyzing the degree of environmental risk: Regular checks should be done for the identification and evaluation of the hazards possessed by the industrial waste generated in any form to the environment (Baas, 1998; Anand and Sibyala, 2017).

8.7 HIERARCHICAL RELATIONSHIP OF WASTE MANAGEMENT

The hierarchical relations for the management of industrial waste is wholly based on the idea of sustainability. Sustainability refers to the ability to work in an integrated manner rather than focussing on the end waste generation parameters. Hierarchical relationship involves a series of waste reduction and managing the waste processes. The basic aim is the extraction of maximal asset and generation minimal extent of waste. This can be advantageous in various processes. It helps in prevention of release

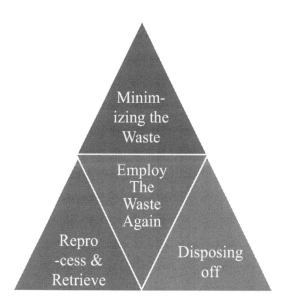

FIGURE 8.4 Hierarchical Classification of Waste Management

of exhaust gases, pollutants from the industrial processes. In addition, it helps in conservation of energy as well as resources. Also help in creating employment and stimulation of green technology in industrial operations for minimal waste generation and reducing the threat (Environmental et al., 1995; Plaut, 1998; Anand and Sibyala, 2017). The hierarchical classification of waste management is shown in Figure 8.4.

8.8 ASSORTMENT AND TREATMENT OF INDUSTRIAL WASTE

8.8.1 SEGREGATION OF WASTE

The waste generated is a mixture of different types of waste such as hazardous and non-hazardous wastes. The first significant step towards the environmental protection is the segregation of waste into different constituent elements. This refers to differentiating the type of waste into solid, liquid, gaseous elements. In addition, segregation of waste from hazardous to non-hazardous.

8.8.2 METHODS FOR THE WASTE TREATMENT

8.8.2.1 Liquid Industrial Waste

The output efficacy of treatment plant is usually evaluated by the functional parameters which include the biochemical oxygen demand (BOD), chemical oxygen demand (COD), and its suspended solids (SS). These functional variables are regulated and maintained for evaluating the efficacy. The output of the plant is determined in terms of the minimization of SS, BOD, and COD of discharge.

The most commonly used methods for disposal of industrial wastewater are:

1. *Waste stabilization ponds*: This secondary treatment methods uses biological processes by daily organic loading in designed earthen ponds.
2. *Activated sludge*: This process involves feeding of industrial waste into an aeration tank, where suspension of microorganisms is placed in the tank. The microorganisms then act on the waste such as organic matter causing sedimentation and are then separated from the discharge. The flocs of microorganisms are then reused in the aeration tank to perform oxidation of the organic matter.
3. *Trickling Filters*: This process includes the waste filtration and filter includes contact media including either a bed of broken stones, or any substance that can be used to screen the effluent. (Lohwongwatana et al., 1990).

8.8.2.2 Solid Industrial Waste

Management of solid industrial waste usually involves the following:

1. Reduction of sources
2. Reuse of the waste material
3. Recycling the materials
4. Incineration
5. Landfilling operations (Onipede and Bolaji, 2004).

8.8.2.2.1 Reduction of Sources

Reducing the wastage at a basic preliminary level by minimizing the source is the most potential outcome for the solid waste management. Thus, it occupies the top most position in the solid waste hierarchy. Source minimization refers to the reduction in the raw material usage. Thus, leading to the decline in energy as well as waste production. Factors responsible for source reduction includes:

1. Modification in equipment
2. Alterations in production flow
3. Alterations in operations
4. Alterations in process.

8.8.2.2.2 Reusing the Materials

Reusing the materials refer to the usage of the material again for the same purpose or using it with certain modifications or its utilization in any other process with an aim of minimizing the waste (Adewumi, 2000; Onipede and Bolaji, 2004).

8.8.2.2.3 Recycling the Materials

In countries which are developing the tradition for recycling helps in the reduction of waste at an utmost level. It refers to accumulating and sortation as well as recovery of the materials which can be used again for the reprocesses involved in the formation of new product or to carry the old process again. Composting differs from recycling

Industrial Waste Management System

in a way that it deals with the accumulation of waste which is organic in nature and is ultimately introduced in the soil (Staniskis et al., 2006).

8.8.2.2.4 Incineration

Incineration refers to the combustion process of the waste generated. It is often referred to as thermal treatment. Because of high temperature being used in this. Conversion of waste in to a form of its by product such as ash, gas, heat or steam. It can be designed and modified according to its utilization for specific industrial waste (Abdulkadir, 2009). The waste is heated at a temperature of about 9,000–10,000 degrees, leaving behind ash material which is disposed of in the man-made landfills. This method has been used as an alternative to landfilling processes (Swain et al., 2018). Factors affecting the process of incineration are:

1. Amount of combustible material more than 60%
2. Number of non-combustible solid less than 5%
3. Content of moisture less than 30%
4. Amount of fine non-combustible material less than 20%.

(Teka, 1984)

A characteristic incinerator usually consists of an input material which is subjected to combustion process liberating large amount of energy. It has been typically used for the disposal of hazardous industrial waste. One demerit of this disposal process is the generation of harmful gaseous pollutants to the environment. In addition to this fractional carbon compounds are also subjected to fractional combustion leaving carbon residues (Ecke and Svensson, 2008). Figure 8.5 shows representation of landfill system.

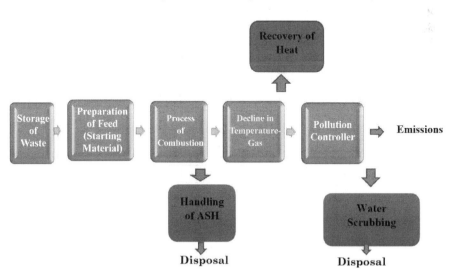

FIGURE 8.5 Representation of Incineration System

Source: Anand (1999).

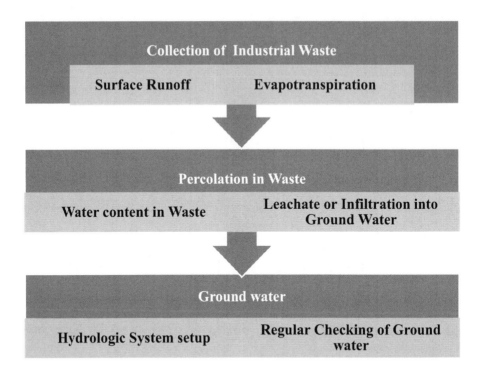

FIGURE 8.6 Landfill Process

Landfilling processes: This method is often described as an elementary method for the disposal of industrial waste because of its lesser cost requirement. In this organic and inorganic waste are subjected to dumping on a are having minimal estate cost. In this method the generated waste is covered with a layer of clay in order to prevent them from animals and insects for breeding. However, one of the major disadvantages of this method is the availability of land. the availability of land has showed a marked decline with the increasing industrialization and population all over the globe. A typical landfill consists of four basic elements which includes a bottom liner; a collection is, cover and a hydrogeologic system setting. The latter is used for the minimization of waste discharged into the groundwater. A typical landfill process is described in Figure 8.6.

8.9 CONCLUSION

Industrial waste management should be regard as significant for the protection of the environment completely. Waste should not be regarded as typical waste, instead it should be regarded as an asset. Education among the population and industry workers will ultimately lead to the working of an efficient management system for the large amount of industrial waste produced. The increase in production and marketing processes have led to the stake of the natural resources to a danger level. Therefore, the idea of sustainability should be adapted by the

Industrial authorities. This idea helps in reducing the natural resources and avoid the risk of environmental deuteration. The disposal techniques employed should be modified timely and analyzed on a regular basis as they also cause serious hazards to the environment directly or indirectly. Hence, proper planning and development techniques should be employed for the reducing the risk of industrial health and promoting the improvement as well as prevention of environment from harmful effects.

REFERENCES

Abdulkadir, K., 2009. General characteristics of waste management: A review. *Energy Education Science and Technology Part A: Energy Science and Research.* 23(1):55–69.

Adewumi, S. O., 2000. Policy guidelines for sewage and solid waste disposal. *Resources and Environmental Development.* 11(2):161–176.

AEVG, 2008. Waste management concept of the city of Graz. Available online at www.aevg. com

Anand, K. and Varma, S., 2017. *Principles of Industrial Waste Management: Waste Prevention,* Lambert Academic Publishing, USA.

Anand, P. B., 1999. Waste management in Madras revisited. *Environment and Urbanization.* 11(2):161–176.

Baas, L., 1998. Cleaner production and industrial ecosystems, a Dutch experience. *Journal of Cleaner Production.* 6(3–4):189–197.

Babu, B. V. and Ramakrishna, V., 2003. Extended studies on mathematical modeling of site sensitivity indices in the site selection criteria for hazardous waste treatment, storage and disposal facility. *Journal of the Institution of Public Health Engineers India.* 11–17.

Babu, S. and Gupta, J. P., 1997. Waste characterization and Treatment. *Chemical Business.* 39–42.

Blanchard, O., 1992. Energy consumption and modes of industrialization: Four developing countries. *Energy Policy.* 20:1174–1185.

Bond, R. G., Straub, C. P. and Prober, H., 1974. *Wastewater Treatment and Disposal of Handbook of Environmental Control.* CRC Press.

Characteristic and Listed Hazardous Wastes, 2013. Kansas Department of Health and Environment, Bureau of Waste management. Kansas. Available online at http://www.kdheks.gov/waste/techguide/HW-2011-G2.pdf

Allen, D. T., 1992. An overview of industrial waste generation. *MRS Bulletin.* 17(3):30–33.

Ecke, H. and Svensson, M., 2008. Mobility of organic carbon from incineration residues. *Waste Management.* 28:1301–1309.

Elnasri, R. A. A., 2003. Assessment of industrial liquid waste management in Omdurman Industrial Area. Available online at www.osti.gov/etdeweb/servlets/purl/20943506

Environmental Management in Developing Countries, 1995. *Waste Management, Institute for Scientific Co-operation,* Environmental Management in Developing Countries, Tubingen, Vol. 2. 214.

Gerbens-Leenes, P. W., Nonhebel, S. and Krol, M. S., 2010. Food consumption patterns and economic growth. Increasing affluence and the use of natural resources. *Appetite.* 55:597–608.

Hand, C. L., 2010. *Waste Management and Minimization-Waste Management in Industry.* Encyclopedia of Life Support Systems (EOLSS) Publication. 1:419.

Koc, J.L.K. and Mazur, T., 1976. Investigations into the fertilizing value of tannery sludges. I. Chemico-physical characteristics of sludges. *RocznikiGleboznawecze.* 27:103–122.

Lohwongwatana, B., Soponkanaporn, T. and Sophonsridsuk, A., 1990. Industrial hazardous waste treatment facilities in Thailand. *Waste Management & Research.* 8(1):129–134.

Metcalf and Eddy, 1991. *Wastewater Engineering: Treatment, Disposal, and Reuse*, 3rd ed., McGraw-Hill, Inc., Singapore.

Munter, R., 2003. *Industrial Wastewater Characteristics*, BaltUniv Program (BUP), Sweden. 195–210.

Nebel, J.B. and Wright, T.R., 1999. Air pollution and its control in: Environmental science. *6'h America*. 371–388.

Noggle, G.R. and Fritz, G.J., 1991. *Introductory Plant Physiology*, Prentice Hall Inc., New Delhi. 688.

Onipede, A.I.M. and Bolaji, B.O., 2004. Management and disposal of industrial wastes in Nigeria. *Nigerian Journal of Mechanical Engineering*. 2(1):49–58.

Plaut, J., 1998. Industry environmental processes: Beyond compliance. *Technology in Society*. 20:469–479.

Pratap, S. K., Biswal, T. and Panda, R.B., 2018. Short review on solid waste generations, recycling and management in the present scenario of India. *Journal of Industrial Pollution Control*. 34(1):2008–2014.

Ramachandra, T.V. and Saira, V., 2004. Exploring possibilities of achieving sustainability in solid waste management. *Indian Journal of Environmental Health*. 45(4):255–264.

Sahai, R. and Neelam, S., 1987. Effect of fertilizer factory and distillery effluent on the seed germination, seedling growth, pigment content and biomass of Phaseolus radiatus Linn. *Indian Journal of Ecology*. 14(1):21–25.

Staniskis, J., Arbačiauskas, V. and Pivoras, T., 2006. Progress in the process of sustainable industrial development in Lithuania. *Environmental Research, Engineering & Management*. 37(3).

Teka, G. E., 1984. *Human Waste Disposal Ethiopia: A Practical Approach to Environmental Health*, MOH, Addis Ababa, Ethiopia. Available online at https://docplayer. net/7520692-Human-wastes-disposal-ethiopia.html

Vaishnavi, A., 2016. Industrial solid waste: Emerging problems, challenges and its solution. *International Journal of Management and Applied Science*. 2(7):68–72.

Webber, J., 1981. The energy and resources institute: Trace metals in agriculture. In Lepp, N.W. (ed.), *Effect of Heavy Metal Pollution on Plants. Metal in the Environment*, Applied Science Publishers, London, Vol. 2. 159–184.

Woodard, F., 2001. *Industrial Waste Treatment Handbook*, Elsevier, Oxford. 455–459.

Zhang, P. et al., 2015. The analysis of obstacles and the promoting strategy of achieving detailed classification of municipal living waste. *Applied Mechanics and Materials*. 768:733–739.

Zhu, D., Asnani, P.U., Zurbrugg, C., Anapolsky, S. and Mani, S., 2008. Improving municipal solid waste management in India. In *A Source Book for Policy Makers and Practitioners*, World Bank, Washington, DC.

9 Role of Microbes in Solid Waste Management
An Insight View

Debajit Borah and Kaushal Sood

CONTENTS

9.1 Introduction 131
9.2 Reduce, Reuse and Recycle 132
    9.2.1 Waste Reduction and Reuse.. 132
    9.2.2 Recycling 133
9.3 Treatment and Disposal 133
    9.3.1 Thermal Treatment 134
        9.3.1.1 Incineration 134
        9.3.1.2 Pyrolysis and Gasification.. 134
        9.3.1.3 Open Burning 134
        9.3.1.4 Landfills 134
9.4 Biological Waste Treatment.. 135
    9.4.1 Vermicomposting and Its Microbiology................................... 135
    9.4.2 Role of Microbes in Aerobic and Anaerobic Digestion of
        Solid Wastes.. 136
    9.4.3 Bioreactor Landfills 138
    9.4.4 Role of Microbes in E-Waste Management 138
9.5 Role of Microbes in Xenobiotics Degradation 139
9.6 Role of Microbial Enzymes in Waste Management 139
9.7 Conclusion 146
References.. 146

9.1 INTRODUCTION

Solid waste is the discarded or unwanted solid materials generated from household activities, industrial or commercial activities or construction and municipal activities in a given area. It has been observed that of the enormous supply of food for human consumption, about one-third gets wasted globally (FAO, 2011). Fruits and vegetables are estimated to generate at least up to 25% to 30% of waste materials, which are not further used (Ajila et al., 2007,2010). In most fruits and vegetables, only the flesh or pulp is consumed, but studies have revealed that significant amounts of phytochemicals and essential nutrients are present in the seeds, peels, and other components of fruits and vegetables that are not usually consumed (Rudra et al., 2015). Some of the bioactive

131

components that can be obtained from vegetable waste are dietary fibres, phenolic compounds, flavouring agents and aroma, enzymes (amylases, cellulases, invertase, pectinases, proteases, lipases etc.), organic acids and proteins (Sagar et al., 2018).

Based on its origin, solid waste may be categorized into municipal, industrial, domestic, commercial, construction or institutional etc. Again, it may be further categorized as biodegradable (e.g.organic) and non-biodegradable (e.g. some of the xenobiotics). Non-biodegradable solid waste may be again sub-divided into recyclable and non-recyclable solid waste. Growing urbanization and industrialization all across the globe have led to generation of massive volumes of solid waste (SW) and developing countries mostly release them into nearby environment. In developing countries, dumping of solid wastes into open areas like wetlands, water bodies, drains, landfills and burning are prevalent forms of waste disposal practices which often results in the pollution of the surroundings (Ihuoma, 2012). Further, dumping of wastes in the open serves as breeding place for flies, insects, and rodents, which increases the potential for the spread of infectious diseases. As a consequence, management of SW nowadays requires modern scientific methods for the management of solid waste for better environment.

Traditionally, the following three strategies are mostly followed for the treatment of solid waste are popularly known as the 3 Rs and is described below.

9.2 REDUCE, REUSE AND RECYCLE

The 3 Rs stands for reduce, reuse and recycle and are the most widely used strategy for the treatment of solid waste all over the world. However, such traditional methods have their own advantages, disadvantages and limits for their employment in waste management.

Besides their disadvantages, they greatly reduce or prevent the emissions of greenhouse gases, conserve resources and also save energy. Therefore, these methods are principally employed for solid waste management.

9.2.1 WASTE REDUCTION AND REUSE

This is possibly the most suitable environment benign technique for the management of solid waste. Waste reduction by reusing the products such as plastic, thermocol, glass, nylon, metallic products and so forth greatly reduces the necessity for the production of such products and also reduces the demands for large scale treatment and disposal facilities. Such strategies also need great public awareness and support. For example, waste reduction may also include using very less packaging materials, encouraging consumers to bring their own reusable bags for packaging, encouraging the public to choose reusable or biodegradable products such as reusable or biodegradable plastic bags, glass or metallic containers, composting the biodegradable products and sharing non-biodegradable and unwanted items rather than discarding them.

Implementation of the above methods not only requires public participation and attention but also requires strict government laws. Educational curriculum on such topics in elementary schools and public awareness camps organized by the government bodies and non-governmental organizations (NGOs) may play a great role in the process.

9.2.2 Recycling

Recycling refers to the removal of used products from the waste materials to be used as raw materials for the manufacturing of new products. To achieve this, it is very important to sort recyclable wastes from other wastes initially before dumping with public participation at home or office level.

Another laborious option is to sort recyclable material from a mixture of nonrecyclable material in the dumping site manually with the help of municipality personal. But such technique has some disadvantages such as possible health impact on the labours, higher cost due to manual sorting, time consumption, and most importantly degradation of value of the recyclable materials in terms of quality as raw materials due to compacted with other garbage with possible adverse impact on the quality of the products to be manufactured. However, one major disadvantage of recycling some of the substrates made of plastics, glass, PVC (polyvinyl chloride) and so forth is the emission of toxic gases into the environment.

9.3 TREATMENT AND DISPOSAL

A number of methods are employed for the treatment of solid wastes which includes thermal treatment, dumps and landfills, biological waste treatment etc. But not all of these techniques are environmentally benign in nature. Based on their applicability in terms of their environmental effect the following pyramid may be proposed (Figure 9.1). The pyramid shows the best to the worst options from its top to the base. The best options for waste treatments (i.e. reduce, reuse and recycle) are already discussed in the previous section.

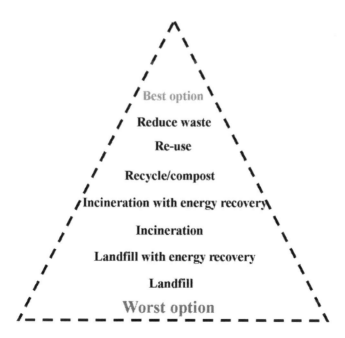

FIGURE 9.1 Solid Waste Management Hierarchy

Apart from these three popular treatment methods, the following treatment process are also widely used in spite of their respective advantages or disadvantages:

- Thermal treatment
- Dumps and landfills
- Biological waste treatment.

9.3.1 Thermal Treatment

These processes refer to the use of heat for the treatment of wastes. The following methods are the most commonly used techniques for the treatment of solid wastes by thermal energy.

9.3.1.1 Incineration

Incineration is one of the universally used thermal treatment processes which involves the combustion of solid wastes in the presence of oxygen. Nowadays, this process is also used for the conversion of heat energy into electric energy. Upon combustion, solid waste gets converted into CO_2, water vapour and ashes, and this technique greatly reduces the volume of solid wastes.

9.3.1.2 Pyrolysis and Gasification

Pyrolysis involves incineration of solid wastes in the absence of free oxygen which in the end produces a very high amount of heat energy. On the other hand, gasification involves incineration of organic wastes in the presence of very little oxygen but allows the recovery of heat energy without air pollution.

9.3.1.3 Open Burning

Open burning may be the most commonly used thermal process for waste management in spite of direct emission of gaseous by products into the atmosphere (EPA, 1997). Uncontrolled open burning of garbage may release many pollutants such as carbon monoxide (CO), volatile organic compounds (VOCs), polycyclic aromatic hydrocarbon (PAH), hexachlorobenzene, dioxins, particulate matters, ash and so forth which poses serious threat to human health. VOCs may cause severe lung and kidney problems and on the other hand dioxins may show adverse effects on reproductive health and also interrupt the hormonal systems. PAHs are mainly responsible for respiratory disease and also known as potential carcinogens. Particulate matters are mostly responsible for most of the respiratory disorders such as asthma or bronchitis. In addition to these, open burning of disposals may also release NO_x which may further contribute to global warming by ozone depletion and also may cause acid rain.

9.3.1.4 Landfills

Landfills may be used as an eco-friendly method of biodegradable waste treatment where boreholes are dug to make cells to fill with wastes for composting. The wells should be dug in such a way so that it doesn't come in contact with underground source of water to prevent water contamination. Biodegradable wastes may be

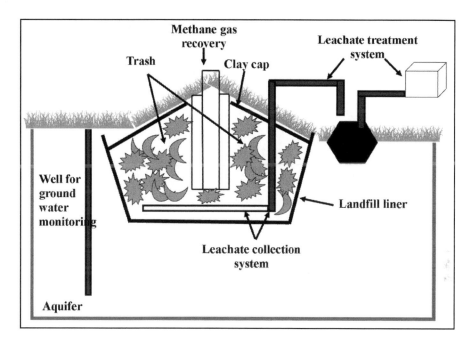

FIGURE 9.2 Main Features of a Modern Landfill

converted into fertile biomass in presence of indigenous methanogenic bacteria with subsequent recovery of methane as combustible gas. A typical schematic diagram of a modern landfill is shown in Figure 9.2.

9.4 BIOLOGICAL WASTE TREATMENT

9.4.1 Vermicomposting and Its Microbiology

Composting is the decomposition of biodegradable wastes with the help of microorganisms or by using earthworms. Decomposition of wastes by earthworms is popularly known as vermicomposting. Household wastes such as vegetable and other organic wastes may be converted into nitrate and phosphate rich fertilizer with the help of vermicomposting. *Eisenia fetida, Lumbricus rubellus, E. andrei, E. hortensis* and *Dendrobaena veneta* are the most popularly used earthworm species for vermicomposting. Maintaining aeration and humidity is very important for achieving best results through this process. Earthworm species such as *E. foetida, L. rubellus, L. castaneus, L. festivus, Eiseniella tetraedra, Bimastus minusculus, B. eiseni, Dendrodrilus rubidus, Dendrobaena veneta* and *Dendrobaena octaedra* are predominantly present in the superficial soil layers and leaf litter which initiates early decomposition of litters (Pathma and Sakthivel, 2012). On the other hand, *Aporrectodea caliginosa, A. trapezoides, A. rosea* and *Millsonia anomala* are the earthworm species which are basically found in the subsoil region and *L. terrestris, L. polyphemus* and *A. longa* are found in the deep soil region (Pathma and Sakthivel, 2012).

TABLE 9.1
Role of Microbes in Vermicomposting (Pathma and Sakthivel, 2012)

Name of Bacteria	Benefits
Pseudomonas oxalaticus	Oxalate degradation
Rhizobium trifolii	Nitrogen fixation and growth of leguminous plants
R. japonicum, P. putida	Plant growth promotion
*P. corrugata*214OR	Suppress *Gaeumannomyces graminis* var. *Tritd* in wheat
*R. melilotiL*5-30R	Increased root nodulation and nitrogen fixation in legumes
Bacillus spp., B. megaterium, B. pumilus, B. subtilis	Antimicrobial activity against *Enterococcus faecalis* DSM 2570, *Staphylococcus aureus* DSM 1104
Fluorescent pseudomonads, Filamentous actinomycetes	Suppress *Fusarium oxysporum, F.* sp. *asparagi* and *F. proliferatum* in asparagus, *Verticillium dahlia* in eggplant and *F. oxysporum* f. sp. *lycopersici* Race 1 in tomato
Free-living N_2 fixers, *Azospirillum, Azotobacter*, Autotrophic *Nitrosomonas, Nitrobacter*, Ammonifying bacteria, Phosphate solubilizers, Fluorescent pseudomonads etc.	Plant growth promotion by nitrification, phosphate solubilization and plant disease suppression
Proteobacteria, Bacteroidetes, Verrucomicrobia, Actinobacteria, Firmicutes	Antifungal activity against Colletotrichum coccodes, *R. solani, P. ultimum, P. capsici* and *F. moliniforme*
Eiseniicola composti YC06271[T]	Antagonistic activity against *F. moniliforme*

Vermicomposting of organic wastes is greatly facilitated by diverge microbial population which are already present in the waste material, soil and also in the gut of earth worms employed for composting. Free living nitrogen-fixing microbes belonging to the genera *Azospirillum, Azotobacter, Autotrophic Nitrosomonas, Nitrobacter* and so forth help in promoting plant growth by nitrifying the soil, and phosphate-solubilizing bacteria such as *Pantoea agglomerans, Microbacterium laevaniformans* and *Pseudomonas putida* help in plant disease suppression (Gopal et al., 2009). On the other hand, *Pseudomonas oxalaticus* helps in oxalate degradation and *Bacillus spp., B. megaterium, B. pumilus* and *B. subtilise* show antimicrobial activities against certain microbial pathogens (Vaz-Moreira et al., 2008). The detailed role of some of the microbes in vermicomposting is shown in Table 9.1.

9.4.2 ROLE OF MICROBES IN AEROBIC AND ANAEROBIC DIGESTION OF SOLID WASTES

Aerobic degradation of organic waste is an exothermic process with the release of heat energy, CO_2 and water vapour along with the formation of stabilized digested product which may have a great value as fertilizer (Insam et al., 2010). Lignin may contribute more than 30% of wood based solid wastes and are efficiently converted by fungi such as *Trametes versicolor, Stereum hirsutum* and *Pleurotus ostreatus* under natural environment (Insam et al., 2010). Fungal species belonging to the genera *Chaetomium, Fusarium, Aspergillus* are also known for cellulose degradation. On the other hand, *Cytophaga hutchinsonii* is a Gram-negative bacterium which is well known for its efficient production of cellulase enzyme which helps in degrading cellulose which is the most abundant natural organic compound. *Cytophaga, Polyangium* and *Sorangium* are some of the widely studied bacterial genera involves in cellulose degradation (Wilson, 2008). With the increase in temperature due to the exothermic process, the thermophiles start degrading the organic matter. Most of such thermophiles belong to the genera *Bacillus, Thermus* and *Stearothermophilus*.

Anaerobic digestion involves the use of anaerobic microbes such as the methanogens for composting organic wastes in absence of free oxygen for the production of biogas and organic fertilizer. This process converts organic wastes into CO_2, water vapour and methane as combustible gas. A schematic diagram of a typical anaerobic digester is shown in Figure 9.3. Humus is also produced from anaerobic digestion of organic wastes.

Most of the methanogenic bacteria are archaea and also considered as extremophiles. *Methanosarcina thermophila, Methanothermobacter sp., Methanobacterium formicicum* and *Methanoculleus thermophilus* are the most popularly reported

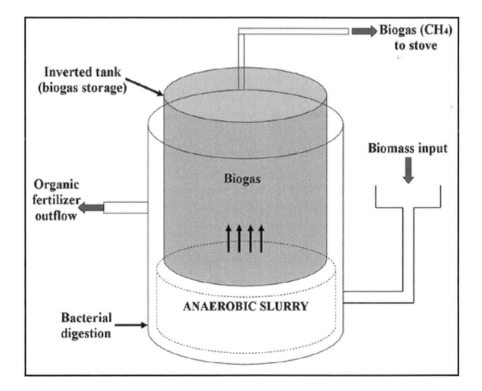

FIGURE 9.3 A Schematic Diagram of Anaerobic Digester

methanogens involved in anaerobic digestion (Thummes et al., 2007). Methanogens are strict anaerobes and do not use free oxygen for respiration; instead their growth may be inhibited in presence of free oxygen. They use CO_2 as an electron acceptor instead of oxygen. Most of the methanogens use H_2 as electron donor with format or secondary alcohols such as isopropanol, ethanol etc. as exceptions in some species (Insam et al., 2010).

9.4.3 BIOREACTOR LANDFILLS

With the advancement of technology, recently bioreactors are employed for organic waste treatment. The major advantage of such technique is that it requires much lesser time for the reduction of biomass. The same is achieved by maintaining optimum conditions such as moisture, temperature and nutritional balance for the microbes involved throughout the process. Impellers are employed occasionally or constantly as per the requirement for the mixing of the biomass with liquid supplement. Such bioreactors may employ both aerobic and anaerobic conditions or may be a combination of both if required. As fermentation process creates acidic environment due to release of a number of organic acids by microorganisms hence the microbes taking part in this process are acid tolerating. The most abundant macromolecule in any fermentation process is carbohydrate (Deshmukh et al., 2017). Various species belonging to the genera *Clostridium*, *Streptococcus* and *Lactobacillus* are commonly reported sugar degraders involved in fermentation process of organic wastes which are anaerobic in nature and yields a number of organic acids, ethanol, CO_2, water and molecular hydrogen (Deshmukh et al., 2017).

9.4.4 ROLE OF MICROBES IN E-WASTE MANAGEMENT

Electronic wastes popularly known as e-wastes are mostly electronic circuits, glass, plastics, heavy metal contents, halogens and so forth. Not all of these components are biodegradable in nature but metallic content of electronic circuits may be recovered by bioleaching with the help of suitable microbes. However, pre-treatment of e-wastes is required for the removal of non-biodegradable components from waste material before subjected to bioleaching.

Gold may be the most abundant metal used for the printing of electronic circuits which may be recovered by bioleaching process. *Chromobacterium violaceum* is reported as the most efficient microbes for the recovery of gold from electronic wastes (Pham and Ting, 2009). *C. violaceum* is also proven for its capability to recover copper from e-wastes. Some of the acidophiles such as *Acidithiobacillus ferrooxidans*, *Acidithiobacillus thiooxidans*, *Leptospirillum ferrooxidans* and *Sulfolobus sp.* are widely known for the recovery of a number of metals such as iron, copper, nickel, zinc and lead (Liang et al., 2010; Pant et al., 2018). Besides, some of the bacterial species includes *Thermothrixthiopara sp.*, *Thiobacillus acidophilus*, *Thiobacillus albertis*, *Thiobacillus capsulatus*, *Thiobacillus concretivorus*, *Thiobacillus rubellus*, *Crenothrix sp.*, *Leptotrix sp.*, archaea species such as *Sulfobacillus thermosulfidooxidans*, *Sulfobacillus thermosulfidooxidans*, *Sulfolobus ambivalens*, *Sulfolobus solfataricus*, fungal species such as *Actinomucor sp.*, *Alternaria sp.*, *Aspergillus amstelodami*, *Aspergillus ficuum*, *Aspergillus fumigatus*, *Aspergillus niger*, *Candida sp., Cerostamella sp., Cladosporium resinae, Coriolus versicolor, Fusarium sp., Mucor sp., Paecilomyces variotii, Penicillium cyclopium, Penicillium ochrochloron, Penicillium oxalicum* are also reported for their bioleaching potential in e-waste management (Pant et al., 2018).

9.5 ROLE OF MICROBES IN XENOBIOTICS DEGRADATION

Xenobiotics are considered as manmade compounds, which may be synthetic in nature and are the outcome of industrial revolution. Xenobiotics such as PAHs, VOCs, alkanes/alkenes, halogenated methane and BTEX are originated from petroleum industries; styrene and phthalic acids are originated from plastic industries; aromatic sulfonates, azo dyes and alkyl ketones are from dye industries; aromatic amines and halogenated heterocyclic from pesticide industries; alkylphenols from surfactant industries; nitroaromatics and heterocyclic from explosives; chlorophenols and dioxins, chloroaromatic hydrocarbon and PCBs from wood processing and pulp bleaching industries.

Bioremediation is the engagement of microorganisms (both bacteria and fungi) and plants for the treatment of contaminants to bring back to its native form. Various species of *Pseudomonas, Acenatobacterium*and *Bacillus* are the most reported microorganisms involved in the degradation of various xenobiotics (Mandri and Lin, 2007; Su et al., 2011; Luo et al., 2012; Yenn et al., 2014; Ameen et al., 2015; Jia et al., 2016). Microbes act on xenobiotics in a number of ways and short chain linear hydrocarbons are more preferred substrate for the microbes as compared to branched chained or cyclic hydrocarbons. Some microbes release biosurfactants which reduces the surface tension of oily substrates and converts them into smaller entities to easily consumable form. On the other hand, some microbes release a cascade of enzymes includes oxidase, lipase and catalase which oxidize organic xenobiotics and convert them into an intermediate metabolite which takes part in the subsequent steps of respective metabolic pathway (Borah, 2018).

Bioremediation may be divided into insitu and exsitu based on the mode of administration. When bioremediation is administered directly in the contaminate site, then the process is known as insitu bioremediation. Prior knowledge on the physicochemical properties such as pH, nutrient balance (C/N ratio) and salinity of the contaminate site may help in achieving better bioremediation rate. Such balance can be maintained by externally adding nutrients and potential microorganisms (Borah, 2018).

Bioremediation strategies that require transportation of contaminants by excavation or by any other mechanical means to treatment sites such as bioreactors, landfills or bio piles are considered exsitu bioremediation. Such technique employs the use of specialized microbes having potential to degrade such xenobiotics. Such technique helps in achieving better results in a shorter duration of time but the volume of substrate that could be treated in a given duration of time in such setups are very limited.

9.6 ROLE OF MICROBIAL ENZYMES IN WASTE MANAGEMENT

Traditional waste management practices involved dumping and composting, which proved to be useful for immediate management but in the longer run has emerged as an environmental disaster as new mountains of garbage keep piling up. The natural processes of waste degradation and recycling are slow and time consuming. Microorganism mediated waste management, popularly referred to as bioremediation has been reported to be useful and efficient (Vidali, 2001; Leung, 2004). This strategy too has its limitations, as most of the microorganisms have performed well under laboratory conditions only (Karigar and Rao, 2011). Environmental and physical parameters such as pH, temperature, moisture, availability of nutrients, interaction with other microorganisms in the field often challenges the potential of microorganisms for effective bioremediation. The bacteria, fungi, algae and plants used in bioremediation exert their effects through enzymes that interact with the pollutants and transform them into non-hazardous forms. These enzymes are versatile and represent the intrinsic bioremediation machinery available in nature that does not involve manipulation of the environmental parameters for its effective implementation. It represents a nature-friendly and cost-effective biotechnology powered by biocatalytic molecules. The detoxification of waste by enzymes is mediated through various chemical modifications that involve reactions like oxidation, reduction and hydrolysis. The microbial enzymes involved in waste management and treatment may be categorized into oxidoreductases, oxygenases, laccases, peroxidases, amylases, cellulases, proteases and lipases.

Oxidoreductases catalyse the biochemical energy-yielding reactions that involve cleavage and transfer of electrons from the donor to the acceptor. These oxidation-reduction reactions transform the contaminants into harmless ones. Some bacteria, fungi and plants of families Solanaceae, Fabaceae and Gramineae have been reported to produce oxidoreductases that take part in the oxidation of environmental contaminants (Karigar and Rao, 2011).

Oxygenases catalyse the transfer of oxygen from molecular oxygen and accomplish the oxidation of substrate using NADH or NADPH or FAD as co-substrate. They have a broad range of substrate and can act on aromatic rings and halogenated aliphatic compounds (Fetzner and Lingens, 1994). Oxygenases are grouped as mono- or di-oxygenases based on the number of oxygen atoms incorporated into the substrate. Mono-oxygenases are further classified based on the accompanying cofactor as flavin-dependent and P_{450} monoxygenases. Their probable mechanisms are shown in Figures 9.4 and 9.5:

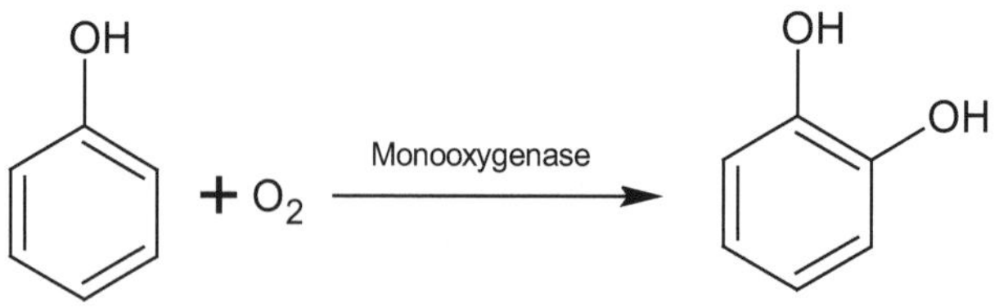

FIGURE 9.4 Degradation of Aromatic Compounds by Monooxygenase

Source: Karigar and Rao (2011).

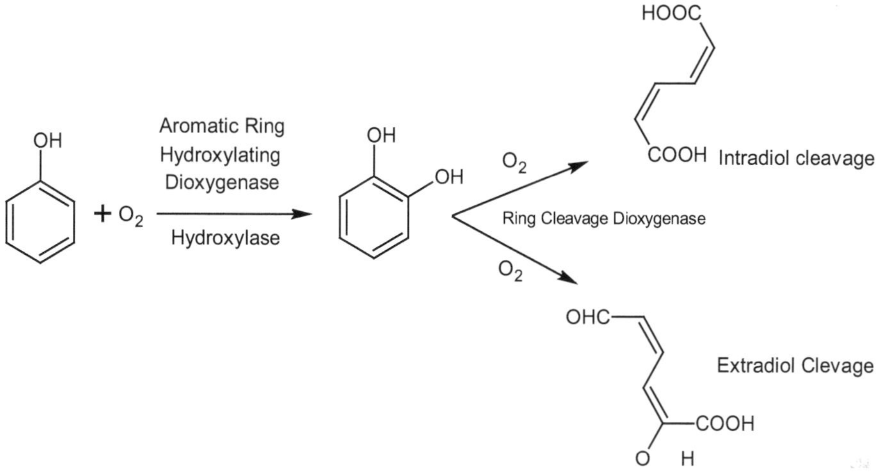

Role of Microbes in Solid Waste Management

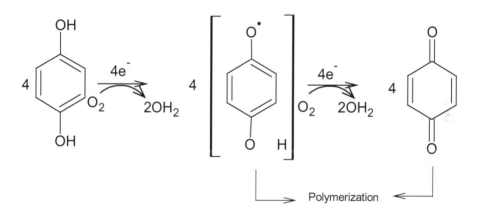

FIGURE 9.5 Degradation of Aromatic Compounds by Dioxygenase
Source: Karigar and Rao (2011).

FIGURE 9.6 General Reaction Mechanism for Phenol Oxidation by Laccase
Source: Karigar and Rao (2011).

Laccases are a family of oxidases that catalyse the oxidation of aromatic and phenolic compounds accompanied by the reduction of molecular oxygen to water (Figure 9.6) (Gianfreda et al., 1999; Mai et al., 2000). These oxidize phenolic and methoxy-phenolic acids and decarboxylate them as well.

Peroxidases are involved in diverse biological reactions and are often categorized as haem and non-haem proteins. These catalyse the oxidation of phenolics and lignins at the expense of hydrogen peroxide. Based on their source and activity, peroxidases have been classified into lignin peroxidase, manganese-dependent peroxidase and versatile peroxidase. Lignin peroxidase participate in the degradation of lignin, a constituent of plant cell wall. It also oxidizes aromatic compounds and plays a key role in aromatic waste management. However, the mechanism of action is still not well understood (Piontek et al., 2001).

Lipases participate in the hydrolysis of lipids. These have been extracted from a variety of sources such as bacteria, fungi, actinomycetes and animal cells. However, microbial lipases have been reported to be versatile and are widely used. Lipases can catalyse a diverse range of reactions including esterification, alcoholysis and aminolysis (Riffaldi et al., 2006) (Figure 9.7).

Cellulases are involved in the degradation of cellulose into glucose and have been a subject of intense scientific research. Cellulases are reported to be a mixture of three major groups of cellulose hydrolyzing enzymes: endoglucanase (attacks regions of low crystallinity in the cellulose fibre and creates free chain ends), exoglucanase or cellobiohydrolase (degrades the cellulose molecule further by removing cellobiose units from the free chain ends) and β-glucosidase which hydrolyzes cellobiose to glucose units (Figure 9.8).

Proteases are known to hydrolyze the peptide bonds in proteins, both of biological and industrial origin, and are extensively used in food, leather, pharmaceutical and detergent industry. They are categorized based on their site of action into endopeptidases and exopeptidases. Endopeptidases are further classified based on the position of active site into serine peptidases, cysteine peptidases, aspartic peptidases and metallopeptidases. The exopeptidases are classified as either aminopeptidases (act on amino terminus) or carboxypeptidases (act on carboxy terminus). The proposed mechanism of protease action is summarized in Figure 9.9.

Table 9.2 summarizes the roles of various enzymes from different sources in waste treatment and management.

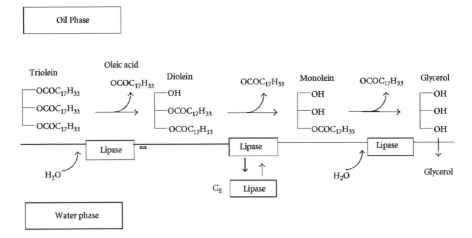

FIGURE 9.7 Proposed Mechanism for Triolein Hydrolysis by *Candida Rugosa* Lipase in Biphasic Oil-Water System. CE Represents the Enzyme Concentration in the Bulk of the Water Phase.

Source: Karigar and Rao (2011); Hermansyah et al. (2007).

Role of Microbes in Solid Waste Management

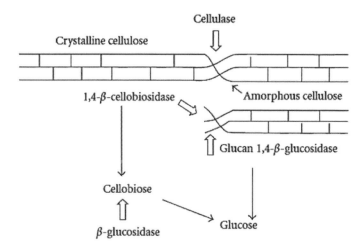

FIGURE 9.8 Proposed Mechanism for the Hydrolysis of Cellulose by the Fungal Cellulase Enzyme System

Source: Karigar and Rao (2011).

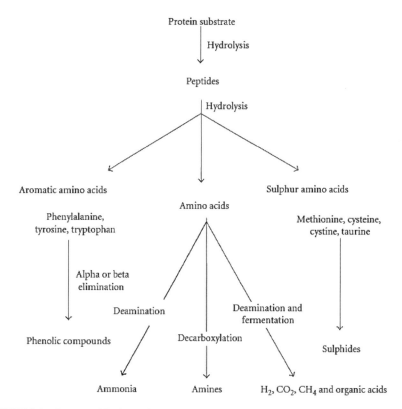

FIGURE 9.9 Proposed Pathway for Protease Hydrolysis

Source: Karigar and Rao (2011).

TABLE 9.2

Applications of Different Enzymes in Waste Management

Enzyme	Source	Applications	Reference
α-amylase (EC 3.2.1.1)	vegetable waste Microorganisms such as *Aspergillus niger, A. awamori, A. oryzae, A. tamarii, Bacillus subtilis, B. licheniformis, Rhizopusoryzae, Candida guilliermondii, Thermomyces lanuginosusetc.*	Hydrolysis of starch	Gangadharan et al. (2008); Said et al. (2014)
Glucoamylase (EC. 3.2.1.3)	Bacteria	Hydrolysis of starch	Blasheck (1992); Shoemaker (1986)
Protease	Wheat bran, soybean meal, cottonseed meal etc.; microbes such as *Bacillus subtilis, Pseudomonas marinoglutinosa, Aspergillus oryzaev* etc.	Solubilization of fish and meat remains, improving sludge dewatering	Castro and Sato (2013); Shoemaker (1986); Dalev (1994)
Oxidoreductases	Bacteria, Fungi and Higher plants	Humification of phenolic substances, detoxification of xenobiotics such as phenolic and anilinic compounds, Degradation of azo dyes	Gianfreda et al. (1999); Bollag and Dec (1998); Vidali (2001); Park et al. (2006); Williams (1977); Husain (2006)
Oxygenases	Bacterial	Degradation of Halogenated organic compounds	Fetzner and Lingens (1994)
Laccases (EC 1.10.3.2)	Plants, insects, and microbes such as *Rhizoctonia praticola, Heterobasidion annosum, Trametes versicolor* etc.	Oxidation of phenolic and aromatic substrates	Gianfreda et al. (1999); Mai et al. (2000); Giardina et al. (1995); Rezende et al. (2005).

Enzyme	Source	Application	References
Peroxidases • Horseradish peroxidase(HRP)EC 1.11.1.7 • *Lignin peroxidase or ligninase or diarylpropaneoxygenase* • *Mn-peroxidase*	Bacteria, Yeast, Plant and Animal Horseradish roots, water hyacinth, tomato, white radish *Phanerochaete chrysosporium* *P. chrysosporium*	Oxidation of lignin and other phenolic compounds Removal of phenols and aromatic compounds, decolourization of Kraft bleaching effluents Oxidation of monoaromatic phenols and aromatic dyes	Hiner et al. (2002); Koua et al.(2009); Aitken and Irvine (1989) Venkatadri and Irvine (1993) Aitken et al. (1994)
Parathion hydrolase	*Pseudomonas sp.* *Flavobacterium* *Streptomyces*	Hydrolyzation of organophosphate pesticides	Smith et al. (1982); Munnecke (1977, 1987); Caldwell and Raushel (1991); Coppella et al. (1990)
Pectin Lyase (EC 4.2.2.10)	*Clostridium beijerinckii*	Pectin degradation	Blasheck (1992)
Lipases EC 3.1.1.3	Bacteria, Plant, Actinomycetes, and Animal cell	Hydrolysis of triacylglycerols, Interesterification, Esterification, Alcoholysis and Aminolysis of various substrates	Prasad and Manjunath (2011)
Cellulases (EC 3.2.1.4)	Microorganisms	Degradation of crystalline cellulose to glucose, Hydrolysis of cellulosic sludges from pulp and paper to produce sugars and alcohol, hydrolysis of cellulose in municipal solid waste to sugars and other energy sources.	Rixon et al. (1992); Duff et al. (1994); Duff et al. (1995); Lagerkvist and Chen (1993); Rivers and Emert (1988); Thomas et al. (1993)
Phosphatase	Citrobacter sp.	Removal of heavy metals	Macaskie and Dean (1984); Macaskie et al. (1987)
Tyrosinase (EC 1.14.18.1)	Mushroom	Removal of phenols	Bollag (1992) Atlow et al. (1984) Wada et al. (1993) Sun et al. (1992) Wada et al. (1992)

9.7 CONCLUSION

Human populations in both developed and developing countries generate huge volumes of heterogeneous solid waste that poses a threat to the environment. Management of such a waste present a great challenge to the agencies due to its diverse composition and potential toxicity to soil, air and water. The situation has been compounded further by the introduction of e-waste, which represents an entirely different and complex matrix in itself. Therefore, the production, commercialization and popularization of biodegradable polymers and such other bio-based products gains significance to minimize the use of traditional polymers. One such commercially available biodegradable polymer is polylactic acid, which is also known as polylactide. Moreover, cellulose and resin based biodegradable polymers are also studied by researchers with promising results.

REFERENCES

Aitken, M.D. and Irvine, R.L., 1989. Stability testing of ligninase and Mn-peroxidase from *Phanerochaete chrysosporium*. *Biotechnology and Bioengineering*. 34:1251–1260.

Aitken, M.D., Massey, I.J., Chen, T. and Heck, P.E., 1994. Characterization of reaction products from the enzyme catalysed oxidation of phenolic pollutants. *Water Research*. 28:1879–1889.

Ajila, C.M., Aalami, M., Leelavathi, K. and Rao, U.P., 2010. Mango peel powder: A potential source of antioxidant and dietary fiber in macaroni preparations. *Innovative Food Science and Emerging Technologies*. 11:219–224.

Ajila, C.M., Bhat, S.G. and Rao, U.P., 2007. Valuable components of raw and ripe peels from two Indian mango varieties. *Food Chemistry*. 102:1006–1011.

Ameen, F., Moslem, M., Hadi, S. and Al-Sabri, A.E., 2015. Biodegradation of diesel fuel hydrocarbons by mangrove fungi from Red Sea Coast of Saudi Arabia. *Saudi Journal of Biological Sciences*. 23(2):211–218.

Atlow, S.C., Bonadonna-Aparo, L. and Klibanov, A.M., 1984. Dephenolization of industrial wastewaters catalysed by polyphenol oxidase. *Biotechnology and Bioengineering*. 26:599–603.

Blasheck, H.P., 1992. Approaches to making the food processing industry more environmentally friendly. *Trends in Food Science and Technology*. 3:107–110.

Bollag, J.M., 1992. Decontaminating soil with enzymes. *Environmental Science and Technology*. 26:1876–1881.

Bollag, J.M. and Dec, J., 1998. Use of Plant material for the removal of pollutants by polymerization and binding to humic substances in *Tech. Rep. R-82092*. Center for Bioremediation and Detoxification Environmental Resources Research Institute the Pennsylvania State University, University Park, PA.

Borah, D., 2018. Microbial bioremediation of petroleum hydrocarbon: An overview. In Kumar, V. et al. (eds.), *Microbial Action on Hydrocarbons*, Springer Nature Singapore Pte Ltd. 321–341. https://doi.org/10.1007/978-981-13-1840-5_13

Caldwell, S.R. and Raushel, F.M., 1991. Detoxification of organophosphate pesticides using an immobilized phosphotriesterase from *Pseudomonas diminuta*. *Biotechnology and Bioengineering*. 37:103–109.

Castro, R.J.S. and Sato, H.H., 2013. Synergistic effects of agro-industrial wastes on simultaneous production of protease and α-amylase under solid state fermentation using a simplex centroid mixture design. *Industrial Crops and Products*. 49:813–821.

Coppella, S.J., Dela Cruz, N., Payne, G.F., Pogell, B., Speedie, M.K., Karns, J.S., Sybert, E.M. and Connor, M.A., 1990. Genetic engineering approach to toxic waste management: Case study for organophosphate waste treatment. *Biotechnology Progress*. 6:76–81.

Role of Microbes in Solid Waste Management

Dalev, P. G., 1994. Utilisation of waste feathers from poultry slaughter for production of a protein concentrate. *Bioresource Technology*. 48:265–267.

Deshmukh, R., Anshuman, A., Khardenavis, H. and Purohit, J., 2017. Bioprocess for solid waste management in book title. In Purohit, H. J., Kalia, V. C. and Vaidya, A. N. and Khardenavis, A. A. (eds.), *Optimization and Applicability of Bioprocesses*, Springer Nature Singapore Pte Ltd. 73–99. https://doi.org/10.1007/978-981-10-6863-8_4

Duff, S. J., Moritz, J. W. and Andersen, K. L., 1994. Simultaneous hydrolysis and fermentation of pulp mill primary clarifier sludge. *Canadian Journal of Chemical Engineering*. 72:1013–1020.

Duff, S. J., Moritz, J. W. and Casavant, T. E., 1995. Effect of surfactant and particle size reduction on hydrolysis of deinking sludge and non-recyclable newsprint. *Biotechnology and Bioengineering*. 45:239–244.

EPA. 1997. *Evaluation of Emissions from the Open Burning of Household Waste in Barrels*. *EPA-600/R-97-134a*. U.S. Environmental Protection Agency, Control Technologies Center, Research Triangle Park, NC.

FAO, 2011. *Global Food Losses and Food Waste: Extent, Causes and Prevention*. FAO, Rome.

Fetzner, S. and Lingens, F., 1994. Bacterial dehalogenases: Biochemistry, genetics, and biotechnological applications. *Microbiological Reviews*. 58(4):641–685.

Gangadharan, D., Sivaramakrishnan, S., Nampoothiri, K. M., Sukumaran, R. K. and Pandey, A., 2008. Response surface methodology for the optimization of alpha amylase production by *Bacillus amyloliquefaciens*. *Bioresource Technology*. 99:4597–4602.

Gianfreda, L., Xu, F. and Bollag, J. M., 1999. Laccases: A useful group of oxidoreductive enzymes. *Bioremediation Journal*. 3(1):1–25.

Giardina, P., Cannio, R., Martirani, L., Marzullo, L., Palmieri, G. and Sannia, G., 1995. Cloning and sequencing of a laccase gene from the lignin-degrading basidiomycete *Pleurotus ostreatus*. *Applied and Environmental Microbiology*. 61(6):2408–2413.

Gopal, M., Gupta, A., Sunil, E. and Thomas, V. G., 2009. Amplification of plant beneficial microbial communities during conversion of coconut leaf substrate to vermicompost by *Eudrilus sp. Curr Microbiol*. 59:15–20.

Hermansyah, H., Wijanarko, A., Gozan, M., Arbianti, R., Utami, T. S., Kubo, M., Shibasaki-Kitakawa, N. and Yonemoto, T., 2007. Consecutive reaction model for triglyceride hydrolysis using lipase. *Jurnal Teknologi*. 2:151–157.

Hiner, N. P., Ruiz, J. H. and Rodri, J. N., 2002. Reactions of the class II peroxidases, lignin peroxidase and *Arthromyces ramosus* peroxidase, with hydrogen peroxide: Catalase-like activity, compound III formation, and enzyme inactivation. *The Journal of Biological Chemistry*. 277(30):26879–26885.

Husain, Q., 2006. Potential applications of the oxidoreductive enzymes in the decolorization and detoxification of textile and other synthetic dyes from polluted water: A review. *Critical Reviews in Biotechnology*. 26(4):201–221.

Ihuoma, S. O., 2012. Characterization and quantification of solid and liquid wastes generated at the University of Ibadan, Ibadan, Nigeria. MSc. thesis Presented to the Department of Agricultural and Environmental Engineering, University of Ibadan, Nigeria.

Insam, H., Franke-Whittle, I. and Goberna, M., 2010. *Microbes in Aerobic and Anaerobic Waste Treatment in Microbes at Work*, edited by Franke-Whittle, H.I.I. and Goberna, M., Springer-Verlag, Berlin Heidelberg. 1–34. doi 10.1007/978-3-642-04043-6_1

Jia, C., Li, X., Allinson, G., Liu, C. and Gong, Z., 2016. Composition and morphology characterization of exopolymeric substances produced by the PAH-degrading fungus of *Mucor mucedo*. *Environmental Science and Pollution Research*. 23(9):8421–3840.

Karigar, C. S. and Rao, S. S., 2011. Role of microbial enzymes in the bioremediation of pollutants: A review. *Enzyme Research*. doi: 10.4061/2011/805187

Koua, D., Cerutti, L., Falquet, L., Sigrist, C. J., Theiler, G., Hulo, N. and Dunand, C., 2009. PeroxiBase: A database with new tools for peroxidase family classification. *Nucleic Acids Research*. 37(1):D261–D266.

Lagerkvist, A. and Chen, H., 1993. Control of two-step anaerobic degradation of municipal solid waste (MSW) by enzyme addition. *Water Science and Technology*. 27:47–56.

Leung, M., 2004. Bioremediation: Techniques for cleaning up a mess. *Journal of Biotechnology*. 2:18–22.

Liang, G., Mo, Y. and Zhou, Q., 2010. Novel strategies of bioleaching metals from printed circuit boards (PCBs) in mixed cultivation of two acidophiles. *Enzyme and Microbial Technology*. 47(7):322–326.

Luo, Q., Zhang, J.G., Shen, X.R., Fan, Z.Q., He, Y. and Hou, D.Y., 2012. Isolation and characterization of marine diesel oil-degrading Acinetobacter sp. strain Y2. *Annals of Microbiology*. 63:633–640.

Macaskie, L.E. and Dean, A.C.R., 1984. Cadmium accumulation by a Citrobacter sp. *Journal of General Microbiology*. 130:53–62.

Macaskie, L.E., Wates, J.M. and Dean, A.C.R., 1987. Cadmium accumulation by a *Citrobacter sp.* immobilized on gel and solid supports: Applicability to the treatment of liquid wastes containing heavy metal cations. *Biotechnology and Bioengineering*. 30:66–73.

Mai, C., Schormann, W., Milstein, O. and Huttermann, A., 2000. Enhanced stability of laccase in the presence of phenolic compounds. *Applied Microbiology and Biotechnology*. 54(4):510–514.

Mandri, T. and Lin, J., 2007. Isolation and characterization of engine oil degrading indigenous microorganisms in Kwazulu-Natal, South Africa. *African Journal of Biotechnology*. 6:23–27.

Munnecke, D.M., 1977. Properties of an immobilized pesticide hydrolysing enzyme. *Applied Environmental Microbiology*. 33:503–507.

Munnecke, D.M., 1987. Detoxification of pesticide using soluble or immobilised enzymes. *Process Biochemistry*. 13:14–17.

Pant, D., Giri, A. and Dhiman, V., 2018. Bioremediation techniques for E-waste management. In Varjani, S.J., Gnansounou, E., Gurunathan, B., Pant, D., Zakaria, Z.A. (eds.), *Waste Bioremediation, Energy, Environment, and Sustainability*, Springer Nature Singapore Pte Ltd, Singapore. https://doi.org/10.1007/978-981-10-7413-4_5

Park, J.W., Park, B.K. and Kim, J.E., 2006. Remediation of soil contaminated with 2,4-dichlorophenol by treatment of minced shepherd's purse roots. *Archives of Environmental Contamination and Toxicology*. 50(2):191–195.

Pathma, J. and Sakthivel, N., 2012. Microbial diversity of vermicompost bacteria that exhibit useful agricultural traits and waste management potential. *SpringerPlus*. 1:26.

Pham, V.A. and Ting, Y.P., 2009. Gold bioleaching of electronic waste by cyanogenic bacteria and its enhancement with bio-oxidation. *Advanced Materials Research. Trans Tech Publications*. 71:661–664.

Piontek, K., Smith, A.T. and Blodig, W., 2001. Lignin peroxidase structure and function. *Biochemical Society Transactions*. 29(2):111–116.

Prasad, M.P. and Manjunath, K., 2011. Comparative study on biodegradation of lipid-rich wastewater using lipase producing bacterial species. *Indian Journal of Biotechnology*. 10(1):121–124.

Rezende, M.I., Barbosa, A.M., Vasconcelos, A.F.D., Haddad, R. and Dekker, R.F.H., 2005. Growth and production of laccases by the ligninolytic fungi, *Pleurotus ostreatus* and *Botryosphaeria rhodina*, cultured on basal medium containing the herbicide, Scepter (imazaquin). *Journal of Basic Microbiology*. 45(6):460–469.

Riffaldi, R., Levi-Minzi, R., Cardelli, R., Palumbo, S. and Saviozzi, A., 2006. Soil biological activities in monitoring the bioremediation of diesel oil-contaminated soil. *Water, Air, and Soil Pollution*. 170(1–4):3–15.

Rivers, D.B. and Emert, G.H., 1988. Factors affecting the enzymatic hydrolysis of municipal-solid-waste components. *Biotechnology and Bioengineering*. 31:278–281.

Role of Microbes in Solid Waste Management

Rixon, J.E., Ferreira, L.M.A., Durrant, A.J., Laurie, J.I., Hazlewood, G.P. and Gilbert, H.J., 1992. Characterization of the genecelD and its encoded product 1,4-β-D-glucanglucohydrolase D from *Pseudomonas fluorescens* sub sp. Cellulose. *Biochemical Journal*. 285(3):947–955.

Rudra, S.G., Nishad, J., Jakhar, N. and Kaur, C., 2015. Food industry waste: Mine of nutraceuticals. *International Journal of Environmental Technology*. 4:205–229.

Sagar, N.A., Pareek, S., Sharma, S., Yahia, E.M. and Lobo, M.G., 2018. Fruit and vegetable waste: Bioactive compounds, their extraction, and possible utilization. *Comprehensive Reviews in Food Science and Food Safety*. 17:512–531.

Said, A., Leila, A., Kaouther, D. and Sadia, B., 2014. Date wastes as substrate for the production of α-amylase and invertase. *Iranian Journal of Biotechnology*. 12:41–49.

Shoemaker, S., 1986. The use of enzymes for water management in the food industry. In Harlander, S.K. and Labuza, T.P. (eds.), *Biotechnology in Food Processing*, Nayes Publications, Park Ridge. 259–267.

Smith, J.M., Payne, G.F. and Lumpkin, J.A., 1982. Enzyme based strategy for toxic waste treatment and waste minimization. *Biotechnology and Bioengineering*. 39:741–752.

Su, W.T., Wu, B.S. and Chen, W.J., 2011. Characterization and biodegradation of motor oil by indigenous Pseudomonas aeruginosa and optimizing medium constituents. *Journal of the Taiwan Institute of Chemical Engineers*. 42:689–695.

Sun, W.Q., Payne, G.F., Moas, M., Chu, J.H. and Wallace, K.K., 1992. Tyrosinase reaction/chitosan adsorption for removing phenols from wastewater. *Biotechnology Progress*. 8:179–186.

Thomas, L., Jungschaffer, G. and Sprossler, B., 1993. Improved sludge dewatering by enzymatic treatment. *Water Science and Technology*. 28(1):189–192.

Thummes, K., Kämpfer, P. and Jäckel, U., 2007. Temporal change of composition and potential activity of the thermophilic archaeal community during the composting of organic material. *Systematic and Applied Microbiology*. 30:418–429.

Vaz-Moreira, I., Maria, E., Maria, C.M., Olga, M. and Nunes, C., 2008. Diversity of bacterial isolates from commercial and homemade composts. *Microbial Ecology*. 55:714–722.

Venkatadri, R. and Irvine, R.L., 1993. Cultivation of *Phanerochaete chrysosporium* and production of lignin peroxidase in novel bio-film reactor systems: Hollow fiber reactor and silicone membrane reactor. *Water Research*. 27:591–596.

Vidali, M., 2001. Bioremediation: An overview. *Pure and Applied Chemistry*. 73(7):1163–1172.

Wada, S., Ichikawa, H. and Tatsumi, K., 1992. Removal of phenols with tyrosinase immobilized on magnetite. *Water Science and Technology*. 26(9–11):2057–2059.

Wada, S., Ichikawa, H. and Tatsumi, K., 1993. Removal of phenols from wastewater by soluble and immobilized tyrosinase. *Biotechnology and Bioengineering*. 42:854–858.

Williams, P.P., 1977. Metabolism of synthetic organic pesticides by anaerobic microorganisms. *Residue Reviews*. 66:63–135.

Wilson, D.B., 2008. Three microbial strategies for plant cell wall degradation. *Annals of the New York Academy of Sciences*. 1125:289–297.

Yenn, R., Borah, M., Boruah, H.P., Roy, A.S., Baruah, R., Saikia, N., Sahu, O.P. and Tamuli, A.K., 2014. Phytobioremediation of abandoned crude oil contaminated drill sites of Assam with the aid of a hydrocarbon-degrading bacterial formulation. *International Journal of Phytoremediation*. 16(7–12):909–925.

10 Management of Solid and Hazardous Waste as per Indian Legislation

Ashok K. Rathoure and Unnati Patel

CONTENTS

10.1 Introduction 152
10.2 Type of Wastes and Their Rules 153
10.3 Plastic Waste (Management and Handling)
     Rules, 2011 153
    10.3.1 Plastic Waste Generation in India 153
    10.3.2 Type of Plastic 154
    10.3.3 Plastic Waste Management (as per Plastic Waste Management
        Rules, 2016) 155
    10.3.4 Technologies for Disposal of Plastic Waste 155
        10.3.4.1 Utilization of Plastic Waste in Road Construction 155
        10.3.4.2 Co-Processing of Plastic Waste in Cement Kilns 156
        10.3.4.3 Conversion of Plastic Waste into Fuel-Oil: Refused-
                Derived Fuel (RDF) 157
        10.3.4.4 Disposal of Plastic Waste through Plasma Pyrolysis
                Technology (PPT) 157
10.4 E-Waste (Management and Handling) Rules, 2011 157
    10.4.1 Indian Scenario for E-Waste Management 157
    10.4.2 E-Waste Treatment 158
        10.4.2.1 1st Level Treatment 158
        10.4.2.2 2nd Level Treatment 158
        10.4.2.3 3rd Level Treatment 159
10.5 Bio-Medical Waste Management Rules, 2016 160
    10.5.1 Indian Scenario of Bio-Medical Waste Management 160
    10.5.2 Bio-Medical Waste Management 160
    10.5.3 Bio-Medical Wastes Categories and Their Segregation,
        Collection, Treatment, Processing and Disposal Options 161
10.6 Construction and Demolition Waste Management Rules, 2016 161
    10.6.1 Indian Scenario for Construction and Demolition Waste 165
    10.6.2 C&D Waste Generation in Cities/Towns 165
    10.6.3 Construction and Demolition Waste Management 166
        10.6.3.1 Recycling and Reuse of Construction and Demolition
                Waste 166

10.6.3.2 Disposal (Landfill)	167

10.7 Hazardous and Other Wastes (Management and Transboundary Movement) Rules, 2016 167
    10.7.1 Indian Scenario for Hazardous Waste 167
    10.7.2 Classification of Hazardous Waste 168
    10.7.3 Hazardous Waste Treatment 168
        10.7.3.1 Physical and Chemical Treatment 168
        10.7.3.2 Thermal Treatment 169
        10.7.3.3 Biological Treatment 170
10.8 Municipal Solid Wastes (Management and Handling) Rules, 2000 170
    10.8.1 Indian Scenario of Municipal Solid Waste 170
    10.8.2 Municipal Solid Waste Management 171
        10.8.2.1 Landfill 172
        10.8.2.2 Incineration 172
        10.8.2.3 Biodegradation processes 172
        10.8.2.4 Composting 172
10.9 Conclusion 172
References 173

10.1 INTRODUCTION

Notification of treatment and management of waste is covered under the Environment (Protection) Act, 1986, regulated by the Ministry of Environment, Forest and Climate Change Government of India by PARIVESH (Pro-Active and Responsive facilitation by Interactive and Virtuous Environmental Single-window Hub) Portal. The Environment (Protection) Act was enacted in 1986 with the objective of providing for the protection and improvement of the environment. It empowers the central government to establish authorities (under section 3(3)) charged with the mandate of preventing environmental pollution in all its forms and to tackle specific environmental problems that are peculiar to different parts of the country. The EPA Act was enacted in the wake of the Bhopal tragedy. The Government of India enacted the Environment Protection Act, 1986 under Article 253 of the Constitution. This act was passed in March 1986; it came into force on 19 November 1986 and it has 26 sections. Wastes means substances or objects which are disposed of or are intended to be disposed of or are required to be disposed of by the provisions of national law define by UNEP. Solid and liquid, hazardous and non-toxic wastes are generated in our households, offices, schools, hospitals and industries. Waste management means the collection, storage, transportation reduction, reuse, recovery, recycling, composting or disposal of plastic waste in an environmentally safe manner (CPCB, 2016f). The 3 R's stands for reduce, reuse and recycle and are the most widely used strategy for the treatment of solid waste in all over the world. However, such traditional methods have their own advantages, disadvantages and limits for their employment in waste management. The treatment and management of different types of waste have their own rules.

10.2 TYPE OF WASTES AND THEIR RULES

Type of waste and their rules as per Environment (Protection) Rule, 1986, are described in Table 10.1.

10.3 PLASTIC WASTE (MANAGEMENT AND HANDLING) RULES, 2011

This rule applies to every waste generator, local body, Gram Panchayat, manufacturer, importer and producer. Plastic means a material which contains an essential ingredient like high polymer such as polyethylene terephthalate, high density polyethylene, low density polyethylene, polypropylene, polystyrene resins, multi-materials like acrylonitrile butadiene styrene, polyphenylene oxide, polycarbonate or polybutylene terephthalate. Plastic waste means any plastic discarded after use or after their use is over (CPCB, 2016f).

10.3.1 PLASTIC WASTE GENERATION IN INDIA

As per the study conducted by Central Pollution Control Board (CPCB) in 60 major cities of India, it has been observed that around 4,059 T/day of plastic waste is generated from these cities. The fraction of plastic waste in total Municipal Solid Waste

TABLE 10.1
Wastes and Their Rules

S. No.	Type of Waste	Rule for the Waste
1.	Plastic waste	Plastic Waste (Management and Handling) Rules, 2011. Amendment G.S.R. 320 (E) (18–03–2016): Plastic Waste Management Rules, 2016
2.	E-waste	E-Waste (Management and Handling) Rules, 2011. Amendment G.S.R. 338 (E) (23–03–2016): E-waste (Management) Rules, 2016
3.	Biomedical waste	The Bio-Medical Waste (Management and Handling) Rules, 1998. Amendment G.S.R. 343(E) (28–03–2016): Bio-Medical Waste Management Rules, 2016
4.	Construction and demolition waste	G.S.R. 317(E) (29–03–2016): Construction and Demolition Waste Management Rules, 2016
5.	Hazardous waste	Hazardous Wastes (Management and Handling) Rules in 1989. Amendment G.S.R No. 395 (E) (04–04–2016): Hazardous and Other Wastes (Management and Transboundary Movement) Rules, 2016
6.	Municipal solid waste	Municipal Solid Wastes (Management and Handling) Rules in 2000. Amendment S.O. 1357(E) (08–04–2016): Solid Waste Management Rules, 2016

(MSW) varies from 3.10% (Chandigarh) to 12.47% (Surat). Average plastic waste generation is around 6.92% of municipal solid waste. With extrapolation of the plastic waste generation data from 60 major cities, it is estimated that around 25,940 T/day of plastic waste is generated in India. As per the results of the study, out of total plastic waste, around 94% waste comprises thermoplastic content, which is recyclable such as PET, LDPE, HDPE and PVC and the remaining 6% belongs to the family of thermoset and other categories of plastics such as sheet moulding compound (SMC), fibre reinforced plastic (FRP), multi-layered and thermocol, which are non-recyclable (CPCB, 2017b).

10.3.2 Type of Plastic

The rapid rate of urbanization and development has led to increase in consumption of plastic products and plastic waste generation. It is a fact that plastics waste constitutes a significant portion of the total municipal solid waste (MSW) generated in India (Bureau of Indian Standards, March 1998. IS 14534: 1998). To identify the raw material of plastic products, the symbols defined by Society of the Plastics Industry (SPI, USA) shall be marked on each product. Different types of plastics and their uses are given in Table 10.2.

TABLE 10.2

Different Types of Plastics and Their Uses (CPCB, 2017b)

S. No.	Symbol	Short Name	Scientific Name	Use
1.		PET	Polyethylene terephthalate	Soft drink bottles, furniture, carpet, paneling etc.
2.		HDPE	High-density polyethylene	Bottles, carry bags, milk pouches, recycling bins, agricultural pipe, base cups, playground equipment etc.
3.		PVC	Polyvinyl chloride	Pipe, window profile, fencing, flooring, shower curtains, lawn chairs, non-food bottles and children's toys etc.
4.		LDPE	Low-density polyethylene	Plastic bags, various containers, dispensing bottles, wash bottles, tubing etc.
5.		PP	Polypropylene	Auto parts, industrial fibres, food containers, dishware etc.

(Continued)

TABLE 10.2 (Continued)
Different Types of Plastics and Their Uses (CPCB, 2017b)

S. No.	Symbol	Short Name	Scientific Name	Use
6.		PS	Polystyrene	Cafeteria trays, plastic utensils, toys, video cassettes and cases, clamshell containers, insulation board etc.
7.		O	Other	Thermoset Plastics, Multilayer and Laminates, Bakelite, Polycarbonate, Nylon SMC, FRP etc.

10.3.3 PLASTIC WASTE MANAGEMENT (AS PER PLASTIC WASTE MANAGEMENT RULES, 2016)

The plastic waste management by the local bodies in their respective jurisdiction shall be as under:

1. Plastic waste, which can be recycled, shall be channelized to registered plastic waste recycler and recycling of plastic shall confirm to the Indian Standard: IS 14534:1998 titled as Guidelines for Recycling of Plastics, as amended from time to time.
2. Local bodies shall encourage the use of plastic waste (preferably the plastic waste which cannot be further recycled) for road construction as per Indian Road Congress guidelines or energy recovery or waste to oil and so forth. The standards and pollution control norms specified by the prescribed authority for these technologies shall be complied with.
3. Thermoset plastic waste shall be processed and disposed of as per the guidelines issued from time to time by the Central Pollution Control Board.

The inert from recycling or processing facilities of plastic waste shall be disposed of in compliance with the Solid Waste Management Rules, 2000 or as amended from time to time.

10.3.4 TECHNOLOGIES FOR DISPOSAL OF PLASTIC WASTE

Provision 5(b) of PWM Rules, 2016, encourages the use of technologies for disposal of plastic waste. The major technologies for the disposal of plastic waste are discussed in Figure 10.1. Different type of technologies for disposal of plastic waste is given below as per CPCB 2017b.

10.3.4.1 Utilization of Plastic Waste in Road Construction

Plastic waste is collected and segregated (except chlorinated/brominated plastic waste) from mixed MSW. The segregated plastic waste is stored and should

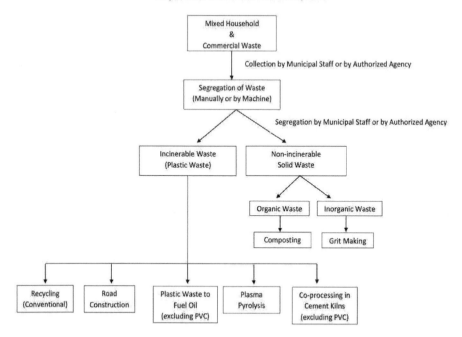

FIGURE 10.1 Flow Diagram for Plastic Waste Management
Source: CPCB (2017b).

be transported to the location working site for drying. The dried plastic waste is shredded to 2–4 mm size and added to heated stone aggregate followed by mixing. Further, the coated aggregate is mixed with hot bitumen, which is used for laying and compaction. The use of plastic waste in road construction shall follow the IRC: SP:98–2013, titled as Guidelines for the use of waste plastic in hot bituminous mix (dry mixing) in wearing courses. Presently, several roads have been constructed by using plastic waste with bitumen in many of the states/UTs, such as Tamil Nadu, Himachal Pradesh, Nagaland, West Bengal, Pondicherry and so forth.

10.3.4.2 Co-Processing of Plastic Waste in Cement Kilns

Co-processing in cement plants, plastic waste is used as Alternate Fuel and Raw-material (AFR), subjected to higher temperature around 1,400–1,500°C. During the process, energy is recovered while burning of plastic waste and its inorganic content get fixed with clinker. This technology is used successfully in some of the states where cement plants (which have facility for co-processing of waste) are present, such as Gujarat, Tamil Nadu, Karnataka, Chhattisgarh, Himachal Pradesh, Madhya Pradesh and Odisha.

10.3.4.3 Conversion of Plastic Waste into Fuel-Oil: Refused-Derived Fuel (RDF)

For converting plastic waste into fuel-oil (RDF), plastic waste is collected and segregated. The segregated plastic waste is then fed into multi fractionalization, where the unwanted material is rejected for better handling & processing. The segregated plastic waste (only HD, LD, PP and multilayer packaging except PVC) is then fed into in-vessel for depolymerization system. The catalytic gasolysis in-vessel is designed to handle polymers. The selection of catalyst depends on the type of raw material used. The reactor operates at high temperature and in absence on Air The vapours produced are condensed in the Condensers and collected as crude oil. This technology is used by few municipalities like Vadodara (Gujarat) and NDMC (New Delhi).

10.3.4.4 Disposal of Plastic Waste through Plasma Pyrolysis Technology (PPT)

In plasma pyrolysis, first the plastics waste is fed into the primary chamber at 850°C through a feeder. The waste material dissociates into carbon monoxide, hydrogen, methane, higher hydrocarbons and so forth. Induced draft fan drains the pyrolysis gases and plastics waste into the secondary chamber where these gases are combusted in the presence of excess air. The inflammable gases are ignited with high voltage spark. The secondary chamber temperature is maintained at 1,050°C. The hydrocarbon, CO and hydrogen are combusted into safe carbon dioxide and water. This process is used by few municipalities and hospitals; however, this can be useful for tourist places, hill stations, pilgrimages, coasts and other remote places.

10.4 E-WASTE (MANAGEMENT AND HANDLING) RULES, 2011

E-waste means electrical and electronic equipment, whole or in part discarded as waste by the consumer or bulk consumer as well as rejects from manufacturing, refurbishment and repair processes (CPCB, 2016a). Common electronic equipment includes televisions and monitors, computers, audio/stereo equipment, VCRs, DVD players, video cameras, telephones, fax and copy machines, cellular phones, wireless devices and video game consoles. The two categories of electrical and electronic equipment are (1) IT and telecommunication equipment and (2) consumer electricals and electronics such as TVs, washing machines, refrigerators and air conditioners covered under these rules. The main feature of this rule is extended producer responsibility (EPR) (CPCB, 2016a).

10.4.1 INDIAN SCENARIO FOR E-WASTE MANAGEMENT

The electronics industry is the world's largest and fastest growing manufacturing industry. The threat of Waste Electrical or Electronic Equipment (WEEE) or e-waste consists of obsolete electronic devices. E-waste is one of the fastest growing waste in the world. In developing countries, WEEE accounts for 1% of total solid waste and is expected to grow to 2% by 2010 (UNEP, 2009). Two major problems are associated with generation of WEEE including the large volume generated and safe environmental disposal of e-waste. Studies conducted have shown that about 330,000 tons of e-waste is generated annually in India and the total generation of e-waste was almost about 4.8 tons by 2011 as predicted (Chaturvedi et al., 2007). TRAI reports suggested about 113.26 million new cellular customers in 2008, with an average 9.5 million customers added every month. The cellular market grew from 168.11 million in 2003–2004 to 261.97 million in 2007–2008 (Research Unit (Larrdis) Rajya Sabha Secretariat, 2011). The total amount of e-waste that is generated only about 19,000 tons of the e-waste is recycled, 95% in the informal sector. E-waste contains valuable constituents such as precious metals like silver, gold and copper and hence is economically viable to recycle. The fraction including iron, copper, aluminium, gold and other metals in e-waste is over 60%, while plastics account for about 30% and the hazardous pollutants comprise only about 2.70% (Widmer et al., 2005).

10.4.2 E-Waste Treatment

E-waste treatment technologies are used at the following three levels:

- 1st level treatment
- 2nd level treatment
- 3rd level treatment.

All the three levels of e-waste treatment are based on material flow. The material flows from the first level to the third level of treatment. After the third level treatment, the residue is disposed of either in TSDF or incinerated. The efficiency of operations at first and second level determines the quantity of residues going to TSDF or incineration (TERI, 2014).

10.4.2.1 1st Level Treatment

Unit operations at first level of e-waste treatment are:

- Decontamination: removal of all liquids and gases
- Dismantling: manual/mechanized breaking segregation.

10.4.2.2 2nd Level Treatment

Unit operation at second level of e-waste treatment includes:

- Hammering
- Shredding
- Special treatment processes comprising CRT treatment.

Hammering and shredding objective is size reduction. Electromagnetic and eddy current separation utilizes properties of different elements like electrical conductivity, magnetic properties and density to separate ferrous, nonferrous metal and precious metal fractions. CRT segregated after first level WEEE/e-waste treatment. The CRT is manually removed from its plastic/wooden casing. In depressurization

Waste Management as per Indian Legislation

and splitting, the picture tube is split and the funnel section is then lifted off the screen section and the internal metal mask can be lifted to facilitate internal phosphor coating.

10.4.2.3 3rd Level Treatment

The 3rd level e-waste treatment is carried out mainly to recover ferrous and non-ferrous metals, plastics and other items of economic value.

10.4.2.3.1 Plastic Recycling

There are three different types of plastic recycling options (i.e. chemical recycling, mechanical recycling and thermal recycling). The two major types of plastic resins, which are used in electronics, are thermosets and thermoplastics.

10.4.2.3.2 Mechanical Recycling Process

Contaminated plastic such laminated or painted plastic are removed by grinding, cryogenic method, abrasion/abrasive technique, solvent stripping method and high temperature aqueous based paint removal method. Magnetic separators are used for ferrous metals separation, while eddy current separators are used for nonferrous metals separation. Air separation system is used to separate light fractions such as paper, labels and films.

10.4.2.3.3 Chemical Recycling Process

This process was developed by the Association of Plastic Manufacturers in Europe (APME). Mixed plastic waste is first depolymerized at about 350–400°C and de-halogenated (Br and Cl). This step also includes removal of metals. In hydrogenation unit 1, the remaining polymer chains from depolymerized unit are cracked at temperatures between 350–400°C and hydrogenated at pressure greater than 100 bar. After hydrogenation, the liquid product is subjected to distillation and left-over inert material is collected in the bottom of distillation column as residue, hydrogenation bitumen. In hydrogenation unit 2, high-quality products like off gases and sync rude are obtained by hydro-treatment, which are sent to petrochemical process.

10.4.2.3.4 Thermal Recycling Process

In thermal recycling process, plastics are used as fuel for energy recovery. Since plastics have high calorific value, which is equivalent or greater than coal, they can be combusted to produce heat energy in cement kilns. APME has found thermal recycling of plastic as the most environmentally sound option for managing WEEE/e-waste plastic fraction.

10.4.2.3.5 Precious Metals Recovery

In this process the anode slime recovered from copper electrolytic process used for precious metal recovery. The process involves the following steps. Anode slime is leached by pressure. The leached residue is then dried and, after the addition of fluxes, smelted in a precious metals furnace. Selenium is recovered during smelting. The remaining material from smelter is caste into anode and undergoes electrolysis to form high purity silver cathode and anode gold slime. The anode gold slime is further leached and high purity gold, palladium and platinum sludge are recovered (TERI, 2014).

10.5   BIO-MEDICAL WASTE MANAGEMENT RULES, 2016

Bio-medical Waste (Management & Handling) Rules, 1998 were notified by the Ministry of Environment, Forests and Climate Change (MoEF&CC) under the Environment (Protection) Act, 1986. In exercise of the powers conferred by Section 6, 8 and 25 of the Environment (Protection) Act, 1986 (29 of 1986), and in supersession of the Bio-Medical Waste (Management and Handling) Rules, 1998 and further amendments made thereof, the Central Government vide G.S.R. 343(E) dated 28 March 2016 published the Bio-medical Waste Management Rules, 2016. This rule apply to all persons who generate, collect, receive, store, transport, treat, dispose or handle bio medical waste in any form including hospitals, nursing homes, clinics, dispensaries, veterinary institutions, animal houses, pathological laboratories, blood banks, clinical establishments, research or educational institutions, health camps, medical or surgical camps, vaccination camps, blood donation camps, first aid rooms of schools, forensic laboratories and research labs (CPCB, 2017a). Biomedical waste means any waste which is generated during the diagnosis, treatment or immunization of human beings or animals or research activities pertaining thereto or in the production or testing of biological or in health camps (CPCB, 2016b).

10.5.1   INDIAN SCENARIO OF BIO-MEDICAL WASTE MANAGEMENT

A nationwide survey performed by the International Clinical Epidemiology Network in 25 districts across 20 states highlighted that only two big cities in India (Chennai and Mumbai) had comparatively better system for BMWM. Improper pre-treatment of BMW at source and improper terminal disposal was the major challenges observed. It was observed that around 82% of primary, 60% of secondary and 54% of tertiary care health facilities were in the red category (i.e. the absence of a credible BMWM in place or ones requiring major improvement). According to the studies conducted by the World Health Organization (WHO) in 22 developing countries showed that the proportion of health-care facility (HCF) that do not use proper waste disposal methods range from 18% to 64%. In India, annually about 0.33 million tons of BMW is generated and rate ranges from 0.5 to 2.0 kg per bed per day. The poor BMWM practices are attributed to lack of awareness and training as was concluded in a recent study. India was one of the first countries to implement BMWM rules in 1998 (amended as draft in 2003, 2011) under Environment Protection Act (EPA), 1986. India was signatory to an international legally binding and environmental treaty, Stockholm Convention, 2004, on persistent organic pollutants (POPs) that aims to eliminate or restrict production of POPs (Capoor and Bhowmik, 2017).

10.5.2   BIO-MEDICAL WASTE MANAGEMENT

The first five steps (segregation, collection, pre-treatment, intramural transportation and storage) are the responsibility of the health care facility. Treatment and disposal

Waste Management as per Indian Legislation

is the primary responsibility of the common biomedical waste treatment facility (CBWTF) operator except for lab and highly infectious waste, which is required to be pre-treated by the HCF (CPCB, 2016h).

Following are the responsibilities of the health care facility (HCF) for management and handling of bio-medical waste:

1. Biomedical waste should be segregated at the point of generation by the person who is generating the waste in designated colour coded bin/container.
2. Biomedical waste and general waste shall not be mixed. Storage time of waste should be as less as possible so that waste storage, transportation and disposal is done within 48 hours.
3. No secondary handling or pilferage of waste shall be done at health care facility. If a CBWTF facility is available at a distance of 75 km from the HCF, biomedical waste should be treated and disposed only through such CBWTF operator.
4. Only laboratory and highly infectious waste shall be pre-treated on-site before sending for final treatment or disposal through a CBWTF operator.
5. Provide bar-code labels on all colour-coded bags or containers containing segregated bio-medical waste before such waste goes for final disposal through a CBWTF.

10.5.3 Bio-Medical Wastes Categories and Their Segregation, Collection, Treatment, Processing and Disposal Options

The Bio-Medical Waste Management Rules, 2016 categorizes the biomedical waste generated from the health care facility into four categories based on the segregation pathway and colour code. Various types of bio medical waste are further assigned to each one of the categories, as detailed in Table 10.3.

1. Yellow category
2. Red category
3. White category
4. Blue category (CPCB, 2016b).

10.6 CONSTRUCTION AND DEMOLITION WASTE MANAGEMENT RULES, 2016

These rules shall apply to every waste resulting from construction, re-modelling, repair and demolition of any civil structure of individual or organization or authority who generates construction and demolition waste such as building materials, debris, rubble. Wastes also include surplus and damaged products and materials arising in the course of construction work or used temporarily during the course of on-site activities. As per Rule construction and demolition waste means waste comprising of building materials, debris and rubble resulting from construction, re-modelling, repair and demolition of any civil structure. (CPCB, 2017c).

TABLE 10.3

Biomedical Waste Categories and Their Segregation, Collection, Treatment, Processing and Disposal (CPCB, 2016b)

Category	Type of Waste	Type of Bag or Container to Be Used	Treatment and Disposal Options
Yellow	(a) Human Anatomical Waste: Human tissues, organs, body parts and fetus below the viability period (as per the Medical Termination of Pregnancy Act 1971, amended from time to time).	Yellow coloured non-chlorinated plastic bags	Incineration or Plasma Pyrolysis or deep burial*
	(b) Animal Anatomical Waste: Experimental animal carcasses, body parts, organs, tissues, including the waste generated from animals used in experiments or testing in veterinary hospitals or colleges or animal houses.		
	(c) Soiled Waste: Items contaminated with blood, body fluids like dressings, plaster casts, cotton swabs and bags containing residual or discarded blood and blood components.		
	(d) Expired or Discarded Medicines: Pharmaceutical waste like antibiotics, cytotoxic drugs including all items contaminated with cytotoxic drugs along with glass or plastic ampoules, vials etc.	Yellow coloured non-chlorinated plastic bags or containers	Expired 'cytotoxic drugs and items contaminated with cytotoxic drugs to be returned back to the manufacturer or supplier for incineration at temperature >1200°C or to common bio-medical waste treatment facility or hazardous waste treatment, storage and disposal facility for incineration at >12000C Or Encapsulation or Plasma Pyrolysis at >12000C. All other discarded medicines shall be either sent back to manufacturer or disposed by incineration.
	(e) Chemical Waste: Chemicals used in production of biological and used or discarded disinfectants	Yellow coloured containers or non-chlorinated plastic bags	Disposed of by incineration or Plasma Pyrolysis or Encapsulation in hazardous waste treatment, storage and disposal facility.

(f) Chemical Liquid Waste: Liquid waste generated due to use of chemicals in production of biological and used or discarded disinfectants, Silver X-ray film developing liquid, discarded Formalin, infected secretions, aspirated body fluids, liquid from laboratories and floor washings, cleaning, house-keeping and disinfecting activities etc.	Separate collection system leading to effluent treatment system	After resource recovery, the chemical liquid waste shall be pre-treated before mixing with another wastewater. The combined discharge shall conform to the discharge norms given in Schedule III.
(g) Discarded linen, mattresses, beddings contaminated with blood or body fluid.	Non-chlorinated yellow plastic bags or suitable packing material	Non- chlorinated chemical disinfection followed by incineration or Plasma Pyrolysis or for energy recovery. In absence of above facilities, shredding or mutilation or combination of sterilization and shredding. Treated waste to be sent for energy recovery or incineration or Plasma Pyrolysis.
h) Microbiology, Biotechnology and other clinical laboratory waste: Blood bags, Laboratory cultures, stocks or specimens of microorganisms, live or attenuated vaccines, human and animal cell cultures used in research, industrial laboratories, production of biological, residual toxins, dishes and devices used for cultures	Autoclave safe plastic bags or containers	Pre-treat to sterilize with no chlorinated chemicals on-site as per National AIDS Control Organization or World Health Organization guidelines thereafter for Incineration.
Red	Contaminated Waste (Recyclable) (a) Wastes generated from disposable items such as tubing, bottles, intravenous tubes and sets, catheters, urine bags, syringes (without needles and fixed needle syringes) and vacutainers with their needles cut) and gloves.	Red coloured non-chlorinated plastic bags or containers	Autoclaving or micro-waving/ hydroclaving followed by shredding or mutilation or combination of sterilization and shredding. Treated waste to be sent to registered or authorized recyclers or for energy recovery or plastics to diesel or fuel oil or for road making, whichever is possible. Plastic waste should not be sent to landfill sites.

(Continued)

TABLE 10.3 (Continued)

Biomedical Waste Categories and Their Segregation, Collection, Treatment, Processing and Disposal (CPCB, 2016b)

Category	Type of Waste	Type of Bag or Container to Be Used	Treatment and Disposal Options
White (Translucent)	Waste sharps including Metals: Needles, syringes with fixed needles, needles from needle tip cutter or burner, scalpels, blades, or any other contaminated sharp object that may cause puncture and cuts. This includes both used, discarded and contaminated metal sharps	Puncture proof, Leak proof, tamper proof containers	Autoclaving or Dry Heat Sterilization followed by shredding or mutilation or encapsulation in metal container or cement concrete; combination of shredding cum autoclaving; and sent for final disposal to iron foundries (having consent to operate from the State Pollution Control Boards or Pollution Control Committees) or sanitary landfill or designated concrete waste sharp pit.
Blue	(a) Glassware: Broken or discarded and contaminated glass including medicine vials and ampoules except those contaminated with cytotoxic wastes	Cardboard boxes with blue coloured marking	Disinfection (by soaking the washed glass waste after cleaning with detergent and Sodium Hypochlorite treatment) or through autoclaving or microwaving or hydro-claving and then sent for recycling.
	(b) Metallic Body Implants	Cardboard boxes with blue coloured marking	

Definitions (Under Construction and Demolition Waste Management Rules, 2016d):

1. Construction means the process of erecting of building or built facility or other structure, or building of infrastructure including alteration in these entities;
2. De-construction means a planned selective demolition in which salvage, reuse and recycling of the demolished structure is maximized;
3. Demolition means breaking down or tearing down buildings and other structures either manually or using mechanical force (by various equipment) or by implosion using explosives. (CPCB, 2016d).

10.6.1 Indian Scenario for Construction and Demolition Waste

Scientific processing and utilization of C&D waste has achieved isolated successes in India. Delhi was the first city to implement a C&D waste management plan through a pilot processing facility developed under a public-private-partnership (PPP) in 2010. After the initial success of the pilot plant processing waste at 500 tonnes per day (TPD), the capacity of the plant has been increased to 2,000 TPD. Ahmedabad was the second city in India to implement C&D waste processing, by adopting a similar PPP model as that in Delhi. A 300 TPD processing facility was launched in 2014, the capacity of which was increased to 600 TPD 2016 after successful operation and now to 1,000 TPD in 2018. In both Delhi and Ahmedabad, the Design Build Operate Finance and Transfer (DBOFT) model is being followed. The municipal corporation contracts a private party and this authorized agency is responsible for both transportation and processing of the C&D waste and develops the necessary infrastructure with its own financing. The municipal corporation offers land to the contracted party for establishing the processing facility and also designates a series of intermediate collection points at favourable locations throughout the city. The authorized agency collects C&D waste from these designated collection points as well as from unauthorized dumps, as directed by the urban authority, and transports it to the processing facility. The municipal corporation pays the authorized agency an agreed fee per tonne of waste that is collected and transported (Ministry of Housing & Urban Affairs, 2018).

10.6.2 C&D Waste Generation in Cities/Towns

Activities which generate C&D waste in cities/towns are mainly from (CPCB, 2017c):

1. Renovation of existing buildings (residential or commercial)
2. Construction of new buildings (residential or commercial or hotel etc.)
3. Excavation/ reconstruction of asphalt/concrete roads
4. Construction of new fly over bridges/under bridges/subways etc.
5. Renovation/ Installation of new water/telephone/internet/sewer pipe lines etc.
6. Present collection and disposal system.

10.6.3 Construction and Demolition Waste Management

C&D waste managed by the concept of 3R which refers to reduce, reuse and recycle particularly in the context of production and consumption is well known today. It is something like using recyclable materials is more than actual practice, reusing of raw materials if possible and reducing use of resources and energy. These can be applied to the entire life cycles of products and services—starting from design and extraction of raw materials to transports, manufacture, use, dismantling and disposal. CPHEEO (2000). Figure 10.2 shows the hierarchy of C&D waste management.

10.6.3.1 Recycling and Reuse of Construction and Demolition Waste

The reuse and recycling of these materials is effective in reducing both cost and environmental impact. The standard point of energy saving and conservation of natural resources, the use of alternative constituents in construction materials is now a global concern. The use of these materials basically depends on their separation and condition of the separated material. A majority of these materials are durable and therefore, have a high potential of reuse. It would, however, be desirable to have quality standards for the recycled materials. Construction and demolition waste can be used in the following manner (CPHEEO, 2000):

- Reuse (at site) of bricks, tiles, stone slabs, timber, conduits, piping railings etc. to the extent possible and depending upon their condition. Among these the ceramic tiles are used in this project.
- Sale/auction of material which cannot be used at the site due to design constraint or change in design.
- Plastics, broken glass, scrap metal etc. can be used by recycling industries.
- Rubble, brick bats, broken plaster/concrete pieces etc. can be used for building activity, such as, levelling, under coat of lanes where the traffic does not constitute of heavy moving loads.

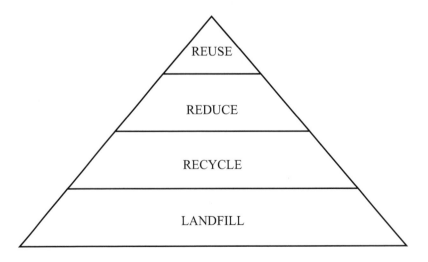

FIGURE 10.2 Construction and Demolition Waste Management Hierarchy

Waste Management as per Indian Legislation 167

- Larger unusable pieces can be sent for filling up low-lying areas.
- Fine material, such as, sand, dust etc. can be used as cover material over sanitary landfill.

10.6.3.2 Disposal (Landfill)

Being predominantly inert in nature, construction and demolition waste does not create chemical or biochemical pollution. Hence maximum effort should be made to reuse and recycle them. The material can be used for filling/levelling of low-lying areas. In the industrialized countries, special landfills are sometimes created for inert waste, which are normally located in abandoned mines and quarries (TERI, 2014). The same can be attempted in our country also for cities, which are located near open mining quarries or mines where normally sand is used as the filling material. However, proper sampling of the material for its physical and chemical characteristics has to be done for evaluating its use under the given circumstances.

10.7 HAZARDOUS AND OTHER WASTES (MANAGEMENT AND TRANSBOUNDARY MOVEMENT) RULES, 2016

Hazardous waste management rules are notified to ensure safe handling, generation, processing, treatment, package, storage, transportation, use reprocessing, collection, conversion, and offering for sale, destruction and disposal of hazardous waste. This rule came into effect in the year 1989 and have been amended in the years 2000, 2003 and with final notification of the Hazardous Waste (Management, Handling and Transboundary Movement) Rules, 2008 in supersession of former notification. The rules lay down corresponding duties of various authorities such as MoEF and CC, CPCB, State/UT Govts., SPCBs/PCCs, DGFT, Port Authority and Custom Authority while State Pollution Control Boards/Pollution Control Committees have been designated with wider responsibilities (CPCB, 2017d). Hazardous waste means any waste which by reason of characteristics such as physical, chemical, biological, reactive, toxic, flammable, explosive or corrosive, causes danger or is likely to cause danger to health or environment, whether alone or in contact with other wastes or substances (CPCB, 2016e).

10.7.1 INDIAN SCENARIO FOR HAZARDOUS WASTE

The hazardous waste generated in the country per annum is estimated to be around 4.4 million tonnes while as per the estimates of Organisation for Economic Co-operation and Development (OECD) derived from correlating hazardous waste generation and economic activities, nearly five million tonnes of hazardous waste are being produced in the country annually. This estimate of around 4.4 million MTA is based on the 18 categories of wastes which appeared in the HWM Rules first published in 1989. Out of this, 38.3% is recyclable, 4.3% is incinerable and the remaining 57.4% is disposable in secured landfills. Thirteen states of the country (Maharashtra, Gujarat, Tamil Nadu, Orissa, Madhya Pradesh, Assam, Uttar Pradesh, West Bengal, Kerala, Andhra Pradesh, Telangana, Karnataka and

Rajasthan) account for 97% of total hazardous waste generation. The top five waste generating states are Maharashtra, Gujarat, Andhra Pradesh, Telangana and Tamil Nadu. On the other hand, states such as Himachal Pradesh, Jammu and Kashmir and all the North Eastern states excepting Assam generate less than 20,000 MT per annum. Hazardous waste generation is maximum in Maharashtra (45.47%) followed by Gujarat (9.73%). Minimum hazardous wastes are reported in Chandigarh (0.0069%). The number of industries that generate hazardous wastes are maximum in Maharashtra (30.38%) followed by Gujarat (22.93%) (Agarwal and Gupta, 2011).

10.7.2 CLASSIFICATION OF HAZARDOUS WASTE

Hazardous wastes are classified into four lists as F, K, P, and U list by Devi et al. (2018). The detail given below:

F-list: The F-list contains hazardous wastes from non-specific source. The list consists of solvents commonly used in degreasing, metal treatment baths and sludges, wastewaters from metal plating operations and dioxin containing chemicals or their precursors.

K-list: The K-list contains hazardous wastes generated by specific industrial processes. Industries which generate K-listed wastes include wood preservation, pigment production, chemical production, petroleum refining, iron and steel production, explosive manufacturing and pesticide production.

P and U lists: The P and U lists contain discarded commercial chemical products, off-specification chemicals, container residues and residues from the spillage of materials. An example of a P or U listed hazardous waste is a pesticide, which is not used during its shelf-life and requires to be disposed in bulk.

10.7.3 HAZARDOUS WASTE TREATMENT

There are various alternative hazardous waste treatment technologies:

* Physical and chemical treatment
* Thermal treatment
* Biological treatment.

The selection of the most effective technology depends upon the wastes being treated (Ramachandra, 2006).

10.7.3.1 Physical and Chemical Treatment

Physical and chemical treatments are an important part of most hazardous waste treatment operations, and the treatments include the following (Freeman, 1988):

1. *Filtration and separation*: The application of filtration for treatment of hazardous waste fall into the following categories: Clarification, this is usually accomplished by depth filtration and cross-flow filtration and the primary aim is to produce a clear aqueous effluent, which can either be discharged

Waste Management as per Indian Legislation

directly, or further processed. The suspended solids are concentrated in a reject stream. Dewatering of slurries of typically 1% to 30% solids by weight. This is usually accomplished by cake filtration.

2. *Chemical precipitation*: In the treatment of hazardous waste, the process has a wide applicability in the removal of toxic metal from aqueous wastes by converting them to an insoluble form. This includes wastes containing arsenic, barium, cadmium, chromium, copper, lead, mercury, nickel, selenium, silver, thallium and zinc. The sources of wastes containing metals are metal plating and polishing, inorganic pigment, mining and the electronic industries. Hazardous wastes containing metals are also generated from clean-up of uncontrolled hazardous waste sites (e.g. leachate or contaminated groundwater).

3. *Chemical oxidation and reduction (redox)*: Such reactions are used in treatment of metal-bearing wastes, sulphides, cyanides and chromium and in the treatment of many organic wastes such as phenols, pesticides and sulphur containing compounds. Since these treatment processes involve chemical reactions, both reactants are generally in solution. However, in some cases, a solution reacts with a slightly soluble solid or gas.

4. *Solidification and stabilization*: In hazardous waste management, solidification and stabilization (S/S) is a term normally used to designate a technology employing activities to reduce the mobility of pollutants, thereby making the waste acceptable under current land disposal requirements. Solidification and stabilization are treatment processes designed to improve waste handling and physical characteristics, decrease surface area across which pollutants can transfer or leach, limit the solubility or detoxify the hazardous constituent.

5. *Evaporation*: this process is used in the treatment of hazardous waste and the process equipment is quite flexible and can handle waste in various forms—aqueous, slurries, sludges and tars. Evaporation is commonly used as a pre-treatment method to decrease quantities of material for final treatment. It is also used in cases where no other treatment method was found to be practical, such as in the concentration of trinitrotoluene (TNT) for subsequent incineration.

10.7.3.2 Thermal Treatment

The two main thermal treatments used with regard to hazardous wastes are:

1. *Incineration*: Incineration can be use as either a pre-treatment of hazardous waste, prior to final disposal or as a means of valorizing waste by recovering energy. The concept of treating hazardous waste is similar to that of municipal solid waste.

2. *Pyrolysis*: The application of pyrolysis to hazardous waste treatment leads to a two-step process for disposal. In the first step, wastes are heated separating the volatile contents (combustible gases, water vapour etc.) from non-volatile char and ash. In the second step volatile components are burned under proper conditions to assure incineration of all hazardous components (Freeman, 1988).

10.7.3.3 Biological Treatment

Following are the list of some techniques that are used for biological treatment of hazardous waste (Freeman, 1988).

1. *Land treatment*: This is a waste treatment and disposal process, where a waste is mixed with or incorporated into the surface soil and is degraded, transformed or immobilized through proper management. Compared to other land disposal options (e.g. landfill and surface impoundments), land treatment has lower long-term monitoring, maintenance and potential clean up liabilities. It is a dynamic, management-intensive process involving waste, site, soil, climate and biological activity as a system to degrade and immobilise waste constituents.
2. *Composting*: The microbiology of hazardous wastes differs from that of composting in the use of inoculums. The reaction is that certain types of hazardous waste molecules can be degraded by only one or a very few microbial species, which may not be widely distributed or abundant in nature. The factors important in composting of hazardous wastes are those that govern all biological reactions.
3. *Aerobic and anaerobic treatment*: Hazardous wastes can be treated using either aerobic or anaerobic treatment methods. In aerobic treatment, under proper conditions, microorganisms grow. They need a carbon and energy source, which many hazardous wastes satisfy, nutrients such as nitrogen, phosphorus and trace metals and a source of oxygen. Some organisms can use oxidized inorganic compounds (e.g. nitrate) as a substitute for oxygen. Anaerobic treatment Hazardous waste streams often consist of hydrocarbons leading to higher concentrations of chemical oxygen demand (COD). Depending upon the nature of waste, the organic constituents may be derived from a single process stream or from a mixture of streams.

10.8 MUNICIPAL SOLID WASTES (MANAGEMENT AND HANDLING) RULES, 2000

Municipal Solid Wastes (Management and Handling) Rules, 2000 (MSW Rules) are applicable to every municipal authority responsible for collection, segregation, storage, transportation, processing and disposal of municipal solid. The Ministry of Environment, Forests and Climate Change (MoEF and CC) recently notified the new S.O. 1357(E) (08–04–2016) Solid Waste Management Rules, 2016. These rules will replace the Municipal Solid Wastes (Management and Handling) Rules, 2000. This rule is the sixth category of waste management rules brought out by the ministry, as it has earlier notified plastic, e-waste, biomedical, hazardous and construction and demolition waste management rules (CPCB, 2016g).

10.8.1 INDIAN SCENARIO OF MUNICIPAL SOLID WASTE

Municipal areas in the country generate 133,760 metric tonnes per day (TPD) of MSW, of which only 91,152 TPD waste is collected and 25,884 TPD treated.

Waste Management as per Indian Legislation

Considering that the volume of waste is expected to increase at the rate of 5% per year on account of increase in the population and change in lifestyle of the people, it is assumed that urban India will generate 276,342 TPD by 2021, 450,132 TPD by 2031 and 1,195,000 TPD by 2050. The challenge is in managing this waste which is projected to be 165 million by 2031 and 436 million by 2050. The CPCB report also reveals that only 68% of the MSW generated in the country is collected of which, 28% is treated by the municipal authorities. Thus, merely 19% of the total waste generated is currently treated. The untapped waste has a potential of generating 439 MW of power from 32,890 tonnes per day of combustible wastes including refuse derived fuel (RDF), 1.3 million cubic metres of biogas per day or 72 MW of electricity from biogas and 5.4 million metric tonnes of compost annually to support agriculture. (CPCB, 2016c). According to Union Minister of State for Environment, Forests and Climate Change, Prakash Javedkar, 62 million tonnes of waste is generated annually in the country at present, out of which 5.6 million tonnes is plastic waste, 0.17 million tonnes is biomedical waste, hazardous waste generation is 7.90 million tonnes per annum and 1.5 million tonnes is e-waste. He added that only about 75%–80% of the municipal waste gets collected and only 22%–28% of this waste is processed and treated.

10.8.2 Municipal Solid Waste Management

The MSW contains organic as well as inorganic matter. The latent energy of its organic fraction can be recovered for gainful utilization through the adoption of suitable waste processing and treatment technologies. Table 10.4 shows the process associated with different components of municipal solid waste (TERI, 2014).

TABLE 10.4
Type and Treatment of Municipal Solid Waste (TERI, 2014)

Type	Components of Waste	Process	
Biodegradable	Kitchen, Garden and Food Waste	Biological Treatment	Aerobic processes, Anaerobic processes
		Thermal Treatment	Incinerations, Pyrolysis systems, Gasification systems
		Transformation	Mechanical Transformation, Thermal Transformation
Recyclable	Plastic	Plasma pyrolysis technology, Alternate fuel as refuse derived fuel RDF	
	Paper	Dissolution, screening, De-inking, Sterilization, and Bleaching process	
	Glass	Vitrification Technology	
Inert	Sand	Landfilling: Jaw & Pulse Crusher	
	Pebbles & Gravels		

A few technologies are listed below:

- Sanitary landfill
- Incineration
- Biodegradation processes
- Composting.

10.8.2.1 Landfill

MSW that is not recycled is typically sent to landfills—engineered areas of land where waste is deposited, compacted, and covered. MSW landfills are designed to protect the environment from contaminants which may be present in the solid waste stream and as such are required to comply with federal Resource Conservation and Recovery Act (RCRA) regulations or equivalent state regulations, which include standards related to location restrictions, composite liners requirements, leachate collection and removal systems, operating practices, groundwater monitoring requirements, closure and post-closure care requirements, corrective action provisions, and financial assurance (USEPA, 2014).

10.8.2.2 Incineration

As per TERI (2014), incineration is a process of direct burning of waste in presence of excess air oxygen, at temperatures of about 800°C and above, liberating heat energy, inert gases and ash. In practice, about 65%–80% of the energy content of the organic matter can be recovered as heat energy, which can be utilized either for direct thermal application or for producing power via steam turbine generation (with conversion efficiency of about 30%).

10.8.2.3 Biodegradation processes

In the aerobic degradation process, organic material is oxidized to give a humus product commonly called compost, which can be used as a fertilizer. As the aerobic degradation process involves decay of the organic material, such as garbage, leaves, manure etc. the process takes a considerable time. Anaerobic digestion leads to a highly marketable product called methane. The first step is to break down the complex organic materials present in the refuse into organic acid and CO_2. The second step involves the action of bacteria known as methane formers on the organic acid to produce methane and carbon dioxide.

10.8.2.4 Composting

Composting is seen as a key process in the waste hierarchy and has an important role in reducing the volume of biodegradable municipal solid waste going to landfill. microorganisms convert organic materials such as manure, sludge, leaves, fruits, vegetables and food wastes into product. Through composting organic waste materials are decomposed and stabilized into a product that can be used as soil conditioner and/or organic fertilizer. Decomposers include bacteria, actinomycetes and fungi that are widespread in nature (Atalial et al., 2017).

10.9 CONCLUSION

Implementation of the above methods not only requires public participation and attention but also requires strict government laws. Despite the fact that solid waste management practices have been improving in recent years, the space of improvement needs to be accelerated. Measures mentioned in MSW rules must be implemented. The time has come to encourage technology-based entrepreneurship to achieve effective solid waste management as per Indian legislations. Public involvement in management of solid waste is of significant importance. The waste should be treated as resource and formal recycling sector/industries be developed to recycle non-biodegradable recyclable component from the waste thereby providing employment to rag-pickers and absorb them in mainstream. Also, a policy, fiscal intensive and development of quality standard for reuse and recycle of hazardous and solid waste be developed and notified so that producers dispose/reuse it as per guidelines, thereby reducing burden on landfill. Manufacturing of non-recyclable polyethylene bags should be banned or research should be initiated to develop biodegradable polyethylene.

REFERENCES

Agarwal, D. and Gupta, A. K., 2011. Hazardous waste management: Analysis of Indian scenario and perspective governance. *VSRD Journal*. 2:484–495. Available online at https://www.researchgate.net/publication/310166596_Hazardous_Waste_Management_Analysis_of_Indian_Scenario_and_Perspective_Governance

Atalia, K. R., Buha, D. M., Bhavsar, K. A. and Shah, N. K., 2017. A review on composting of municipal solid waste. *Journal of Environmental Science, Toxicology and Food Technology*. 9(5):20–29. Available online at https://pdfs.semanticscholar.org/69d3/625 bf8871f0cd7b7f6c4de8650bcab8a9b2a.pdf

Bureau of Indian Standards, 1998. IS 14534: 1998 guidelines for recycling of plastics. Available online at www.questin.org/sites/default/files/standards/is.14534.1998_0.pdf

Capoor, M. R. and Bhowmik, K. T., 2017. Current perspectives on biomedical waste management: Rules, conventions and treatment technologies. *Indian Journal of Medical Microbiology*. 35:157–164. Available online at www.ijmm.org/article.asp?issn=025508 57;year=2017;volume=35;issue=2;spage=157;epage=164;aulast=Capoor

Chaturvedi, A., Arora, R., Khatter, V. and Kaur, J., 2007. E-waste assessment in India: specific focus on Delhi. Manufacturer's association for information technology (MAIT) and German technical cooperation organization (GTZ) New Delhi, India: BIRD. *Society Policy Science*. 2007:127–143. Available online at https://www.researchgate.net/publication/304351685_E-waste_assessment_in_India_specific_focus_on_Delhi_Manufacturer's_association_for_information_technology_MAIT_and_German_technical_cooperation_organization_GTZ_New_Delhi_India_BIRD

CPCB, 2016a. E-waste (management) Rules, 2016, G.S.R. 338 (E): Available online at http://cpcb.nic.in/displaypdf.php?id=aHdtZC9HVUlERUxJTkVTX0VXQVNURV9SVUxFU18yMDE2LnBkZg)

CPCB, 2016b. Bio-medical waste (management and handling) Rules, 2016, G.S.R. 343(E): Available online at http://cpcb.nic.in/displaypdf.php?id=aHdtZC9CaW8tbWVkaWNhbbF9XYXN0ZV9NYW5hZ2VtZW50X1J1bGVzXzIwMTYucGRm

CPCB, 2016c. Annual report 2015–2016. Available online at http://cpcb.nic.in/displaypdf.php?id=aHdtZC9NU1dfQW5udWFsUmVwb3J0XzIwMTEtMTIucGRm

CPCB, 2016d. Construction and demolition waste management rules, 2016. Available online at http://cpcb.nic.in/c-d-waste-rules/

CPCB, 2016e. Hazardous and other wastes (management and transboundary movement) Rules, 2016. Available online at http://cpcb.nic.in/displaypdf.php?id=aHdtZC9IV01fUnVsZXNfMjAxNi5wZGY=

CPCB, 2016f. Plastic waste (management and handling) rules, 2011 G.S.R. 320 (E). Available online at http://cpcb.nic.in/displaypdf.php?id=cGxhc3RpY3dhc3RlL1BXTV9HYXpld HRlLnBkZg==

CPCB, 2016g. Solid waste management rules, S.O. 1357(E). Available online at http://cpcb.nic.in/uploads/MSW/SWM_2016.pdf

CPCB, 2016h. Guidelines for management of healthcare waste as per biomedical waste management rules. Available online at http://cpcb.nic.in/uploads/hwmd/Guidelines_healthcare_June_2018.pdf

CPCB, 2017a. Available online at http://cpcb.nic.in/bio-medical-waste-rules/ (Accessed on 14 September 2017).

CPCB, 2017b. Consolidated guidelines for segregation, collection and disposal of plastic waste. Available online at http://cpcb.nic.in/uploads/plasticwaste/Consolidate_Guidelines_for_disposal_of_PW.pdf

CPCB, 2017c. Guidelines on environmental management of Construction & Demolition (C & D) wastes.

CPCB, 2017d. Available online at http://cpcb.nic.in/hazardous-waste-rules/.

CPHEEO, 2000. Solid waste management manual, Construction and Demolition waste. Available online at http://mohua.gov.in/upload/uploadfiles/files/chap4.pdf

Devi, K. S., Sujana, O. and Singh, T. C., 2018. Hazardous waste management in India—A review. *International Journal of Creative Research Thoughts.* 6(1):1547–1555. Available online at www.researchgate.net/publication/323028874_Hazardous_Waste_Management_in_India_-_A_Review

Freeman, M. H., 1988. *Standard Handbook of Hazardous Waste Treatment and Disposal,* McGraw-Hill Book Company, New York.

Ministry of Housing & Urban Affairs, 2018. Strategy for promoting processing of Construction and Demolition (C&D) waste and utilisation of recycled products| 1–31. Available online at http://niti.gov.in/writereaddata/files/CDW_Strategy_Draft%20Final_011118.pdf

Ramachandra, T. V., 2006. Municipal solid waste management book. Available online at https://nptel.ac.in/courses/120108005/module9/lecture9.pdf

Research Unit (Larrdis) Rajya Sabha Secretariat, 2011. *Report on E-waste in India,* New Delhi. Available online at https://rajyasabha.nic.in/rsnew/publication_electronic/E-Waste_in_india.pdf

TERI (The Energy and Resources Institute), 2014. Waste to resources: A waste management handbook. Available online at http://cbs.teriin.org/pdf/Waste_Management_Handbook.pdf

UNEP, 2009. Inventory assessment manual. Available online at www.unep.or.jp/ietc/Publications/spc/EWasteManual_Vol1.pdf

USEPA, 2014. Municipal solid waste landfills manual. Available online at https://www3.epa.gov/ttn/ecas/docs/eia_ip/solid-waste_eia_nsps_proposal_07-014.pdf

Widmer, R, Heidi, O. K., Khetriwal, D. S., Schnellmann, M., Heinz, B., 2005. Global perspectives on e-waste. *Environmental Impact Assessment Review.* 25:436–458.

11 Good Practices of Hazardous Waste Management

Vidushi Abrol, Manoj Kushwaha, Nisha Sharma, Sundeep Jaglan and Sharada Mallubhotla

CONTENTS

11.1 Background of Hazardous Waste Management..175
11.2 Identification of Hazardous Waste Material..176
11.3 Classification and Segregation of Hazardous Waste Material.....................177
    11.3.1 Other Measures for Hazardous Waste Characterizations.................178
11.4 Consideration of Regulatory Groups for Hazardous Waste Management....178
    11.4.1 Generators of Hazardous Waste Material..179
    11.4.2 Carriers for the Transportation of Hazardous Waste.......................180
    11.4.3 Receivers of Hazardous Waste Material..180
    11.4.4 Others Regulatory Bodies Involved for Hazardous Waste
        Management 180
        11.4.4.1 Department of Infrastructure, GNWT..............................180
        11.4.4.2 Department of Lands, GNWT 180
        11.4.4.3 Fire Marshal Office, GNWT 180
        11.4.4.4 Officer of Chief Public Health Authority, GNWT 180
        11.4.4.5 Environment and Climate Change Canada (ECCC) 180
        11.4.4.6 Worker's Safety and Compensation
                Commission (WSCC) 181
        11.4.4.7 Local Governments...181
11.5 Choosing Right Treatment, Storage and Disposal of Hazardous Wastes.....181
11.6 Hazardous Waste Minimization and Prevention 182
    11.6.1 Source Reduction Strategy 182
    11.6.2 On-Site/Off-Site Recycling..182
11.7 Conclusion 183
References...184

11.1 BACKGROUND OF HAZARDOUS WASTE MANAGEMENT

Hazardous waste is one the major challenges that is affecting public health and environment nowadays. The massive expansion of industries, urban areas, agriculture and natural resources leads to the deterioration and exploitation of natural habitat and resources generates large amount of hazardous wastes materials of all varieties (Hester and Harrison, 2002). This includes explosive, flammable liquids/solids, poisonous, toxic and other infectious substances (Agrawal et al., 2004). Other waste material may also possess hazardous components, like clinical and pharmaceutical waste material, industrial or other production wastes, formulations and adhesives, latex and plasticizers. Toxic materials may also arise from surface treatment and disposal of metals and plastics substances, residues generates from the incineration of household waste, industrial residues like copper, zinc, cadmium, mercury, lead and so forth (Cheng et al., 2019). Therefore, reducing the production of hazardous substances and treating waste toxic material is the need of the hour (Wang et al., 2004). So the best practice for disposing of hazardous waste—any toxic material or residues which are dangerous to human or to the environment other than radioactive wastes—should be identified, categorized, handled, stored, transported and treated or discarded with special precautions (Misra and Pandey, 2005). The Resource Conservation and Recovery Act (RCRA) gave the EPA authorization to determine what type of hazardous waste is and how to dispose of it carefully (Grasso, 2009).

11.2 IDENTIFICATION OF HAZARDOUS WASTE MATERIAL

For the classification of hazardous waste material, the initial phase of identification procedures is to identify that the waste material is solid (Smith and Mccauley, 2006). Then, the secondary phase is to determine is the waste is particularly eliminated from regulation as a hazardous or solid waste material. As the producer analyzed and investigated

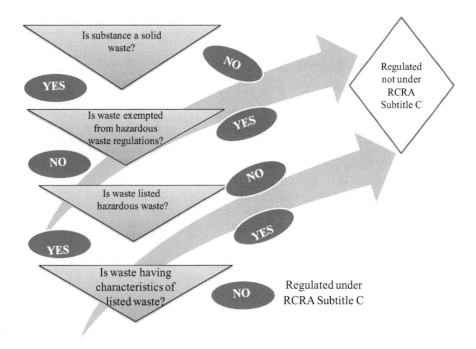

FIGURE 11.1 Outlay of Identification of Hazardous Waste Process

that their waste product justifies the example of solid waste and listed as hazardous waste material (Shah, 2000). Finally, it is also important to note that some authorities appeal EPA to remove their wastes category from RCRA Subtitle C regulation. So, considering all the facts, hazardous waste material is identified and further categorized (Figure 11.1).

11.3 CLASSIFICATION AND SEGREGATION OF HAZARDOUS WASTE MATERIAL

This phase involves the identification of the toxic material based on their characteristic feature on which basis it is listed as hazardous waste. Ignitability, toxicity, reactivity, corrosiveness and other features those are responsible for making a particular substance as hazardous (Basden, 2002) (Figure 11.2).

To handle these wastes safely, one must be aware of what type of waste is, how it may act and what all its resources. While managing individual must know the level of toxicity, what biosafety measures must be taken, the waste compatibility and incompatibility (Ramachandra, 2006). Based on this fact, generators are required to separate hazardous waste accordingly. Hazardous waste segregation can be performed in two ways:

1. During sampling and examination of the waste
2. Based on the process knowledge the waste can be categorized.

This has been observed that EPA has analyzed and listed as hazardous wastes substances under four different categories that are mentioned in the regulations at Part 261, (D) (Greenberg, 2017). These four categories are:

1. The *F-list*: This designates specifically for solid wastes material that are generated from common manufacturing industries of different sectors as toxic components and are known as non-specific waste source material.

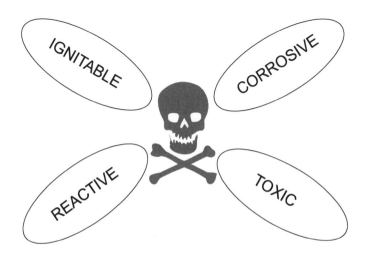

FIGURE 11.2 Hazardous Waste Classification

178   Zero Waste

2. The *K-list*: This defines specific solid wastes or hazardous material from some specific industries and is called as wastes from particular waste source material.
3. The *P-list and the U-list*: Both the P and U lists are similar in that both lists have commercial grade formulations for some specific unused chemicals as toxic wastes.

These mentioned lists assigned anywhere from 30 to hundreds of waste streams as hazardous substances. Each waste designates with a waste code having alphabet along with three numbers (LaGrega et al., 2010). These codes are very significant part of the RCRA regulatory system. Providing the accurate waste code to a waste material has important involvement for managing the hazardous wastes.

11.3.1 OTHER MEASURES FOR HAZARDOUS WASTE CHARACTERIZATIONS

1. Individuals must be aware of the material safety data sheet (MSDS) for the information like physical and chemical property, explosion of hazard, chemical reactivity, fire and special protection information if it is available.
2. It should be mentioned that if any changes in a production procedure is possible that may alter the composition of the waste generated.
3. Always be very particular in tagging/labelling of the wastes container that may symbolizes that a substance is flammable, toxic etc.
4. Based on the waste characterization, if a waste is incompatible or reactive with other substances. It is mandatory to segregate as per their properties. This is important that hazardous wastes must be kept in a way to avoid explosions.
5. Always keep the waste away from becoming too hot, try to keep the substances that are producing toxic materials and flammable mists, fumes, gases in separate section that will prevent from being in direct contact to the waste and will avoid explosions.
6. Always be aware, that mixing of incompatible wastes won't rupture or damage the container and be safe for the individual and the environment in all ways.

11.4 CONSIDERATION OF REGULATORY GROUPS FOR HAZARDOUS WASTE MANAGEMENT

The Department of Environment and Natural Resources (ENR) is the Government of the Northwest Territories (GNWT) regulatory body that is answerable for control and prevention of the release of toxic contaminants. They are responsible for certifying the environmentally friendly acceptable protocols for management and handling, their level of release and discarding steps are managed and maintained. Environmental Protection Act (EPA) and the Pesticide Act administered the legislative authority. The Environment Division (ED) of ENR observes the transportation of hazardous waste material from the producers to destined disposal and known as hazardous waste movement document. The EPA is not the only regulatory body who is having the authority over the issues related to hazardous waste (Pichtel, 2005).

Good Practices of Hazardous Waste Management

There are many reputed agencies which regulate the handling and disposal of waste material, like the Department of Transportation, Drug Enforcement Authority and OSHA, and our local government may also have more demanding regulations (Blackman, 2016).

It is important to note that if the hazardous waste is to be migrated from the originating site, the originator must be enrolled with ED. Once enrolled, the ID number will be generated and assigned to through with the document movement. The basic outlay for the migration of hazardous waste substances is outlined in Figure 11.3.

11.4.1 Generators of Hazardous Waste Material

- Generators are responsible for the packaging, classification, characterization, quantification, marking/labelling and storing of hazardous waste carefully.
- Producers must ensure that the disposal of hazardous waste must be in acceptable form.
- Always make sure that the workers who are handling these waste materials are well trained in the practices of hazardous waste management.
- The generator must register as a generator of hazardous waste when the hazardous waste is to be migrated.
- Always ensure the waste is transported by an authorized hazardous waste transport to a receiver who is having the authority to receive the type of hazardous waste along with the document of movement.

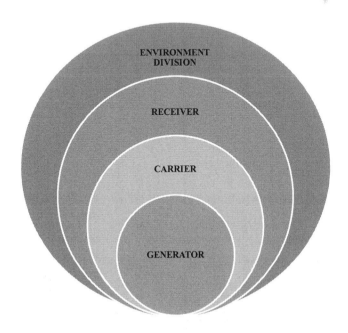

FIGURE 11.3 Movement of Hazardous Waste Materials

11.4.2 Carriers for the Transportation of Hazardous Waste

Carriers must be authorized and registered with ED before the transportation of hazardous waste (Hassett, 1994; Dutta and Janakiraman, 2002). Hazardous waste material should be carried in suitable transport authority according to the requirement (i.e. Air International Civil Aviation Organization (ICAO), Marine International Maritime Dangerous Goods Code (IMDG) Road, Rail Transportation of Dangerous Goods Regulations (TDGR)). Carriers are responsible in completing the movement document of hazardous waste material also or maintaining the record of disposal of wastes. Always make sure that the staff must be well trained in the transport for any emergency situation, and having knowledge and awareness for safely transportation hazardous wastes.

11.4.3 Receivers of Hazardous Waste Material

For best hazardous waste management practices with the updated facilities that treat and manage hazardous waste substances from different generators are registered/authorized as receivers. The work person that can handle and operate the hazardous waste management authority in the NWT is enforced to login the facility with ED to handle particular type of hazardous waste (Federal Hazardous Waste, 2018).

11.4.4 Others Regulatory Bodies Involved for Hazardous Waste Management

11.4.4.1 Department of Infrastructure, GNWT

For the transportation of hazardous goods act and regulations (GNWT) the road licensing and safety division is having the authority for monitoring (Kent et al., 2003). This division is also possessing head supervision of drivers, vehicles and road security/safety under add-on transport legislation. The carrier for transportation of hazardous wastes by rail (TDGR), marine (IMDG) or by air (ICAO) is regulated by Transport Canada.

11.4.4.2 Department of Lands, GNWT

This division issues lands and administers different authorizations related to land and its public use (Sandlos and Keeling, 2016). In this, public land is leased by the GNWT with a condition require proper management of hazardous waste.

11.4.4.3 Fire Marshal Office, GNWT

This division has authority for keeping the flammable, combustible and hazardous substances under the NFC (National Fire Code) (Hayden, 2010).

11.4.4.4 Officer of Chief Public Health Authority, GNWT

This division should be communicated regarding compulsion when waste handling activities may affect environment and public health.

11.4.4.5 Environment and Climate Change Canada (ECCC)

This is responsible for hazardous waste management regulations under the Canadian Environmental Protection Act (CEPA) (Environmental Board, 2014). Off-site shipments/transportation of hazardous waste is managed under the import and export of hazardous waste and hazardous recyclable material regulations.

11.4.4.6 Worker's Safety and Compensation Commission (WSCC)

The WSCC is authorized for monitoring the NWT Safety Act and the Occupational Health and Safety (OHS) Regulations, that defines the safety of employees, workers and the working area (Zimolong and Elke, 2006). This act signifies that the worker shall maintain their working area and take all required and necessary precautions. In this work area of hazardous materials information system guidelines were approved to ensure worker training and safety purposes.

11.4.4.7 Local Governments

Governments of local municipal corporations are involved in a multiple way. Below different legislation local municipal governments possess full command for taking decisions about public foundations and community precautions including disposal systems and so forth.

11.5 CHOOSING RIGHT TREATMENT, STORAGE AND DISPOSAL OF HAZARDOUS WASTES

Based on the above factors, choosing the right treatment with all of these factors plays an important role, specifically with the accurate authenticated organization. It is required that a dedicated and experienced waste management authority be employed for the dispose of hazardous waste material promptly and properly (Federal Hazardous Waste, 2018).

As mentioned earlier different best practices for the hazardous waste management like identification, generation, processing, storage and transport, treatment and disposal (LaGrega et al., 2010). But before the disposal of hazardous wastes, appropriate treatment is mandatory depending on the type and category of the wastes. Different type of hazardous waste treatment can be sectioned under physical, chemical, thermal and biological treatments (Figure 11.4). There is different treatment,

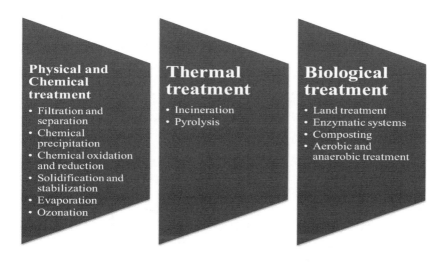

FIGURE 11.4 Hazardous Waste Treatment

storage and disposable facilities are available for the hazardous waste material. The government of a state, operator of a facility, any organization individually or in collaboration with, is solely responsible for establishing the facility for treatment, storage and disposal of the hazardous waste material. Establishment of facilities for treatment and disposal was directed as per technicalities and guidelines that were approved by the Central Pollution Control Board (CPCB) and obtained from the State Pollution Control Board office. The State Pollution Control Board shall examine the complete setup and working environment for the treatment, storage and disposal facilities regularly based on the records that is maintained by the operator for the hazardous waste management.

11.6 HAZARDOUS WASTE MINIMIZATION AND PREVENTION

Waste minimization is a strategy or a technique that reduces the amount of waste material. This waste minimization supports the economic strategies along with the protection of the environment (Cheremisinoff, 2003). It is very much necessary that every single member of the society must be aware of the hazardous waste material and its dangerous impacts on the environmental and society and should follow necessary measure for its reduction for its future production (Jin et al., 2006). It is important that accurate hazardous waste management practices are essential part for operating processes. In addition to the environment protection, waste minimization also improves the quality of the product and also provides economic benefits (Higgins, 2017). Hazardous waste minimization revolves around few processes i.e. reduce, reuse, recycle, recover and treat.

11.6.1 SOURCE REDUCTION STRATEGY

Alteration of strategies and the procedures for the reduction of production of hazardous waste material is known as source reduction strategy. Strategies involved process alteration or modification, chemical substitution or deletion, improvement of handling and operating processes etc. (Demirbas, 2011). Some other processes for reduction of hazardous wastes material include; proper handling and operation of materials, accurate segregation of hazardous and non-hazardous waste, clean performance of reduced waste products, always take proper precautions for packaging and sealing to prevent leakage to the environment, always mention the manufacturing date and expiry date of the products to avoid harmful effect of the material, always use proper labelled product to prevent accumulation of unwanted products, always use products after analyzing complete demonstration of the material, wherever possible.

11.6.2 ON-SITE/OFF-SITE RECYCLING

Those resources that attained their plateau of life or precipitated at the end of the processes are advised to utilize and recycled for further usages. This technique is one of the most significant techniques that are referred as recycling. Some examples include re-distillation of solvents are used and recovered, recycling of used oils,

Good Practices of Hazardous Waste Management 183

batteries, electronic devices like computers and so forth. Other material like medical equipment may be used again after a sterilization process; example are scalpels, forceps, glass bottles, glass containers, syringes. These are easy to reuse after collected separately in aseptic conditions (Osmani et al., 2006). But this also depends on the product and the type of usage has been performed. If the equipment has been used with highly toxic components than it should be discarded directly, else can be reused or recycled after considering complete precautions and sterilization strategies (Ajayi et al., 2015). On-site recycling is the process of reusing waste material for the same purpose or for another use within the industry whereas off-site recycling is the improvement and modification of waste material to transform into a substance for its another application within or outside the industry (Waite, 2013).

The major significance of hazardous waste minimization is cost saving that reduces raw material consumption; minimize its dangerous impact to the ecosystem and environment, production and development of new or more sustainable materials, provides competitive benefits and improves its social impact on to the society (Nair and Abraham, 2018).

The Government of India promotes the Environment Protection Act in 1986 to promulgate the protection and improvement of the environment and management and handling the hazardous waste substances. It was evaluated that almost 9.3 million metric tons of hazardous waste material is produced in India. Out of 1.35 million metric tons are recycled wastes and 0.11 million metric tons are incinerable waste, and 0.49 million metric tons waste for the disposal purposes. There are only two engineered landfill sites that are available in India, and both are in Gujarat (Grasso et al., 2017). A study in 2000 also reveals that identification of hazardous waste material, its quantification analysis and segregation of varieties of hazardous waste material according to the Indian classification system carried out in five major districts of Gujarat. On the basis of this classification sludge waste material from plant effluent therapy and waste from dye materials observed the most dominant in respect to the other waste categories (Haq and Chakrabarti, 2000). A case study reveals that economic development, industrialization advancements and changing lifestyles of the public and society, hazardous waste material in India is increasing notably. This is specifically in those areas where population is increasing and growing progressively. The major factors that are involved are inadequate treatment and disposal facilities, untrained and unskilled stakeholders, less governance authorities; the management of hazardous waste material is very poor except in Gujarat. This is one of the states that has better management amenities and majorly adaptive the advanced technologies for the best hazardous waste management practices (Karthikeyan et al., 2018).

11.7 CONCLUSION

Based on the study it is revealed that hazardous waste substances are generated by every sector of the society from large industries to health care, to the business, Farms and also from household wastages. Disposal of hazardous wastes in pits, ponds and so forth possess warning to human health and the social environment. Therefore, depending on the level of risks it is very much important to treat and manage the waste material judiciously and appropriately in accordance with the legal authorities.

It has been also observed that many public and private organizations and government regulatory bodies also support the best practices for the hazardous waste management and minimization as a strategy for reducing waste generation for years to come. As the actual present generators are well aware with their material of work, and they are accurate source of ideas to prevent against their respective hazardous waste materials and their minimization. Henceforth, the EPA has analyzed and prepared a hazardous waste list for evaluation of hazards and risks considering their physical, chemical, environmental, health and other hazard exposures. So, that keeping in view of all the practices and necessary measures, depending on the willingness and active participation management of hazardous waste material must be successfully attained and prevent against waste generation.

REFERENCES

Agrawal, A., Sahu, K. K. and Pandey, B. D., 2004. Solid waste management in non-ferrous industries in India. *Resources, Conservation and Recycling.* 42(2):99–120.

Ajayi, S. O., Oyedele, L. O., Muhammad, B., Akinade, O. O., Alaka, H. A., Owolabi, H. A. and Kadiri, K. O., 2015. Waste effectiveness of the construction industry: Understanding the impediments and requisites for improvements. *Resources, Conservation and Recycling.* 102:101–112.

Basden, R. A., 2002. Hazardous material classification system. Google Patents.

Blackman, W. C. Jr., 2016. *Basic Hazardous Waste Management*, CRC Press.

Cheng, K., Yu, P., Yu, W., Chris, W., Laura-lee, I. and Kaksonen, A. H., 2019. A new method for ranking potential hazards and risks from wastes. *Journal of Hazardous Materials.* 365:778–788.

Cheremisinoff, N. P., 2003. *Handbook of Solid Waste Management and Waste Minimization Technologies*, Butterworth-Heinemann.

Demirbas, A., 2011. Waste management, waste resource facilities and waste conversion processes. *Energy Conversion and Management.* 52(2):1280–1287.

Dutta, R. and Janani, J., 2002. Package tracking system. *International Business Machines Corp.* United States patent US 6,433,732.

Environmental, Canadian, and Treasury Board, 2014. *Chemicals Management Plan Progress Report*, Government of Canada, 7:1–13.

Federal Hazardous Waste, 2018. *Hazardous Waste Management*, Alabama Department of Environmental Management, Montgomery, AL.

Grasso, D., 2017. *Hazardous Waste Site Remediation*, Routledge, New York.

Grasso, D., Kahn, D., Kaseva, M. E. and Mbuligwe, S. E., 2009. Hazardous waste, In Domenico, G., Smith, S. R., Nath, C. Y. and Nath, B., (eds.), *Natural and Human Induced Hazards and Environmental Waste Management, Volume 1, Encyclopedia of Life Support Systems (EOLSS)*, UNESCO. Chapter 1, 1–54.

Greenberg, M. R., 2017. *Hazardous Waste Sites: The Credibility Gap*, Routledge, New York.

Haq, I. and Chakrabarti, S. P., 2000. Management of hazardous waste: A case study in India. *International Journal of Environmental Studies.* 57(6):735–752.

Hassett, J. J., 1994. Hazardous waste transport management system. Google Patents.

Hayden, S., Mackenzie Valley Land, and Water Board. 2010. *Operation and Maintenance Manual for Sewage and Solid Waste Facility*, Department of Environment and Natural Resources, GNWT. 1–21.

Hester, R. E. and Harrison, R. M., 2002. *Environmental and Health Impact of Solid Waste Management Activities*, Vol. 18, Royal Society of Chemistry.

Higgins, T. E., 2017. *Hazardous Waste Minimization Handbook*, CRC Press.

Jin, J., Wang, Z. and Shenghong, R., 2006. Solid waste management in Macao: Practices and challenges. *Waste Management.* 26(9):1045–1051.

Karthikeyan, L., Venkatesan, S., Krishnan, V., Tudor, T. and Varshini, V., 2018. The management of hazardous solid waste in India: An overview. *Environments.* 5(9):103.

Kent, R., Marshall, P. and Hawke, L., 2003. Guidelines for the planning, design, operations and maintenance of modified solid waste sites, in the Northwest Territories. In *Report for the Department Municipal and Community Affairs Government of the Northwest Territories,* Canada, Ferguson Simek Clark Architects & Engineers. Project No. 2001-1330.

LaGrega, M. D., Buckingham, P.L. and Evans, J.C., 2010. *Hazardous Waste Management,* Waveland Press.

Misra, V. and Pandey, S.D., 2005. Hazardous waste, impact on health and environment for development of better waste management strategies in future in India. *Environment International.* 31(3):417–431.

Nair, S. and Abraham, J., 2018. Hazardous waste management with special reference to biological treatment. *Handbook of Environmental Materials Management.* 1–27.

Osmani, M., Jacqueline, G. and Price, A.D.F., 2006. Architect and contractor attitudes to waste minimisation, In *Proceedings of the Institution of Civil Engineers: Waste and Resource Management.* Thomas Telford Publishing. 159: 65–72.

Pichtel, J., 2005. *Waste Management Practices: Municipal, Hazardous, and Industrial,* CRC Press.

Ramachandra, T. V., 2006. *Management of Municipal Solid Waste,* The Energy and Resources Institute (TERI), New Delhi, ISBN: 978-81-7993-188-2.

Sandlos, J. and Keeling, A., 2016. Aboriginal communities, traditional knowledge, and the environmental legacies of extractive development in Canada. *The Extractive Industries and Society.* 3(2):278–287.

Shah, K.L., 2000. Basics of solid and hazardous waste management technology. *Technology.* 550:50787.

Smith, C. A, and Mccauley, J.R., 2006. Pharmaceutical hazardous waste identification and management system. *Pharm Ecology Associates LLC, assignee.* United States Patent US 7,096,161.

Waite, R., 2013. *Household Waste Recycling,* Routledge, New York.

Wang, L. K., Hung, Y-T., Howard, H. L. and Yapijakis, C., 2004. *Handbook of Industrial and Hazardous Wastes Treatment,* CRC Press.

Zimolong, B.M. and Elke, G., 2006. Occupational health and safety management. In *Handbook of Human Factors and Ergonomics.* John Wiley & Sons, Inc, New York, Chapter 26, 673–707.

12 Hydroponic Treatment System Plant for Canteen Wastewater Treatment in Park College of Technology

Shailendra Kumar Yadav and Kanagaraj Rajagopal

CONTENTS

12.1 Introduction 187
12.2 Literature Review 188
12.3 Hydroponic Wastewater Treatment for Canteen Wastewater in Park
College of Technology 189
    12.3.1 Quantity and Quality of Wastewater to Be Treated......................... 189
        12.3.1.1 Flow Rate Measurement 189
        12.3.1.2 Quality of Wastewater to Be Treated................................. 189
        12.3.1.3 Removal of Suspended Matter.. 190
        12.3.1.4 Removal of Oil and Grease.. 191
12.4 Treatment Plant Layout and Design Details.. 191
    12.4.1 Design for Treatment System for Canteen Wastewater 191
12.5 Selection of Plants 193
    12.5.1 Sludge Formation.. 193
12.6 Economic Aspect.. 193
12.7 Performance of the Modified Hydroponic Treatment System Unit............. 193
12.8 Results and Discussion 195
12.9 Conclusion 202
References.. 202

12.1 INTRODUCTION

The increasing global demand for water in combination with mounting levels of domestic and industrial pollution of rivers and lakes make reclamation and reuse of water a necessity (UNEP, 2007). In many regions like tropical and subtropical areas where plenty of sunlight and land is available for usage, treating of domestic wastewater is highly recommended using macrophytes. This hydroponic method of treating wastewater uses simple construction in the ground and uses macrophytes as a tool for treating the domestic wastewater. The hydroponic method of treatment has very good advantages and also one of the efficient methods for treating the wastewater in household as well as industrial kitchen wastewater sources.

12.2 LITERATURE REVIEW

J. von Sachs and W. Knop, two scientists in the 19th century, showed that terrestrial plants could grow independently without soil in nutrient solution. They performed experiments and demonstrated that plants can grow with the help of sunlight and inorganic elements. This technique is called the hydroponics technique (Stanhill and Enoch, 1999). Hydroponics is not a new idea; it is deep rooted in history. Egyptians, Chinese and Indians used it almost 4,000 years ago. This may include the Hanging Gardens of Babylon about 660 BC. The earliest published work on growing terrestrial plants without soil was in 1627. The discoveries of German botanists, in the years 1859–1865, resulted in a development of the technique of soilless cultivation. In 1929, the use of solution culture for agricultural crop production was publicly promoted. The term hydroponics was introduced in 1937 and hydroponics developed from the 1960s.

The literature study shows growing plants in a defined nutrient solution (i.e. hydroponic) is sometimes different from cultivating in a natural solution, such as domestic wastewater which is also called as biophonic or organic hydroponics. Wastewater hydroponics, which uses wastewater as the nutrient source, is gaining attention as a bio-integrated food production system, as it helps meet the goals of sustainability by following certain principles; the waste produced from one source is recycled and at the same time it is used as a feed or food for another biological system. Ohtani et al. (2000) reviewed the importance of reusing the drainage nutrient water from the traditional hydroponic system, which helps in preventing the environment as well as reducing water demand. The water from the pond can be again used for various other domestic applications as well as gardening purposes.

The term hydroponics is used for defining the method of growing the plants by making it float in the wastewater by submerging its roots into it. Compared to the traditional or conventional ways of wastewater treatment, the hydroponic system is quite different because it looks like a green pond with plants floating on it, which is actually a reactor for treating the kitchen wastewater. Plants take up nutrients and water by their roots. In a soil substrate, the roots usually obtain sufficient oxygen; but when plants are grown in a liquid medium, they might experience diffusion limitations. Oxygen deficiency in the reactor may inhibit cellular respiration, which prevents a supply of metabolic energy to drive absorption processes (Taiz and Zeiger, 1991). As a result, nutrient absorption and biomass production are reduced. Hence aeration of the reactor especially for the ponds is considered one of the important parameters. Other important parameters in hydroponics are pH and the nutrient concentration. The uptake rate changes according to the growth stage, the growing conditions and the ion source. Theoretically, the composition of the nutrient solution should be changed continuously, in order to be adjusted to the plant uptake (Stanhill and Enoch, 1999).

12.3 HYDROPONIC WASTEWATER TREATMENT FOR CANTEEN WASTEWATER IN PARK COLLEGE OF TECHNOLOGY

Hydroponic system of wastewater treatment construction, operation, analysis of performance is done in Park College of technology, Environmental Engineering Department, which is situated in Coimbatore district of Tamil Nadu state. The college canteen generates 1.206 m^3/hr of wastewater per day. It is composed mainly of food particles, oil and grease, starch water, wash water and cleansing agents. There was no technical treatment system for treating the effluent. It is made to pass through manhole and from the manhole it was diverted to a pond almost 4.6 m away behind the canteen premise. The main reason for having manhole is to serve as a settling tank and to retain the settable solid waste before entering in to the pond. The water is not flowing continuously into the pond and in between at manhole the wastewater was seen overflowing and spreading around the manhole and making the place less aesthetic. The pond in which the untreated wastewater was being dumped into was not able to assimilate the wastewater and was found to have become chocked and eutrophic. To overcome these aesthetically unpleasant defects and to provide treatment of the canteen wastewater prior to the disposal in to the pond, a hydroponic treatment system is constructed at the premise of the canteen.

12.3.1 QUANTITY AND QUALITY OF WASTEWATER TO BE TREATED

12.3.1.1 Flow Rate Measurement

The average flow rates are determined at one-hour intervals, from 10:00 a.m. to 6:00 p.m., Monday to Saturday. The flow rate measurement is done by diverting the flow from the seventh manhole through a ball valve. The measurement was taken using a 20-liter plastic bucket. The average flow rate is measured to be 1,206.59 L/hr. The maximum and minimum flow rate is at 5:00 p.m. and 4:00 p.m., respectively. The variation in the flow rates are shown in Figure 12.1.

12.3.1.2 Quality of Wastewater to Be Treated

The samples were collected at the inlet of the site for five days. To have a better and more accurate value inlet samples were collected at 10:00 a.m., 1:00 p.m. and 5:00 p.m. The raw wastewater had an average of 600 mg/L of COD, 265 mg/L of TSS,

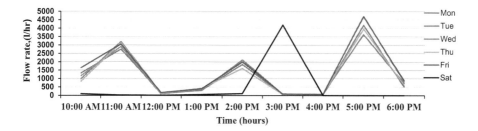

FIGURE 12.1A Variation of Flow Rate in a Week

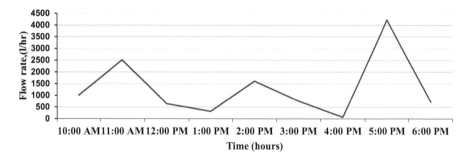

FIGURE 12.1B Hourly Variations in Flow Rate

TABLE 12.1
Quality of Raw Wastewater

Date	Time	pH	EC (μmhos/cm)	TS (mg/l)	TSS (mg/l)	TDS (mg/l)	COD (mg/l)
16.03.15	10:00 a.m.	5.2	671	900	400	560	780
	01:00 p.m.	5.4	573	710	194	450	540
	05:00 p.m.	5.1	481	540	162	374	460
17.03.15	10:00 a.m.	6.3	679	970	380	570	765
	01:00 p.m.	5.9	604	624	206	489	598
	05:00 p.m.	5.3	474	498	147	336	483
23.03.15	10:00 a.m.	5.6	628	995	420	540	772
	01:00 p.m.	6.1	536	632	253	417	552
	05:00 p.m.	6.23	413	472	184	298	511
24.03.15	10:00 a.m.	5.67	665	967	390	560	783
	01:00 p.m.	5.78	572	653	212	431	525
	05:00 p.m.	6.1	441	512	181	314	424
25.03.15	10:00 a.m.	5.43	646	876	410	530	789
	01:00 p.m.	6.2	598	728	242	472	581
	05:00 p.m.	5.4	430	486	199	304	477

443 mg/L of TDS, 704 mg/L of TS, 561 μmhos/cm of EC and pH value of 5.4. The characteristic and the variation of raw canteen wastewater is shown in Table 12.1.

12.3.1.3 Removal of Suspended Matter

The canteen wastewater is composed of food waste including vegetable waste, chilies, fine food waste, oil and grease, starch water, wash water and cleansing agents. The settable solids can be removed from the wastewater by the following method.

Placing a sedimentation bucket at the outlet of the manhole so that the settable solids can be retained at the bottom of bucket with specific inlet design (Figure 12.2)

FIGURE 12.2 Hydroponic Wastewater Treatment Site

and almost all heavy settable solids is removed and the remaining fine food particles are removed from the next settling tank. All settled materials are removed by the sludge outlet pipe at bottom of settling bucket and tank. Plastic or iron mesh can be placed inside the canteen wastewater outlet pipe so that large settable will be retained and can be removed manually by the canteen people and the remaining solids will be removed as they pass over the manhole.

12.3.1.4 Removal of Oil and Grease

There is an entry of oil and grease into the treatment system. In the current site there is set up for the removal of oil and grease (Figure 12.2). There are many options for the removal of oil and grease like solvent extraction, floatation, using traps, skimmers etc. But all these are costly and we have to go in for economically and technically feasible methods. Researchers have shown that several biomasses like kapok fibre, cattail fibre, salvinia sp. (*Salvinia cucullata* Roxb), wood chip, rice husk, coconut husk, and bagasse can be used as sorbents for the oil and grease removal. On this site bricks, stone and jute cover mess are used in oil and grease trapping tank with specific outlet and inlet position design. This method is tried in lab scale to remove oil and grease from canteen wastewater in a cheaper and eco-friendly ways and then implanted on site.

12.4 TREATMENT PLANT LAYOUT AND DESIGN DETAILS

12.4.1 Design for Treatment System for Canteen Wastewater

To treat the wastewater before discharging in to the pond a hydroponic treatment system is constructed at the premise of the canteen following the manhole. The design is done based on the flow rate and the physio-chemical parameters measured at the inlet of the manhole. The average flow rate is measured to be 800 l/hr. and the COD was 900 mg/l. Based on these data a hydroponic treatment system facility is set up. It is designed with a zigzag pattern of water flow with three flow channels of dimension 4.33 m × 1 m × 0.3 m (length × width × depth). The plan of the design is shown in Figure 12.3. Water hyacinth and *Assesslis*, an aquatic free-floating plant, are used for treating the effluent. The design parameter of the existing system is shown in Table 12.2 and the features of existing site is shown in Figure 12.4.

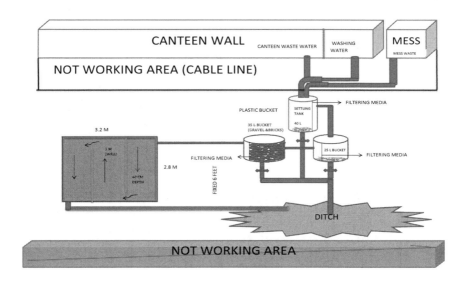

FIGURE 12.3 Flow Chart of Canteen Wastewater Treatment Site

TABLE 12.2
Design Parameters of the Treatment Site

	Parameters
HRT	3 hrs
Flow rate	1,206.59 l/hr
Total number of trenches	3
Volume of 1 channel	1.3 m³
Depth of the trench	0.60 m
Dimensions of the channel	4.33 m × 1 m × 0.3 m (length × width × depth) × 3 channels
Depth of water level	0.3 m
Distance between the three buckets, tanks screens	45 cm
Number of baffles	2
Dimension of settling tank	120 l bucket
Dimension of oil baffle	As per availability
Dimension on 2nd settling tank	As per availability
Size of fine screen with jute	0.02 mm

FIGURE 12.4 Canteen Wastewater Treatment Site at Park College of Technology

12.5 SELECTION OF PLANTS

Since the depth of the water level is 30 cm, plants with roots growing up to that length and more are chosen for the treatment. One such deep rooted plant, water hyacinth is used in first two trenches as they are easily available. *Assesslis* is used in last trench to improve the treatment system. There is lot of literature and experiment available about water hyacinth and *Assesslis* for wastewater treatment efficiency and capability. The plants are collected from Premaravi Nagar Karumathampatti, Coimbatore. Since the plants were taken from fresh water, it took two weeks for the plant to acclimatize to the canteen wastewater (Figure 12.5).

12.5.1 SLUDGE FORMATION

The filtered solids form a thick layer of sludge of about 10 cm thick and float on the water. Sludge formation was found in three weeks after the filtration has started. The floating sludge was skimmed off manually and disposed.

12.6 ECONOMIC ASPECT

The capital cost incurred in setting up the site including the cost involved for troubleshooting are discussed in Table 12.3.

From Table 12.3 it's proved that the hydroponic treatment system is a cost-effective treatment system.

12.7 PERFORMANCE OF THE MODIFIED HYDROPONIC TREATMENT SYSTEM UNIT

The performance of the site was assessed based on the physio chemical parameters like BOD, COD, TS, TDS, TSS, EC, pH and total Kjeldahl nitrogen. Initially samples were collected at 11:30 a.m., 2:30 p.m. and 5:30 p.m. for five days. But to study the variation in the parameters more accurately, samples were collected for three days at 8:30 a.m., 11:30 p.m., 2:30 p.m. and 6:30 p.m., since the canteen's working time is from 8:30 a.m. to 6:30 p.m. For the analysis of COD and nitrogen, filtered samples were collected and for the other parameters unfiltered samples were used.

FIGURE 12.5 (a) Trench with Plant. (b) Inlet Sample Collection. (c) Outlet Sample Collection.

Canteen Wastewater Treatment in Park College

TABLE 12.3
The Overall Expenditure and Total Cost of the Project

Materials	Cost in INR
HDPE Sheet	1,000
Screen	150
mesh	100
Pipe and Solution	2,320
Valve	180
Super Bond	60
Labour Cost	4,350
Operation cost and maintenance cost (Only for emptying the outlet pond when filled)	500
Total cost	**8,660**

12.8 RESULTS AND DISCUSSION

The performance of the hydroponic treatment system reactor was assessed based on the removal efficiencies of BOD, COD, TS, TDS, TSS, EC and total Kjeldahl nitrogen. The experiments were done based on the APHA standard procedures. The samples were collected at 8:30 a.m., 11:00 a.m., 11:30 a.m., 2:30 p.m. and at 6:30 p.m. to analyse the fluctuations in a day. The results from the experiments are discussed below (Figures 12.6–12.15; Tables 12.4–12.6).

- The EC of the raw canteen wastewater was in the range of 324 µmhos/cm to 679 µmhos/cm. The average was 537 µmhos/cm. The peak value was at 11:00 a.m. The average drop in EC was 41% and the maximum removal was at 2:30 p.m.
- The wastewater was acidic having an average inlet pH value of 5.4 and the outlet pH value of 5.5.
- The COD was in the range of 424 mg/l to 812 mg/l. The average was 630 mg/l higher value of COD is due to the oil and grease in the canteen wastewater. The average removal efficiency was 52%.
- The TS was in the range of 320 mg/l to 1,500 mg/l. The average was 717 mg/l. The average removal efficiency was 45%.
- The TDS was in the range of 180 mg/l to 560 mg/l. The average value was 256 mg/l. The average removal efficiency was 42%.
- The TSS was in the range of 110 mg/l to 901 mg/l. The average value was 295 mg/l. The average removal efficiency was 49%.
- The BOD was in the range of 280 mg/l to 375 mg/l. The average value was 324 mg/l. The average removal efficiency was 45%.
- The total Kjeldahl nitrogen was in the range of 20.5 mg/l to 22.4 mg/l. The average value was 21.3 mg/l. The removal efficiency was 69%.
- The capital cost involved for the construction, operation and maintenance and the labour cost was about INR 16,554, thus proving the cost effectiveness of the treatment system.

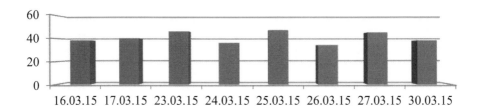

FIGURE 12.6A Performance of Hydroponic Wastewater Treatment System in Terms of EC

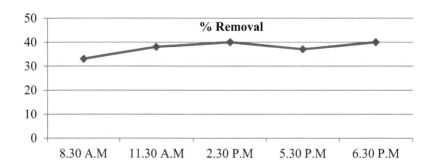

FIGURE 12.6B Performance of Hydroponic Wastewater Treatment System in Terms of EC (in a Day)

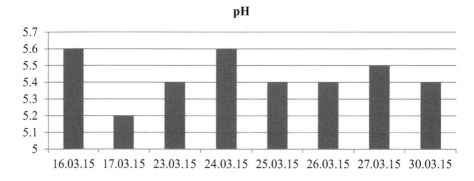

FIGURE 12.7A Performance of Hydroponic Wastewater Treatment System in Terms of pH

Canteen Wastewater Treatment in Park College

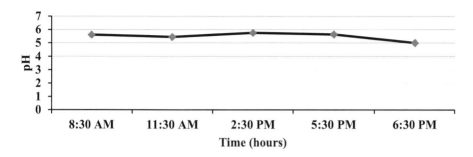

FIGURE 12.7B Performance of Hydroponic Wastewater Treatment System in Terms of pH (in a Day)

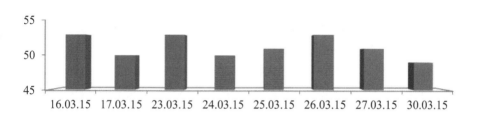

FIGURE 12.8A Performance of Hydroponic Wastewater Treatment System in Terms of COD Removal

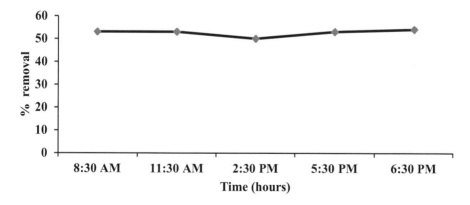

FIGURE 12.8B Performance of Hydroponic Wastewater Treatment System in Terms of COD (in a Day)

FIGURE 12.9A Performance of Hydroponic Wastewater Treatment System in Terms of TS Removal

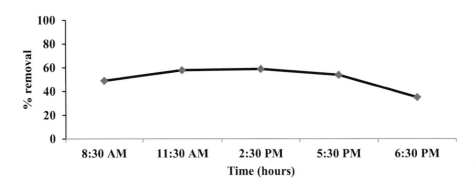

FIGURE 12.9B Performance of Hydroponic Wastewater Treatment System in Terms of TS Removal (in a Day)

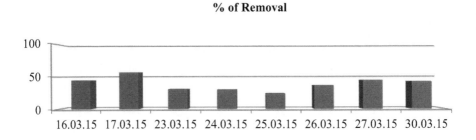

FIGURE 12.10A Performance of Hydroponic Wastewater Treatment System in Terms of TDS Removal

Canteen Wastewater Treatment in Park College

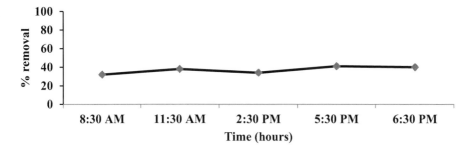

FIGURE 12.10B Performance of Hydroponic Wastewater Treatment System in Terms of TDS Removal (in a Day)

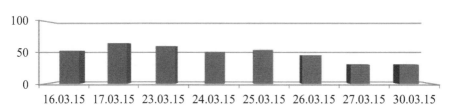

FIGURE 12.11A Performance of Hydroponic Wastewater Treatment System in Terms of TSS Removal

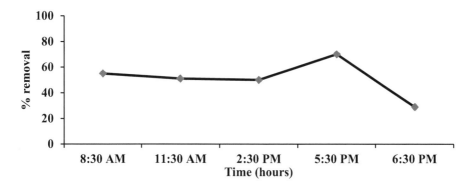

FIGURE 12.11B Performance of Hydroponic Wastewater Treatment System in Terms of TSS Removal (in a Day)

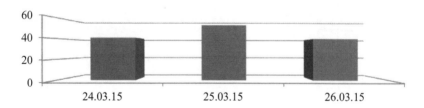

FIGURE 12.12A Performance of Hydroponic Wastewater Treatment System in Terms of BOD Removal

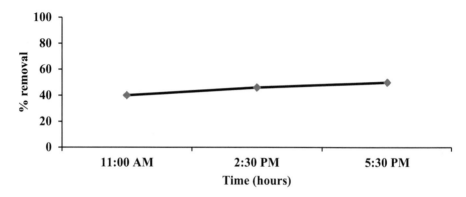

FIGURE 12.12B Performance of Hydroponic Wastewater Treatment System in Terms of BOD Removal (in a Day)

TABLE 12.4
Performance of Hydroponic Wastewater Treatment System in Terms of TKN Removal

Date	Time	TKN mg/l Inlet	TKN mg/l Outlet	% Removal
30.03.15	11:30 a.m.	22.4	5.6	75
	02:30 p.m.	21	7.6	64
	06:30 p.m.	20.5	6.3	69

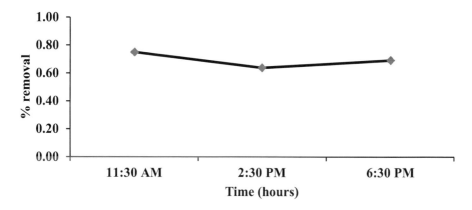

FIGURE 12.13 Performance of Hydroponic Wastewater Treatment System in Terms of TKN Removal

TABLE 12.5
Overall Performance of the Treatment Unit on Day Basis

Date	% Removal				
	COD	TS	TSS	TDS	EC
16.03.15	53	47	53	45	38
17.03.15	51	52	66	57	40
23.03.15	53	61	60	32	46
24.03.15	50	60	51	30	36
25.03.15	51	58	54	24	47
26.03.15	53	40	46	38	34
27.03.15	52	44	39	45	45
30.03.15	49	46	31	44	33

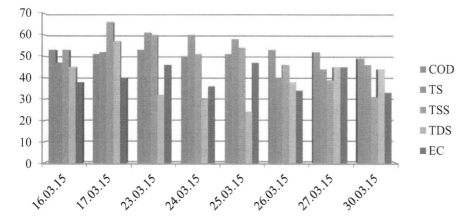

FIGURE 12.14 Overall Performance of the Treatment Unit on Day Basis

TABLE 12.6
Overall Performance of the Treatment Unit at Different Times of the Day

Time	% Removal				
	COD	TS	TSS	TDS	EC
8:30 a.m.	53	49	55	32	33
11:30 a.m.	53	58	51	38	40
2:30 p.m.	50	59	50	34	44
5:30 p.m.	53	54	70	41	40
6:30 p.m.	54	35	29	40	43

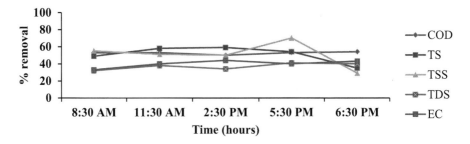

FIGURE 12.15 Overall Performance of Hydroponic Wastewater Treatment System at Different Times of the Day

12.9 CONCLUSION

From the above results and from the economic aspects discussed, the hydroponic treatment system has proved to be an efficient, low-cost, economically important system suitable for small communities for treating the domestic wastewaters. The treatment efficiency can be still increased to more than 90% by increasing the HRT of the reactor by adding on additional trenches to the existing modified system and by trying out with different plant species. The cost of the project can be still optimized. The food wastes filtered and settled at the bottom of the trench and the harvested plant biomass can be tried for composting or vermicomposting and used as manure.

REFERENCES

Ohtani, T., Kaneko, A., Fukuda, N., Hagiwara, S. and Sase, S., 2000. Development of a membrane disinfection system for closed hydroponics in a greenhouse. *Journal of Agriculture Engineering Research.* 77(2):227–232.

Stanhill, G. and Enoch, H. Z., 1999. *Ecosystems of the World 20: Greenhouse Ecosystems.* Elsevier, Amsterdam.

Taiz, L. and Zeiger, E., 1991. *Plant Physiology: Mineral Nutrition.* The Benjamin Cummings Publishing Co., Inc. Redwood City.

UNEP, 2007. *Global Environment Outlook (GEO–4) Environment for Development.* Available online at https://na.unep.net/atlas/datlas/sites/default/files/GEO-4_Report_Full_en.pdf

13 Ecological and Economic Importance of Wetlands and Their Vulnerability
A Review

Sudipto Bhowmik

CONTENTS

13.1 Introduction 203
13.2 Importance of Wetlands 204
    13.2.1 Biodiversity...204
    13.2.2 Carbon Sequestration..205
    13.2.3 Nutrient Removal..206
    13.2.4 Water Quality..207
    13.2.5 Economic Importance 208
13.3 Major Causes of Vulnerability of Wetlands 209
    13.3.1 Climate Change 209
    13.3.2 Land Use Change and Biodiversity Loss...210
    13.3.3 Pollution...211
    13.3.4 Eutrophication..213
13.4 Conclusion 213
References...214

13.1 INTRODUCTION

Each of the wetlands can be an individual ecosystem. They are different by species composition, geologic location, exposure to different types of landscapes, climate and many other factors. Their surroundings, water sources, atmosphere and so forth can regulate them, and wetlands can regulate them also. Brinson (1993) pointed out that though wetness of wetlands is usually studied for understanding their function, these functions can be potentially effected by drainage network, size of wetland, source of water, biogeochemical inflow and outflow and so forth. Hence to understand the function of wetlands, each wetland can be studied while considering these and many other factors, according to the uniqueness of the wetland. The biodiversity, climate, water and hydrological regime, ecosystem services, soil and sediments, anthropogenic dependence and activity and pollution status in and around the wetlands are interdependent. To identify a particular cause of degradation of wetland ecosystem and applying a restoration strategy may or may not result in complete success depending on the other regulating factors as well as factors that were remained unstudied. The understanding of total biotic and abiotic setup is necessary for the conservation strategy development. This particular study is a review of the outcomes of the efforts of researchers to understand the ecological and economic importance of the wetlands and the major causes of their vulnerability. Separate headings and subheadings are used to focus on individual factors, however interconnection of these factors are well established by researchers, and in this literature review some of those connections are mentioned. The examples mentioned here to establish the importance and role of different factors in regulation of the wetland ecosystem are only some of the vast research that has already been done, and yet some factors may remain unmentioned here which may be of equivalent importance to develop strategies for conservation and restoration of the wetlands.

13.2 IMPORTANCE OF WETLANDS

13.2.1 BIODIVERSITY

Wetlands are unique ecosystems, and they host a range of diverse floral, faunal and microbial species around the world. The members of the biotic communities are dependent on each other, where the diversities are also dependent on wetland size, shape, location, climate, water and nutrient resources. Biodiversity consisting of plants, animals and microbes regulates the biogeochemistry of the wetland, at the same time changes in physical and chemical nature of the wetlands are regulators of the community structure of the wetlands.

Several studies are done and being done around the world to understand the community structure and their interdependence. Some of these studies are discussed here. In a study by Hansson et al. (2005), it was shown that in 32 constructed wetland areas in southern Sweden, benthic invertebrate species diversity was positively related to age and surface area of the wetland. Bird species richness was positively related to surface area for up to 4% ha, whereas high shoreline complexity and age of the wetlands was positively related to establishment of high macrophytes. Small Sanjiang plain in China is an important habitat of red crown cranes, oriental white storks and the white-tailed sea eagle (Liu et al. 2004). In India, some of the important indigenous diversity of species and migratory bird species find habitats at the wetlands of Western Ghat, Loktak lake of Manipur, Bharatpur Wildlife Sanctuary, Little Rann of Kutch, and the coastal wetlands of Saurastra (Bassi et al. 2014). Kantrud and Stewart (1997), in a study at the natural basin wetlands of North Dakota, had shown that the breeding of waterfowl in the wetlands is dependent on several factors like water permanance, nutrient availability, wetland size, land use etc. Siwakoti (2006) mentions about 318 wetland-dependent species in the Terai region of Nepal (6 climbers, 287 herbs, 9 shrubs and 16 trees). The Banganga wetlands of Uttarakhand, India, hosts a rich biodiversity. There are 19 aquatic plant species, 25 in inundated shores and 27 in uplands, among which *Polygonum barbaratum, Ipomoea carnea, Phragmites karka,*

Importance of Wetlands

and *Typha elephantina* are common in all habitats, and the region also has animals like swamp deer, hogs and different birds (Adhikari et al., 2008).

The studies reviewed here are some of the important researches on wetland biodiversity around the world. To understand the importance of regional biodiversity of each type of wetland is necessary to develop conservation strategies. Some of the threats to the biodiversity are discussed later in this review. The studies show a glimpse of the rich diversity found or yet to be found in the wetlands. Biodiversity not only regulates the ecosystem but also carries importance in an economic way, the biodiversity of wetlands can support fisheries and eco-tourism, which can help the development of regional economy. So, policy makers may consider the economic importance of some special biodiversity of particular wetlands, but the ecological importance of natural diversity of the usually neglected or unmonitored wetlands is no less important, and these should be protected in order to have a sustainable future.

13.2.2 Carbon Sequestration

Different studies and reviews by researchers around the world indicate that wetlands are one of the important carbon sinks. Comparative analytical studies are also done on carbon loss from the wetland. More in-depth studies on carbon sequestration are required to understand the formation of organic matter, release of carbon dioxide and methane, the associated microbes, role of vegetation, climate, sediment chemistry, geology, anthropogenic impact etc.

Sequestration may vary according to the types of wetland and several other factors. Stallard (1998) mentioned that fertilization by agricultural runoff, atmospheric fallout and increased CO_2 can increase carbon sequestration, whereas the process is affected by wetland drainage, peat mining and accelerated oxidation. In a study at the wetlands of northwest Florida by Choi et al. (2001), it was shown that carbon sequestration is greater in landwards expansions of coastal wetlands. In contradiction to Stallard (1998), through a study in a study on alpine wetland of Qinghai at Tibet plateau, Junhong et al. (2010) showed that wetland drainage can increase carbon loss due to increased decomposition of soil organic matter in these landscapes. Alongi et al. (2016) gave an estimation that approximately 17% of the total blue carbon of the world is sequestered by estuarine and marine wetlands of Indonesia, which mainly consists of mangroves and marshes. In the coastal wetland of Louisiana, about 63% of the total carbon sequestered was dependent on vertical growth of marshes, as shown by Delaune and White (2012). Studies by Mitsch et al. (2016) at mangroves of southern Florida and by Loomis et al. (2010) at tidal marshes of Georgia, USA, indicated that increase in salinity decreases the capacity of carbon sequestration by wetlands. Mitsch et al. (2016) attributed the reduction to different physiological changes in the vegetation due to increased salinity. The study also showed that the sequestration is greatest in riverine mangroves, then in fringe mangroves and lowest in basin mangroves. Bassi et al. (2014) mention that soil of wetland can sequester 200 times more carbon than their vegetation. It was also mentioned that Indian mangroves sequester the largest amount of carbon among Indian wetlands due to their greater size, diversity and complicated network of canals and tidal creeks. McCarty et al. (2009) mentioned that carbon sequestration in wetlands can be increased by downward mixing or diffusion or leaching during sedimentation, and the mechanism performs better if there is a series of smaller sedimentations or a steady state rate of sedimentation occurs.

However along with sequestration, there occurs the loss of carbon also by biological pathways in the form of CO_2, CH_4 and dissolved organic carbon (DOC). A study in Veracruz, Mexico, by Marín-Muñiz et al. (2014) showed swamp soil sequesters more carbon than marsh soil due to differences in vegetation. Increase of water level decreases CO_2 production and favours the production of methane (Lloyd, 2006). Mitsch et al. (2013) mentioned that in most of the wetlands, sequestration is usually greater than carbon loss in form of methane. Methane emission from wetland is approximately 19% of the total carbon sequestered by them (Bassi et al., 2014). Purvaja and Ramesh (2001) showed that in unpolluted coastal wetlands, methane emission is inversely related to salinity and sulphate concentration, but the situation may not remain the same in polluted wetlands. Climate change may increase the loss of carbon from wetlands; drier summers can increase the CO_2 production, and greater spring runoff can increase removal of DOC (Clair et al., 2002).

These examples are not exhaustive and there are several researches being performed on efficiency of wetlands in carbon sequestration. However, it is clear that several factors as vegetation, climate, geologic position, biogeochemistry, nutrients etc. regulate the sequestration process as well as the carbon loss. These shows the uniqueness of each wetland, and hence it can be implied that there cannot be a generalized methodology to restore and/or maintain these wetlands at the peak of their performance, but individual research effort is required around the world to cumulatively reduce carbon from atmosphere with wetlands.

13.2.3 NUTRIENT REMOVAL

Nutrient removal is one of the important ecosystem services of the wetlands. The removal of nutrients added to water from landscapes or other resources reduces the nutrient load in rivers and oceans, thus the natural habitats of the river and ocean are protected. Protection of natural biodiversity takes place and their utility for ecological and economic purposes is also preserved. The removed nutrients can follow several changes in form or stage of removal, and they can be taken up by living organisms also, or in extreme cases may lead to eutrophication which may lead to death of the existing organism, and change the community structure in different ways and dimensions.

Saunders and Kalff (2001) mentioned nitrogen retention capacity is highest in wetlands followed by lakes and rivers respectively. According to Hemond and Benoit (1988), nitrogen that usually enters the wetland in the form of nitrate and ammonium ions is removed effectively in three different ways. Short-term removal is done by plant uptake; long-term removal is done by accumulation in sediment and peat, whereas nitrogen is permanently removed from water by denitrification. Morris (1991) mentions that nitrate, ammonium, and dissolved organic nitrogen are the major forms of nitrogen in North American wetlands. Phosphorus in the form of inorganic and organic phosphate enters the wetlands and plant uptake, and accumulation in

Importance of Wetlands

peat and sediments may play important roles in their removal (Hemond and Benoit 1988). According to Lucassen et al. (2005), groundwater that is affected by events like dam construction, drought and water harvest for agriculture is the major source of orthophosphate in wetlands, whereas an important source of nitrate is leaching from agricultural land and forest soils. Turner et al. (1999) mentioned that Baltic wetlands are capable of removing 100,000 t/year of nitrogen from water before reaching the sea. The performance is better when they are nearer to the source of nitrogen. However, it was mentioned that the nitrogen removal is dependent on biogeochemical coupling of C, N, P and Si circulation. Jansson et al. (1998) showed that existing wetlands of the Baltic coast can retain 5%–10% of total nitrogen emission to the Baltic Sea, whereas restoration of the drained wetlands can increase the potential up to 18%–24%. Hydrological fluctuation can increase P concentration and export from restored wetlands (Ardón et al., 2010). Fisher and Acreman (2004) described in a study that riparian wetlands perform better than swamps and marshes in removal of total N and total P, whereas swamps and marshes perform better during removal of ammonium N and soluble P. It was also mentioned that the regulating factors of N and P removal in wetlands are oxygen concentration in sediment, measure of redox, degree of waterlogged condition, hydraulic retention and vegetation processes.

The studies indicate that wetlands around the world actively remove nitrogen, but their rate of removal and process may vary depending upon several factors. The active role is taken by plants, microbes and biogeochemical processes in sediments, whereas the permanence of water, the area, hydrology and many other factors can indirectly regulate the whole process. Further researches are required on individual types of wetlands to understand in detail the nutrient removal process so that their efficiency might be increased and preventing measures can be taken before events like eutrophication occur.

13.2.4 Water Quality

Besides nutrient removal, the wetlands have the ability to reduce the contamination of water from heavy metals, excess sediments, faecal microbes, xenobiotics etc. They maintain the water quality at a certain level that the biodiversity is unaffected, human health is conserved and the water becomes reusable for several purposes including agriculture.

Wetlands can effectively remove heavy metals from water through chemical, physical and biological mechanisms. Sheoran and Sheoran (2006) described that heavy metals can be removed by chemical processes like adsorption to clay and organic matter, hydrolysis and/or oxidation, precipitation and co-precipitation, and carbonate and sulphide formation and biological processes like plant uptake, bacterial metabolism of sulphate and iron reduction. These processes are not independent, but they rely on the pH, ions, organic matter and many other factors of the wetlands. Most of the heavy metals present in the wetlands are Fe, Al, Mn, Cu, Ni, Zn, Mn, Pb, Cd etc. Similar processes can remove many of them or they may go through special removal mechanisms according to external conditions. Cu, Mn, Mo, Ni, V and Zn co-precipitate with oxides of iron, whereas Co, Fe, Ni, Pb, Zn and Co co-precipitate with oxides of manganese, and organic matter like humic, tannic and fulvic acids remove heavy metals by forming complexes (Matagi, 1998). Plants can effectively metals from wetland. *Equisetum rasmosisti* is efficient for removal of Pb and Zn, and *Equisetum ralleculosa* for Cu (Deng et al., 2004).

According to Gambrell (1994), near pH condition, mineral abundance in flooded wetland facilitates metal immobilization. But the study also mentioned other possible ways of heavy metal removal. Soluble metals that can be taken up by the plants and metal ions bound to soil by cation exchange were described as easiest to mobilize, whereas metals bound in a crystal lattice of clay are described as most immobile. Oxidation and reduction were also mentioned to play important role. Microbial respiration was mentioned to be releasing Fe^{2+} and Mn^{2+} from oxides, making them available to plants. Reduction of metal ions into sulphides are said to be immobilizing metals in coastal marshes whereas oxidation releases them, but state of oxidation is said to be non-effective on metals like Cd, Zn, Cu, Pb, Ni; runoff is said to be more effective in these cases. Matagi (1998) mentioned clay mineral and hydrous oxides absorbs minerals in the order Pb>Cu>Zn>Ni>Cd and peats in the order Pb>Cu>Cd>Zn>Ca. Phytostabilization by lignification and humification, phytovolatilization of highly volatile elements like mercury and selenium and rhyzofiltration by accumulating heavy metals on roots and shoots or precipitation through plant exudates are mentioned as important mechanisms of metal removal from wetlands by plants (Prasad, 2003).

Hemond and Benoit (1988) described several possible ways through which wetlands can change the water quality. According to them, wetlands can remove suspended solid material from water by slowing the flow of water, thus allowing more time to settle. The performance of wetlands with higher residence time is better for removal of suspended solids. Decomposition of organic matter by bacteria can decrease the biochemical oxygen demand (BOD) and chemical oxygen demand (COD). Wetlands can reduce the pathogenic microbes in water by detaining them for a long time till they fail to survive in absence of a host organism. They also mentioned that removal of xenobiotic organic compounds after reaching wetlands can be initiated by slowing down the flow of water. Then these compounds are removed by sorption on sediments, hydrolysis and anaerobic degradation. Yu et al. (2005) showed in their study that autochthonous microorganisms in mangrove sediment can effectively remove polycyclic aromatic hydrocarbons (PAHs).

These studies showed some examples that can explain the removal of several pollutants and contaminants from the water by the wetlands. There are also studies on the positive and negative effects of these processes. Researches around the world are being done to identify the sources of these pollutants. Reducing pollutants at the source may reduce the anthropogenic pollutant stress on the wetland so that the natural conditions can be maintained with receiving greater ecosystem services.

13.2.5 Economic Importance

Fisheries are one of the important uses of wetlands, and there are other uses like fuel wood collection, agricultural water supply, ecotourism etc. The coastal wetland of the Gulf of Mexico provides habitat to juveniles of three species of penaeid shrimps upon which the local shrimp-fishery industry is totally dependent (Engle, 2010). In

India, wetland-dependent fisheries are prevalent in Kerala, Goa, Odisha, Jammu and Kashmir, Assam, Bihar, Andhra Pradesh and Gujarat (Bassi et al., 2014).

In India, several wetlands are tourist attractions. Chilka Lake of Odisha, the backwaters of Kerala and the Small Run of Kutch are some of the examples of many of the wetland-based attraction spots for national and international tourists, and in India tourism is one of the major income sources for the common people (Bassi et al., 2014).

Wetlands, besides proving resources for economic development, may also effectively protect life and property of coastal communities. Bassi et al. (2014) mentioned the efficiency of the Bhitarkanika mangrove of Odisha, India, against the cyclones and protecting the local communities. Through a study by Hashim et al. (2013) it was shown that mangroves can reduce wave height and impact on shorelines and protect the soil from washing away. Wetland can supply required freshwater for agricultural purposes. According to Greenway (2005), the artificial wetlands can effectively remove nutrients from water, and the water can be applied for irrigation purposes in agriculture and at golf courses and many other places.

Though there are several important studies performed around the world the world in the form of actual economic services with factors like cost-effective maintenance, actual income in terms of money etc., this review only tried to touch upon some of the economic fields which can be supported by the wetlands. There are several other ways by which wetlands can support the economy, like maintaining soil fertility by periodic flooding, reducing human health risk from contamination thus saving hospital expenses, conserving biodiversity, keeping the balance of the groundwater table etc.

13.3 MAJOR CAUSES OF VULNERABILITY OF WETLANDS

Several factors, both natural and anthropogenic, are threatening the existence of the wetlands, reducing their ecosystem services, affecting their biodiversity. While climate change is slowly but continuously being an emerging factor of threat, there are existing threats like pollution, eutrophication etc. The causes and the effects can be interdependent or independent. Here some of the threats from studies of the researchers around the world have been discussed.

13.3.1 CLIMATE CHANGE

Climate change is one of the major concerns of the researchers around the world. It is already showing impact temperature fluctuation, melting of snow, increases in frequency and intensity of catastrophic events, outbreak of diseases and in many other ways. Most of the climate change are caused by anthropogenic emission of greenhouse gases (GHGs), and global warming can itself induce more global warming, for example increasing microbial decomposition of organic matter, releasing more GHGs. Wetlands are important carbon sinks, and hence have an important role in regulating global climate. But they are also vulnerable to the impacts of climate change.

Climate change can affect the biogeochemistry, ecosystem and physical nature of the wetlands in many ways. Changes in one or many of these aspects can eventually show effects on the others. In a study at Northern Prairie wetlands by Johnson et al. (2005), it was estimated due to global warming that these wetlands may dry up in future, which will potentially reduce the habitat of the waterfowls that were an integral part of this ecosystem. Erwin (2009) described several possible effects of climate change on the wetlands. Mangroves are to be affected by sea level rise, sea grass/ marine aquatic beds will be vulnerable to increased salinity, salt marshes/intertidal marshes are to be threatened by sea level rise and coastal squeezing, Arctic/Tundra wetlands by prolonged summer and higher rate of evaporation and peatlands by melting of permafrost and desertification. Mortsch (1998) predicted that climate change can alter the seasonal cycle and mean level of water in the Great Lakes, which in turn can change the shoreline water balance and affect the productivity, wildlife, water quality etc. in these lakes. Bohn et al. (2007) mentioned that the methane emission in the Vasyugan wetlands of western Siberia can increase up to 100% due to lowering of the water table under the climate change scenario. Increase in sea level can affect the sensitive biotic communities, and change in precipitation, evaporation and transpiration can alter the size of the wetlands (Brander et al., 2012a).

These studies represent some of the predictions made on impact of global warming and vice versa. It is of urgent necessity to control global warming to save the wetlands as well as saving the wetlands to strengthen the endeavours to counter global warming. The global warming can affect the wetlands by reducing size, permanence, increasing salinity, disturbing the biodiversity and many other ways, at the same time can increase the loss of sequestered carbon which can be challenge to tackling global warming.

13.3.2 Land Use Change and Biodiversity Loss

Land use change is one of the anthropogenic causes that threatens the existence of the wetlands, and it is common around the world. Land use changes can be performed by reclamation of wetlands for agriculture, urbanization and other economic purposes, overuse of the wetland resources, disturbing biodiversity and in many other ways. These processes affect the wetland ecosystem services, soil fertility, hydrological balance and biodiversity in one or multiple ways.

In a study by Liu et al. (2004), it was shown that in small Shanjiang plains of China, change in land use by agricultural and industrial activities caused fragmentation and land reclamation. These resulted in sharp declines in wetland and forest area and increases in paddy land and other land use, which caused loss of nearly 73.6% of the initial wetland and decline in the rich bird communities dependent on the wetland. The flora community has changed from marshes to small-leaved reed species, where remaining marshes are just associated species, and the reason was water loss due to drainage. Jenkins et al. (2003), through a study in wet prairie of Illinois, gave the hypothesis that land conversion in this region has probably reduced crustacean species from 80 to 75, and the remaining species are still vulnerable to further habitat loss. According to Froyd et al. (2014), on Santa Cruz Island, giant tortoises are the most important species: they maintain the vegetation diversity in this region, probably by wallowing and seed dispersal. A reduction in the number of tortoises and coprophilic fungi by direct and indirect human activities has caused

Importance of Wetlands

the conversion of freshwater wetland into sphagnum bogs. Draining of wetlands and pollution from agricultural runoff are mentioned as most important causes of global decline of amphibians and reptile species (Whitfield et al., 2000). Prasad et al. (2002) mentioned that Indian mangroves are suffering from organic and inorganic pollutants from shrimp farms. In this paper, introduction of invasive species have also been mentioned to the very harmful to local biodiversity, as introduction of *Eichhornia crassipes* and *Salvinia moletsa* are threatening indigenous aquatic flora of India. In West Bengal, excessive utilization of groundwater has reduced the water supply by aquifers to the regional wetlands, which in turn is negatively affecting the existence of the wetlands (Bassi et al., 2014). Water withdrawal from the wetlands and artificially elevating the land around wetlands can prevent the wetlands from flooding the surroundings naturally. This can eventually lead towards reduction of soil fertility and increased salinity (Prasad et al., 2002). Road infrastructure development and increased accessibility to the mangroves can increase the fragmentation in mangrove ecosystem and hence can reduce the protection they give to the coastal area (Brander et al., 2002b). Disturbances, moisture and nutrient accumulation can regulate species invasion in wetland, and major invasive plants are grasses, graminoids, forbs, shrubs and trees (Zedler et al., 2004). In coastal wetlands of Ghana, human activities causing increase in grass species and decrease in tree species, and events such as bushfires, herbivory, dispersal and rainfall can regulate the process (Wuver et al., 2003). Olubode (2011) mentioned farming and fishing activities as anthropogenic disturbances in floral communities in wetlands of Ibadan in Nigeria. Water level regulation in Rainy and Namkan lakes has caused notable difference in macrophyte communities with respect to unregulated lakes (Wilcox and Meeker, 1991).

These studies are representing the concerns of the researchers to protect the wetlands. While several similar and diverse researches are also present and being performed, the situation cannot be tackled unless policy makers make strict impositions on any activity that may affect wetlands existence and ecosystem, and people are aware of the upcoming difficulties and losses to society in absence of the wetlands.

13.3.3 POLLUTION

Anthropogenic pollution is something that might have started the day that humans discovered fire. The society relies today on intense agriculture, animal husbandry, industry, vehicles and many other factors, each of which are being responsible for pollution around the world. Wetlands are mostly polluted by untreated sewages, waste dumping, industrial wastewater, runoffs from agricultural field, roads, overuse of synthetic chemicals in household and industry etc. The pollution source can be both point and non-point, though most abundant are the non-point sources. Here some of the concerns of the researchers on wetland pollution are discussed.

According to Hemond and Benoit (1988), BOD and COD in wetlands can be changed by external flow of sewage effluent and organic matter from surface runoff. This can effectively change the biogeochemistry of the wetland, and in turn can be destructive for the biotic communities thriving in the wetland. This can lead to reducing the ecosystem services provided by the wetlands and its biotic communities. Bassi et al. (2014) mentioned that due to eutrophication by untreated sewage water in the wetlands of the Himalayas, there is a resulting steep fall in dissolved oxygen (DO) and increase in BOD.

According to Hemond and Benoit (1988), aerial application of pesticides or their volatilization can carry them through atmosphere to the wetlands, and also the runoff from treated land can add pesticide to the wetlands. This can be extremely harmful for the invertebrates in wetland ecosystem and may disrupt the food chain. In a study by Boone et al. (2003) it was shown that insecticides like carbaryl can destroy salamander larvae and also cause reproductive failure and decline in population carnivorous salamanders, whereas the herbicide atrazine decreased the chlorophyll of algal species, prolonged larval phase of small mouth larvae, and reduced the mass of toads and leopard frogs. Relyea (2009) studied impact of different insecticides and showed, carbaryl and malathion affects cladocerans, chlorpyrifos reduces the population of cladocerans and periphyton, diazinon directly affected the periphytons and indirectly reduced growth of the leopard frog tadpoles and endosulfan reduces population of copepods and periphytons.

In a study by Kumar et al. (2011) at the eastern Kolkata wetlands, it was seen that Mn concentration in the muscles of fishes cultivated in these wetlands was notably higher than the permissible limits of FAO/WHO. Whereas, though within the limits the concentration, Hg(II), Ni, Cd etc. were showing varying concentrations in different species. These heavy metals are probably sourced from the urban sewages, wastewater from tannery, road runoff and local agricultural grounds. In the Nyagugogo wetlands of Rwanda, high concentration of Cr, Pb and Cd is seen in fishes, and in the roots of papyrus metals are accumulated in the order Cu>Zn>Pb>Cr>Cd (Sekomo et al., 2011). Runoff from roads can pollute wetlands and bring change in the biotic communities dwelling in those ecosystems (Spellerberg, 1998). In a study of South Indian coastal wetlands by Purvaja and Ramesh (2001), it was shown that pollution of coastal wetlands by human settlements can turn them into a larger source of methane which can potentially alter the global carbon budget. In a study by Harikumar et al. (2009), it was seen that heavy metal pollution in Vembanad Lake was moderate for Cu, moderate to high for Zn and Ni, and non-polluted to moderate for Pb. Significant contamination of Cd, Zn, and Ni has made the mangroves and mudflats of the Pearl River in China unsuitable for activities like nature reserves, agriculture and reclamation (Li et al., 2007).

PAHs from industrial and domestic wastewater, runoff from road surface and other sources can end up in the wetlands and pollute their ecosystem (Hemond and Benoit, 1988). Xiao et al. (2014) mentions that industrial and shipping activities are causing increase of PAHs in the Pearl River estuary, where different types of PAH are found to be strongly bound to organic matter in soil. Concentration of PAHs in wetland soils of Sundarban (India) shows low to moderate level of contamination, which can be attributed to rapid urbanization and economic development in the coastal areas (Domínguez et al., 2010). Presence of PAHs facilitates growth of some plants (e.g. *Brassica juncea*) and decreases growth of some other (e.g. *Juncus subsecundus*) (Zhang et al., 2011).

These studies show pollutants can be from several sources, affecting the wetlands in several ways, and results are harmful to many organisms. Humans can be potential victims of contaminations like heavy metal pollution. While climate change is

Importance of Wetlands

a global concern, direct and indirect pollution are regional in nature, similar to land use changes, and can be tackled by awareness and strict policies. The sources should be identified and mitigation measures must be taken accordingly. Artificial wetlands can be effective measures for such controls.

13.3.4 EUTROPHICATION

As discussed earlier, it is natural for wetlands to remove nutrients from the water. Major sources of nitrogen and phosphorus in wetlands are urbanized and agricultural lands (Hemond and Benoit 1988). Whereas internal eutrophication can occur in wetlands due to decomposition of organic matter under anaerobic condition with alkaline environment and abundance of SO_4^{2-} (Smolders et al., 2006). According to Lucassen et al. (2005) eutrophication is likely to occur in wetlands if concentration of nitrate is greater than sulphate or sulphate is mobilized by FeS_x. Houlahen et al. (2006) mentions agricultural fertilizer application as one of the important causes of eutrophication, which can decrease the plant species richness and cause a shift in community structure. Increased nitrogen can alter the community structure in wetlands by supporting the nitrophilic plants and endangering the plants adapted to low nitrogen infertile soil conditions, and can also alter rates of microbial nitrogen fixation, decomposition, nitrification and nitrate reduction (Morris, 1991). Eutrophication indirectly increases the risk of miracidial infection in snails and amphibian larvae, particularly the increase in the parasite infection increases severe limb deformity and mortality in amphibians (Johnson et al., 2007). Nyenje et al. (2010) mentioned that disposal of wastewater to wetlands is leading to eutrophication, which is resulting into extinction of fishes, depletion of dissolved oxygen and increasing the abundance of cyanobacteria and *C. botulinum* in sub-Saharan wetlands. Increase in phosphorus causes increase in algae and lemnids (Lucassen et al., 2005).

As the study shows, the eutrophication disturbs the biodiversity of the wetlands, and this can eventually lead to extinctions of some species from a particular wetland, along with the ecosystem services provided by them. The sources of excess nutrients are mainly from agricultural, whereas other possible sources like household wastewater, horticulture, sewage etc. should not be underestimated. Controlled use of fertilizers and artificial wetlands or other dumping sites can be potential solutions. Harvesting, recycling and reusing runoff using proper channels can be a good option.

13.4 CONCLUSION

It can be well understood from the research outcomes from around the world that wetlands are one of the important ecosystems with ecological and economic importance. The biodiversity of wetland is unique, and each of the members of the system is dependent on the others for their survival. Any reduction or extinction of a species may significantly lead to loss or overgrowth of the others, resulting in collapse of the total ecosystem in the wetlands. The loss of the biodiversity by anthropogenic disturbances or overharvest can initiate such problems. The economic support like fisheries, water, soil fertility etc. can be lost in the long term. The process can also alter regional biogeochemistry and also affect human health, other water resources, agriculture etc. in several ways. Similarly, pollution with chemicals, unplanned land use, eutrophication, climate change etc. can themselves impact each other as well as alter biodiversity and biogeochemistry. So wetland biodiversity must be conserved with proper attention, policy and awareness. The alteration in biogeochemistry by excess sedimentation, eutrophication, pollution, unplanned land use, climate change etc. can affect the ecosystem as well as the water and soil quality of the region leading to similar losses.

It will be incomplete to work to protect the individual component of the wetland ecosystem like biodiversity, water quality etc., but a totalitarian approach can be more successful. Any planning to protect a wetland should consider it as a singular entity. The plan should focus on biodiversity, water quality, sources of pollution, types of pollutants, and the required biological, chemical or physical methodologies needed for restoration. Introducing any new species for such purposes should be verified for their invasiveness, and at the same time chemical or physical alterations should be done carefully so that the existing situation is not harmed in any other way. But the restoration planning is not sufficient for protection of wetlands unless people are made aware of the importance of the importance of the wetlands. Both ecological and economic awareness should be developed through school education as well as adult education. Moreover, it can be very helpful to control pollution at the source. Controlled use of the agricultural chemicals, proper drainage from industries and urbanized areas, recycling and reuse of the used water can be potential solutions, whereas the future impact of climate change can only be mitigated by a global coalition to tackle greenhouse gas emissions.

REFERENCES

Adhikari, B. S. and Babu, M. M., 2008. Floral diversity of Baanganga Wetland, Uttarakhand, India. *Check List.* 4(3):279–290.

Alongi, D. M., Murdiyarso, D., Fourqurean, J. W., Kauffman, J. B., Hutahaean, A., Crooks, S., Lovelock, C. E., Howard, J., Herr, D., Fortes, M., Pidgeon, E. and Wagey, T., 2016. Indonesia's blue carbon: A globally significant and vulnerable sink for seagrass and mangrove carbon. Springer *Wetlands Ecol Manage.* 24:3–13.

Ardón, M., Montanari, S., Morse, J. L., Doyle, M. W. and Bernhardt, E. S., 2010. Phosphorus export from a restored wetland ecosystem in response to natural and experimental hydrologic fluctuations, American Geophysical Union. *Journal of Geophysical Research.* 115:G04031. doi:10.1029/2009JG001169

Bassi, N., Kumar, M. D., Sharma, A. and Pardha-Saradhi, P., 2014. Status of Wetlands in India: A review of extent, ecosystem benefits, threats and management strategies, Elsevier. *Journal of Hydrology: Regional Studies.* 2:1–19.

Bohn, T. J., Lettenmaier, D. P., Sathulur, K., Bowling, L. C., Podest, E., McDonald, K. C. and Friborg, T., 2007. Methane emissions from western Siberian Wetlands: Heterogeneity and sensitivity to climate change, IOP Publishing Ltd. *Environ. Res. Lett.* 2:9.

Boone, M. D. and James, S. M., 2003. Interactions of an insecticide, herbicide and natural stressors in amphibian community mesocosms. *Ecological Society of America, Ecological Applications.* 13(3):829–841.

Brander, L. M., Bräuer, I., Gerdes, H., Ghermandi, A., Kuik, O., Markandya, A., Ståle, N., Paulo, A.L.D.N., Schaafsma, M., Wagtendonk, Vos, H. and Wagtendonk, A., 2012a. Using meta-analysis and GIS for value transfer and scaling up: Valuing climate change induced losses of European Wetlands. *Environ Resource Econ.* 52:395–413.

Importance of Wetlands 215

Brander, L. M., Wagtendonk, A. J., Hussain, S. S., McVittie, A., Verburg, P. H., de Groot, R. S. and van der Ploeg, S., 2012b. Ecosystem service values for mangroves in Southeast Asia: A meta-analysis and value transfer application. *ELSEVIER, Ecosystem Services*. 1:62–69.

Brinson, M. M., 1993. Changes in function of Wetlands along environmental gradients. *Wetlands*. 13(2):65–74.

Choi, Y., Yang, W., Yuch-Ping, H. and Robinson, L., 2001. Vegetation succession and carbon sequestration in a coastal wetland in northwest Florida: Evidence from carbon isotopes. *Global Biogeochemical Cycles*. 15(2):311–319.

Clair, T. A., Arp, P., Moore, T. R., Dalva, M. and Meng, F. R., 2002. Gaseous carbon dioxide and methane, as well as dissolved organic carbon losses from a small temperate wetland under a changing climate. *ELSEVIER, Environmental Pollution*. 116:S143–S148.

DeLaune, R. D. and White, J. R., 2012. Will coastal Wetlands continue to sequester carbon in response to an increase in global sea level? a case study of the rapidly subsiding Mississippi river deltaic plain. *Springer, Climatic Change*. 110:297–314.

Deng, H., Ye, Z. H. and Wong, M. H., 2004. Accumulation of lead, zinc, copper and cadmium by12 wetland plant species thriving in metal-contaminated sites in China, Elsevier Ltd. *Environmental Pollution*. 132:29–40.

Domínguez, C., Sarkar, S. K., Bhattacharya, A., Chatterjee, M., Bhattacharya, B. D., Jover, E., Albaigés, J., Albaigés, J. M., Bayona, A.M.A. and Satpathy, K. K., 2010. Quantification and source identification of polycyclic aromatic hydrocarbons in core sediments from Sundarbans Mangrove Wetland, India, Springer. *Arch Environ Contam Toxicol*. 59:49–61.

Engle, V. D., 2011. Estimating the provision of ecosystem services by Gulf of Mexico Coastal Wetlands. *Springer, Wetlands*. 31:179–193.

Erwin, K. L., 2009. Wetlands and global climate change: The role of wetland restoration in a changing world. *Wetlands Ecol Manage*. 17:71–84.

Fisher, J. and Acreman, M. C., 2004. Wetland nutrient removal: A review of the evidence. *Hydrology and Earth System Sciences Discussions, European Geosciences Union*. 8(4):673–685.

Froyd, C. A., Coffey, E.E.D., van der Knaap, W. O., Van Leeuwen, J.F.N., Alan, T. and Katherine, J. W., 2014. The ecological consequences of megafaunal loss: Giant tortoises and wetland biodiversity. *Ecology Letters*. 17:144–154.

Gambrell, R. P., 1994. Trace and toxic metals in Wetlands-a review. *J. Environ. Qual*. 23:883–891.

Greenway, M., 2005. The role of constructed Wetlands in secondary effluent treatment and water reuse in subtropical and arid Australia. *ELSEVIER, Ecological Engineering*. 25:501–509.

Hansson, L., Christer, B. N., Anders, N. P., Kajsa, A. B. R., 2005. Conflicting demands on wetland ecosystem services: Nutrient retention, biodiversity or both? *Freshwater Biology*. 50:705–714.

Harikumar, P. S., Nasir, U. P. and MujeebuRahman, M. P., 2009. Distribution of heavy metals in the core sediments of a tropical wetland system, IRSEN, CEERS, IAU. *Int. J. Environ. Sci. Tech*. 6(2):225–232.

Hashim, A. M. and Catherine, S. M. P., 2013. A laboratory study on wave reduction by mangrove forests. *ELSEVIER, APCBEE Procedia*. 5:27–32.

Hemond, H. F. and Benoit, J., 1988. Cumulative impacts on water quality functions of Wetlands, Springer-Verlag New York Inc. *Environmental Management*. 12(5):639–653.

Houlahan, J. E., Keddy, P. A., Makkay, K. and Scott, F. C., 2006. The effects of adjacent land use on Wetland species richness and community composition. *The Society of Wetland Scientists, Wetlands*. 26(1):79–96.

Jansson, Å., Folke, C. and Langaas, S., 1998. Quantifying the nitrogen retention capacity of natural Wetlands in the large-scale drainage basin of the Baltic Sea. *Kluwer Academic Publishers, Landscape Ecology*. 13:249–262.

Jenkins, D. G., Scott, G. and Keith, M., 2003. Consequences of prairie wetland drainage for crustacean biodiversity and metapopulations. *Conservation Biology.* 17(1):158–167.

Johnson, C. W., Millett, B. V., Gilmanov, T., Voldseth, R. A., Guntenspergen, G. R. and Naugle, D. E., 2005. Vulnerability of Northern Prairie Wetlands to climate change. *Bioscience.* 55(10):863–872.

Johnson, P.T.J., Chase, J. M., Dosch, K. L., Hartson, R. B., Gross, J. A., Larson, D. J., Sutherland, D. R. and Carpenter, S. R., 2007. Aquatic eutrophication promotes pathogenic infection in amphibians, PNAS. *Environmental Sciences.* 104(40):15781–15786.

Junhong, B., Ouyang, H., Rong, X., Gao, J., Gao, H., Cui, B. and Huang, L., 2010. Spatial variability of soil carbon, nitrogen, and phosphorus content and storage in an alpine wetland in the Qinghai—Tibet Plateau, China, CSIRO Publishing. *Australian Journal of Soil Research.* 48:730–736.

Kantrud, H. A. and Stewart, R. E., 1977. Use of natural basin Wetlands by breeding waterfowl in North Dakota. *J. Wildl. Manage.* 41(2):243–253.

Kumar, B., Mukherjee, D. P., Kumar, S., Mishra, M., Prakash, D., Singhl, S. K. and Sharma, C. S., 2011. Bioaccumulation of heavy metals in muscle tissue of fishes from selected aquaculture ponds in east Kolkata Wetlands. *Annals of Biological Research.* 2(5):125–134.

Li, Q., Wu, Z., Chu, B., Na, Z., Cai, S. and Hong Fang, J., 2007. Heavy metals in the coastal wetland sediments of the Pearl River Estuary, China, Elsevier Ltd., *Environmental Pollution.* 149:158–164.

Liu, H., Zhang, S., Li, Z., Lu, X. and Qing, Y., 2004. Impacts on Wetlands of large-scale land-use changes by agricultural development: The small Sanjiang Plain, China. *Royal Swedish Academy of Sciences.* 33(6):307–310.

Lloyd, C. R., 2006. Annual carbon balance of a managed wetland meadow in the Somerset Levels, UK, ELSEVIER. *Agricultural and Forest Meteorology.* 138:168–179.

Loomis, F., Mark, J. and Christopher, Craft, B., 2010. Carbon sequestration and nutrient (Nitrogen, Phosphorus) accumulation in river-dominated tidal marshes, Georgia, USA. *Soil Sci. Soc. Am. J.* 74(3):1028–1036.

Lucassen, E.C.H.E.T., Smolders, A.J.P., Lamers, L.P.M. and Roelofs, J.G.M., 2005. Water table fluctuations and groundwater supply are important in preventing phosphate-eutrophication in sulphate-rich fens: Consequences for wetland restoration. *Springer, Plant and Soil.* 269:109–115.

Marchio, D. A., Savarese, Jr. M., Bovard, B. and Mitsch, W. J., 2016. Carbon sequestration and sedimentation in mangrove swamps influenced by hydro geomorphic conditions and Urbanization in Southwest Florida. *Forests.* 7:116.

Marín-Muñiz, J. L., María, E. H. and Moreno-Casasola, P., 2014. Comparing soil carbon sequestration in coastal freshwater Wetlands with various geomorphic features and plant communities in Veracruz. *Mexico, Plant Soil.* 378:189–203.

Matagi, S. V., Swai, D. and Mugabe, R., 1998. A review of heavy metal removal mechanisms in Wetlands. *Afr. J. Trop. Hydrobiol. Fish.* 8:23–35.

McCarty, G., Pachepsky, Y. and Ritchie, J., 2009. Impact of sedimentation on Wetland carbon sequestration in an agricultural watershed. *Journal of Environmental Quality.* 38:804–813.

Mitsch, W. J., Blanca, B., Nahlik, A. M., loMander, U., Li, Z., Anderson, C. J., Jørgensen Sven, E. and Brix, H., 2016. Wetlands, carbon, and climate change. *Landscape Ecol.* 28:583–597.

Morris, J. T., 1991. Effects of nitrogen loading on wetland ecosystems with particular reference to atmospheric deposition, annual reviews Inc. *Annu. Rev. Ecol. Syst.* 22:257–279.

Mortsch, L. D., 1998. Assessing the impact of climate change on the great Lakes shoreline Wetlands. *Kluwer Academic Publishers, Climate Change.* 40:391–416.

Nyenje, P. M., Foppen, J. W., Uhlenbrook, S., Kulabako, R. and Muwanga, A., 2010. Eutrophication and nutrient release in urban areas of sub-Saharan Africa—a review. *ELSEVIER, Science of the Total Environment.* 408:447–455.

Olubodel, O. S., Awodoyin, R. O. and Ogunyemi, S., 2011. Floral diversity in the Wetlands of Apete River, Eleyele Lake and Oba Dam in Ibadan, Nigeria: Its implication for biodiversity erosion. *West African Journal of Applied Ecology.* 18.

Prasad, M.N.V., 2003. Phytoremediation of metal-polluted ecosystems: Hype for commercialization. *Russian Journal of Plant Physiology.* 50(5):686–700, *From FiziologiyaRastenii.* 50(5):764–780.

Prasad, S. N., Ramachandra, T. V., Ahalya, N., Sengupta, T., Kumar, Alok, Tiwari, A. K., Vijayan, V. S. and Lalitha, V., 2002. International society for tropical ecology, conservation of Wetlands of India—a review. *Tropical Ecology.* 43(1):173–186.

Purvaja, R. and Ramesh, R., 2001. Natural and anthropogenic methane emission from coastal Wetlands of South India, Springer-Verlag New York. *Environmental Management.* 27(4):547–557.

Relyea, R. A., 2009. A cocktail of contaminants: How mixtures of pesticides at low concentrations affect aquatic communities. *Oecologia.* doi:10.1007/s00442-008-1213-9.

Saunders, D. L. and Kalff, J., 2001. Nitrogen retention in Wetlands, lakes and rivers. *Kluwer Academic Publishers, Hydrobiologia.* 443:205–212.

Sekomo, C. B., Nkuranga, E., Rousseau, D.P.L. and Lens, P.N.L., 2011. Fate of heavy metals in an urban natural wetland: The Nyabugogo swamp (Rwanda). *Springer, Water Air Soil Pollution.* 214(1–4):321–333.

Sheoran, A. S. and Sheoran, V., 2006. Heavy metal removal mechanism of acid mine drainage in Wetlands: A critical review. *Elsevier, Minerals Engineering.* 19:105–116.

Siwakoti, M., 2006. An overview of floral diversity in Wetlands of Terai region of Nepal. *Our Nature.* 4:83–90.

Smolders, A.J.P., Lamers, L.P.M., Lucassen, E.C.H.E.T., Van Der Velde, G. and Roelofs, J.G.M., 2006. Internal eutrophication: How it works and what to do about it—a review. *Taylor and Francis, Chemistry and Ecology.* 22(2):93–111.

Spellerberg, I. F., 1998. Ecological effects of roads and traffic: A literature review. *JSTOR, Global Ecology and Biogeography Letters.* 7(5):317–333.

Stallard, R. F., 1998. Terrestrial sedimentation and the carbon cycle: Coupling weathering and erosion to carbon burial. *Global Biogeochemical Cycles.* 12(2):231–257.

Turner, R. K., Georgiou, S., Ing-Marie, G., Wulff, F., Scott, B., Soderqvist, T., Bateman, I. J., Folke, C., Langaas, S., Zylicz, T., Karl-Göran, M. and Markowska, A., 1999. Managing nutrient fluxes and pollution in the Baltic: An interdisciplinary simulation study. *ELSEVIER, Ecological Economics.* 30:333–352.

Whitfield, G. J., Scott, D. E., Ryan, T. J., Buhlmann, K. A., Tuberville, T. D., Metts, B. S., Greene, J. L., Mills, T., Leiden, Y., Poppy, S. and Winne, C. T., 2000. The global decline of reptiles, Déjà Vu Amphibians. *BioScience.* 50(8):653–666.

Wilcox, D. A. and Meeker, J. E., 1991. Disturbance effects on aquatic vegetation in regulated and unregulated lakes in northern Minnesota. *Can. J. Bot.* 69:1542–1551.

Wuver, A. M., Attuquayefio, D. K. and Enu-Kwesi, L., 2003. A study of bushfires in a Ghanaian Coastal Wetland. II. Impact on floral diversity and soil seed bank. *West African Journal of Applied Sociology.* 4:13–26.

Xiao, R., Bai, J., Wang, J., Lu, Q., Zhao, Q., Cui, B. and Liu, X., 2014. Polycyclic aromatic hydrocarbons (PAHs) in wetland soils under different land uses in a coastal estuary: Toxic levels, sources and relationships with soil organic matter and water-stable aggregates. *ELSEVIER, Chemosphere.* 110:8–16.

Yu, K.S.H., Wong, A.H.Y., Yau, K.W.Y., Wong, Y. S. and Tam, N.F.Y., 2005. Natural attenuation, biostimulation and bioaugmentation on biodegradation of polycyclic aromatic hydrocarbons (PAHs) in mangrove sediments. *Mar Pollut Bull.* 51(8–12).

Zedler, J.B. and Kercher, S., 2004. Causes and consequences of invasive plants in Wetlands: Opportunities, opportunists, and outcomes. *Taylor & Francis, Critical Reviews in Plant Sciences.* 23(5):431–452.

Zhang, Z., Rengel, Z., Meney, K., Pantelic, L. and Tomanovic, R., 2011. Polynuclear aromatic hydrocarbons (PAHs) mediate cadmium toxicity to an emergent wetland species. *ELSEVIER, Journal of Hazardous Materials.* 189:119–126.

14 Methylene Blue Dye Degradation by Silver Nanoparticles Biosynthesized Using Seed Extracts of *Foeniculum vulgare*

S. Anjum Mobeen and K. Riazunnisa

CONTENTS

14.1 Introduction 219
14.2 Materials and Methods 220
    14.2.1 Preparation of Seed Extract 220
    14.2.2 Phytochemical Screening 221
    14.2.3 Synthesis of Silver Nanoparticles (AgNPs) 221
    14.2.4 UV-Vis Spectral Analysis of Nanoparticles of Seed Extract 221
14.3 Preparation of MB Dye Solution 221
14.4 Results and Discussion 221
    14.4.1 Phytochemical Screening 221
    14.4.2 UV-Vis Analysis of AgNPs 222
    14.4.3 Catalytic Activity of Synthesized AgNPs 222
    14.4.4 Effect of Contact Time on Adsorption of MB Dye 224
14.5 Conclusion 224
14.6 Acknowledgements 224
References 224

14.1 INTRODUCTION

Synthetic dyes are extensively used in a variety of industries such as textiles, paper, polymers, adhesives, ceramics, construction, cosmetics, food, glass, paints, ink, soap and pharmaceuticals (Zalikha et al., 2012). Dye residues form a group of glaring and grave contaminants since they impart colour to the wastewater even at a very low concentration level. The catalytic reduction studies of these dyes assume greater significance in the present context, as most of these dyes are known to be toxic when inhaled or ingested orally and pose health hazards such as skin and eye irritation in humans. The adverse effects of these synthetic dyes to the environment include the ability to deplete oxygen in the surface waters and streams, thereby affecting the very sustenance of aquatic flora and fauna and causing an inhibitory effect on the photosynthetic activity of plants. The anaerobic degradation products of azo dyes are amines which are very toxic, carcinogenic and mutagenic in nature. The effluents from industries containing these hazardous dyes need to be removed before they are let out for on-land disposal purposes. The treatment technologies such as adsorption, ion exchange (Karcher et al., 2002), filtration, electrode deposition and chemical precipitation are being widely used to remove the dye contaminants from water.

The development of non-toxic methods of synthesizing nanoparticles is a major step in nanotechnology to allow their application in various fields (Stary and Hans, 1998). Among the various categories of compounds synthesized in plants, phytochemicals (antioxidants, flavonoids, flavones, isoflavones, catechins, anthocyanidins, isothiocyanates, carotenoids and polyphenols) are the most potent materials for biological activities and are known as important natural resources for the synthesis of metallic nanoparticles (Park et al., 2011). The recovery of these nanoparticles from plant tissues is tedious and expensive and needs enzymes to degrade the cellulosic materials that surround it (Marshall et al., 2007). Thus, the synthesis of various metal nanoparticles using plant extracts is easy in downstream processing and in scaling up of nanoparticles (Rai et al., 2006; Mukunthan et al., 2011). An attempt has been made to alleviate the detrimental effects of the dyes on the surrounding environment by trying to reduce the dye concentration in the treated effluents such as methylene blue (MB) and silver nanoparticles as a nano-catalyst. The previous studies on the reduction of MB and Congo red by using gold and silver nanoparticles (Wanyonyi et al., 2014; Ganapuram et al., 2015), as nano-catalysts were also reported.

Foeniculum vulgare (common fennel), belonging to the Apiaceae family, is one of the most common spices, known for its highly aromatic nature and flavour in culinary and traditional applications. Various pharmacological properties of *F. vulgare* have been reported in the literature, such as antioxidant, hepatoprotective, antimicrobial oestrogenic, acaricidal, anti-hirsutism, anti-diabetic, anti-inflammatory and anti-thrombotic properties (Nandini and Sreemoyee, 2014). The use of environmentally benign materials such as silver nanoparticles offers numerous benefits of eco-friendliness and compatibility for catalytic degradation application. Keeping in mind the above fact, the present aim of this study was to investigate the bioactive components present in the seed broth of *F. vulgare* which help the biosynthesis of silver nanoparticles (AgNPs) and to analyze their effect on MB dye with respect to both seed extract and AgNPs.

14.2 MATERIALS AND METHODS

14.2.1 Preparation of Seed Extract

Seeds of *Foeniculum vulgare* were collected and the authenticity was confirmed by the Department of Botany, Yogi Vemana University, Kadapa. The material was

Meth-Blue Dye Degradation by Nanoparticles 221

washed first with running tap water later with distilled water in order to remove debris and was completely air-dried at room temperature. They were kept away from direct sunlight to avoid destruction of active compounds. The dried seeds were pounded to powder using a grinder. Sterile distilled water was added to 2.5 g of powder and boiled for 20 minutes at 80°C. The seed infusion was collected through Whatman filter paper No. 25 and stored at 4°C for further analysis.

14.2.2 PHYTOCHEMICAL SCREENING

Prepared extract was subjected to preliminary phytochemical screening for qualitatively determination of secondary metabolites as described by Vani et al. (2018).

14.2.3 SYNTHESIS OF SILVER NANOPARTICLES (AGNPS)

A 1 mM $AgNO_3$ (Sigma-Aldrich) stock solution was prepared freshly. The seed broth of *Foeniculum vulgare* and the $AgNO_3$ solution were mixed in the ratio of 1:9 and kept continuously stirred and a change in colour was noted. The transformation of light creamy to reddish-brown confirms the visual formation of silver nanoparticles.

14.2.4 UV-VIS SPECTRAL ANALYSIS OF NANOPARTICLES OF SEED EXTRACT

The reduction of Ag^+ ions was monitored by measuring the optical density of the reaction mixture at regular intervals with the wavelength range of 350–700 nm by using a UV-Vis spectrophotometer (UV Thermo Scientific Evolution 201 series).

14.3 PREPARATION OF MB DYE SOLUTION

A 1 mM MB dye was prepared and used as a stock solution. The catalytic activity of synthesized AgNPs was carried out in a 3.5 ml quartz cuvette and absorbance values were monitored using the UV-Vis spectrophotometer at room temperature. The value of absorption maxima of the reaction mixture was compared with that of MB. Concentration of dye during degradation was calculated by the absorbance value at 664 nm.

Percentage of dye degradation was estimated by the following formula:

$$\% \text{Decolouruzation} = 100 \times \frac{\left(C_0 - C\right)}{C_0}$$

where C_0 is the initial concentration of dye solution and C is the concentration of dye solution after photocatalytic degradation.

14.4 RESULTS AND DISCUSSION

14.4.1 PHYTOCHEMICAL SCREENING

Phytochemicals in medicinal plants have been reported to be the active principles responsible for the pharmacological potentials of plants. Phytochemical screening of

TABLE 14.1

Qualitative Phytochemical Screening of Aqueous Seed Extract of *Foeniculum vulgare*

Chemical Constituents	Observation
Alkaloids	+
Glycosides	+
Tannins	+
Steroids and Terpenoids	++
Carbohydrates	++
Flavonoids	+
Proteins	+
Phenols	+++

+ = present, ++ = more quantity, +++ = more than quantity

Foeniculum vulgare showed the presence of phenols, carbohydrates, proteins, tannins, alkaloids, flavonoids, steroids, glycosides and diterpenes (Table 14.1). The bioactivity of plant extracts is attributed to phytochemical constituents. Flavonoids are a major group of phenolic compounds reported for their antiviral and antibacterial properties. Phenolic compounds possess hydroxyl and carboxyl groups, and plants with high content of phenolic compounds are one of the best candidates for nanoparticle synthesis.

14.4.2 UV-Vis Analysis of AgNPs

The colour change of the reaction mixture from light creamy to dark brown was observed within 60 minutes, which indicated the formation of silver nanoparticles. This change is due to the excitation of free electrons in nanoparticles, which gives the surface plasmon resonance (SPR) absorption band by the combined vibration of electrons of metal nanoparticles in resonance with light wave. Metal nanoparticles display different colours in solution due to their optical properties (Mobeen et al., 2017).

In the present work, the colour of the freshly prepared aqueous seed extract obtained from *Foeniculum vulgare* changed when silver nitrate solution was added. The reduction of pure Ag^+ ions was monitored by measuring the UV visible spectrum of the reaction medium at regular intervals. The appearance of reddish-brown colour at 60 minutes indicates the formation of silver nanoparticles. A stable maximum absorption peak throughout the experiment was recorded at 420 nm (Figure 14.1).

14.4.3 Catalytic Activity of Synthesized AgNPs

It is a well-known fact that AgNPs and their composites show greater catalytic activity in the area of dye reduction and removes the reduction of MB by arsine in the presence of silver nanoparticles. The present study aims at the reduction of MB by

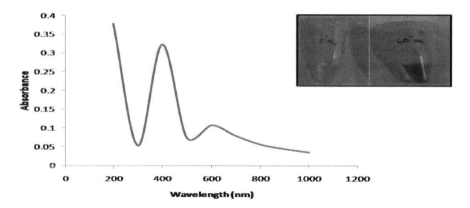

FIGURE 14.1 UV Vis Absorption Spectra of Silver Nitrate with Aqueous *Foeniculum vulgare* Seed Extract (Inset: as Biosynthesized Silver Nanoparticles)

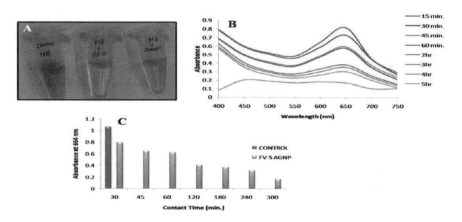

FIGURE 14.2 Catalytic Action of AgNPs A. Visual Identification of Reduction of MB Dye, B. UV -Vis. Spectrum of MB Dye Reduction, C. Effect of Contact Time on Adsorption of MB Dye at 664 nm by *F. Vulgare* Seed Extract.

natural green aqueous extract of *Foeniculum vulgare* containing AgNPs. Pure MB dye has a λ_{max} value of 664 nm (Edison and Sethuraman, 2012). Degradation of dye was visually observed by the change in colour from deep blue to light blue. Finally, the degradation process was completed at the end of the fifth hour and was identified by the change of reaction mixture from colour to colourless (Figure 14.2A). This reveals AgNPs can act as an electron transfer mediator between the extract and MB by acting as a redox catalyst, which is often termed as electron relay effect. Catalytic activity of AgNPs on reduction of MB by *Terminalia chebula* fruit extract was reported by Edison and Sethuraman (2012).

Photocatalytic activity of AgNPs on degradation of dye was demonstrated by using the MB. The degradation of MB was carried out in the presence of AgNPs at different time intervals in the visible region. The absorption spectrum exhibited the decreased peaks for MB at different time periods. The completion of the photo-catalytic degradation of the dyes is known from the gradual decrease of the absorbance value of dye approaching the base line and increased peak for AgNPs. While decreasing the concentration of dye, UV spectra show a typical SPR band for silver AgNPs at five hours of exposure time (Figure 14.2B). The percentage of degradation efficiency of silver AgNPs was calculated as 70% at the fifth hour. The degradation percentage increased with increasing exposure time of dye to AgNPs. Absorption peak for MB dye was centred at 664 nm in visible region which diminished and finally it disappeared while increasing the reaction time, which indicates that the dye had been degraded.

14.4.4 Effect of Contact Time on Adsorption of MB Dye

The adsorption of MB dye onto AgNPs was studied as a function of contact time (Figure 14.2C) revealed that the dye removal percentage increased on increasing the contact time and reached equilibrium after five hours. Karimi et al., 2012 reported the adsorption of methyl orange using silver nanoparticles loaded on activated carbon. They observed that the contact time for methyl orange solutions to reach equilibrium were 15 and 18 minutes for 10 and 20 mg/L of concentration, respectively.

14.5 CONCLUSION

In the present work, silver nanoparticles were effectively green synthesized using the aqueous extract of *Foeniculum vulgare* seeds. The catalytic activity of biologically developed silver nanoparticles was evaluated by choosing methylene blue dye. The produced AgNPs exhibited good catalytic activity in MB dye reduction process. Degradation of MB dye signifies the utilization of the synthesized AgNPs in effluent treatment (dye degradation) of textile industries, pharmaceutical, paper, paints, plastics, cosmetics and chemical industries. Further research is required to evaluate the mechanism of dye reduction by aqueous leaf extract of *F. vulgare* using AgNPs as a nano-catalyst.

14.6 ACKNOWLEDGEMENTS

The author S. Anjum Mobeen is grateful to Maulana Azad National Fellowship (MANF), New Delhi for awarding MANF-JRF fellowship to carry out this study.

REFERENCES

Anjum, M. S., Habeeb, K. C. and Riazunnisa, K., 2017. Green synthesis of silver nanoparticles by *Catharanthus roseus* and their catalytic activity. In Reddy, N. (ed.), *Proceedings of National Seminar on Advanced Trends in Material Science*, Roshan Publishers, New Delhi. 64–69.

Edison, T., N., J., I. and Sethuraman, M. G., 2012. Instant green synthesis of silver nanoparticles using *Terminalia chebula* fruit extract and evaluation of their catalytic activity on reduction of Methylene Blue. *Process Biochemistry*. 47:1351–1357.

Ganapuram, B. R., Alle, M. and Dadigala, R., 2015. Catalytic reduction of methylene blue and Congo red dyes using green synthesized gold nanoparticles capped by *Salmalia malabarica* gum. *International Nano Letters*. 5:215–222. doi: 10.1007/s40089-015-0158-3

Karcher, S., Kornmüller, A. and Jekel, M., 2002. Anion exchange resins for removal of reactive dyes from textile wastewaters. *Water Research*. 36:4717–4724.

Karimi, H., Mousavi, S. and Sadeghian, B., 2012. Silver nanoparticles loaded on activated carbon as efficient adsorbent for removal of methyl orange. *Indian Journal of Science and Technology*. 5:2346–2353.

Marshall, A. T., Haverkamp, R. G., Davies, C. E., Parsons, J. G., Gardea-Torresdey, J. L. and Van-Agterveld, D., 2007. Accumulation of gold nanoparticles in Brassic juncea. *International Journal Phytoremediation*. 9:197–206.

Mukunthan, K. S., Elumalai, E. K., Patel, T. N. and Murty, V. R., 2011. Catharanthus roseus: A natural source for the synthesis of silver nanoparticles. *Asian Pacific Journal Tropical Biomedicine*. 1:270–274.

Nandini, G. and Sreemoyee, C., 2014. Assessment of free radical scavenging potential and oxidative DNA damage preventive activity of trachyspermum ammi L. (Carom) and Foeniculum vulgare Mill. (Fennel) Seed Extracts. *BioMedical Research International*. 8. Available online at http://dx.doi.org/10.1155/2014/582767.

Park, S. Y., Murphy, S. P., Wilkens, L. R., Henderson, B. E. and Kolonel, L. N., 2011. Multivitamin use and the risk of mortality and cancer incidence: The multiethnic cohort study. *American Journal Epidemiology*. 173:906–914.

Rai, A., Singh, A., Ahmad, A. and Sastry, M., 2006. Role of halide ions and temperature on the morphology of biologically synthesized gold nanotriangles. *Langmuir*. 22:736–741.

Stary, F. and Hans, S., 1998. *The National Guides to Medical Herbs and Plants Tiger Books*. International Plc., London.

Vidya, V. M., Anjum, M. S. and Riazunnisa, K., 2018. Phytochemical screening, antioxidant activities and antibacterial potential of leaf extracts of *Buchanania axillaris* L. *Journal of Pharmaceutical International*. 21(3):1–9.

Wanyonyi, W. C., Onyari, J. M. and Shiundu, P. M., 2014. Adsorption of Congo red dye from aqueous solutions using roots of Eichhornia crassipes: Kinetic and equilibrium studies. *Energy Procedia*. 50:862–869. doi: 10.1016/j.egypro.2014.06.105

Zalikha, N., Azlina, W. and Ab, W., 2012. Adsorption of methylene blue by agricultural solid waste of pyrolyzed EFB biochar. *J Purity, Utility Reaction Environment*. 1:376–390.

15 A Green and Sustainable Tool for Physicochemical Analysis of Liquid Solutions
Survismeter

Parth Malik and Rakesh Kumar Ameta

CONTENTS

15.1 Introduction 227
15.2 Survismeter: Conceptualization and Emergence 230
15.3 Working Mechanism for η and γ Measurement 231
15.4 Significance of Survismeter as Cutting
     Edge of Routines 236
     15.4.1 Pharmaceutical Sciences 237
     15.4.2 Quality Control 237
     15.4.3 Working Fascinations 238
15.5 Analytical Tool for Robust Physicochemical Sensing 240
15.6 Outcome Concepts or Mechanisms 241
15.7 Conclusion 242
15.8 Acknowledgements 242
References 243

15.1 INTRODUCTION

Interactions are the cornerstones of several biological, biochemical, pharmaceutical and industrial processes, characterized by energizing of inherent molecular functionalities. Physically and thermodynamically, these are strikingly different from reactions, with no formation of any new species (Malik et al., 2016; Jangid et al., 2018). A typical scenario for better understanding the subtle distinguishing factors of reactions and interactions could be understood through a discussion on solutions and dispersions. Unlike solutions, where a solute loses its chemical and physical identity after mixing with the solvent, which could involve release or absorption of energy, dispersion does not involve the formation of any new substance (Jones et al., 2009; Stefaniak et al., 2017). The best example to understand the terminological distinctions is the mixing pattern of salt and surfactant in water. It is well-known that sodium chloride (common table salt) acts as a potent structure breaker and disintegrates the intermolecular hydrogen bonding (HB) of water molecules through engaging the water molecules with its Na+ and Cl-. Thereby, the native structural networking of water molecules is disrupted, which is more robustly understood by our feeling thirsty after we have taken a salt rich or spicy diet. The surfactants, on the other hand are very different in their chemical behaviour, owing to the fact that these are not dissociated like salts. Dispersion is attractive interaction between any pair of molecules, including non-polar atoms, arising from the interactions of instantaneous multipoles. Figure 15.1 distinguishes the interaction mechanism of NaCl and sodium dodecyl sulphate (SDS). The hydrophobic chain in surfactants is the primary source of Brownian Motions (BM), owing to which the interaction strength of surfactants remains weaker than a corresponding salt (Avramov et al., 1995; Schmitt and Stark, 2016). Nowadays, industrial utility of surfactants is witnessing mounting interest, especially in pharmaceutical and cosmetic domains, since the molecules requiring dissolution are mostly hydrophobic and are therefore poorly water soluble. Some befitting instances making use of dispersion in everyday life are the scientific trials to enhance the bioavailability of hydrophobic molecules, the remedial measures to dispose of the organic residues of petrochemical and textile industries, recalcitrants of waste streams and several others (Walstra, 1983; Mandal and Bera, 2012; Parthasarathy et al., 2013;Gupta et al., 2016). The sole aim is to convert this entire hydrophobicity into distributed hydrophilicity and hydrophobicity which could compensate the enhanced CF between hydrophobic moieties. Surfactants, with

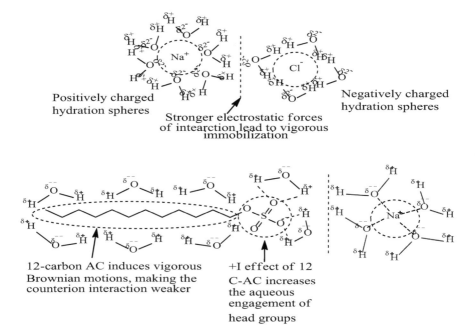

FIGURE 15.1 Pictorial Depiction of Sodium Chloride and Sodium Dodecyl Sulphate (SDS) Aqueous Interaction Mechanisms

Physicochemical Analysis of Liquid Solutions

their amphipathic nature, are able to tap the hydrophobicity of organic molecules and enable their enhanced structural expression through their distributed philicophobic force gradients (Behrens, 1964; Yuan et al., 2014). These differences in the chemical nature of salts and surfactants facilitates an energizing of native structural aspects through non-covalent forces, which is why the interactions with surfactants are moderately driven. The non-covalent nature of surfactant chemical activities is primarily responsible for a retainment of dispersed phase (the solute-like component, analogous to solutions) native structure. Such a scenario leads to a progressive enhancement in the vibrational activities of native structure, as a result of which the structural expression of dispersed phase is markedly improved (Conboy et al., 1996; Sando et al., 2005).

The essence of such molecular interactions generates an interface of workable functional coordinates along the physical and chemical interface, favoured in terminology as physicochemical properties (PCPs) (Gómez-Díaz et al., 2001; Moreno et al., 2015). These properties are swiftly gaining significance as environment sensitive probes to determine the distribution of the solute-like component. The suitability of these PCPs to liquid mixtures is of great significance, since the liquid flow nature varies with need and origin. For instance, the flow of water in a river is quite different from the one undergoneby blood within narrow arteries or the kind of flow possessed by the flowing water in the xylem tissue of a plant. Thus, PCPs are invaluable to conduct the implicit study of these distinct flow regimes, where the flowing liquid undergoes differing levels of shear and thereby the constitutional interactions are highly distinct (Lee, 1996). Survismeter is a device to efficiently track the PCPs of liquid mixtures that are functions of temperature, composition, pressure, local redox environment and the physical barriers/provisions. The study of such properties assumes significance since as per Bernoulli's theorem, the flow regime needs to be specific for characteristic functional activities (Jelali and Kroll, 2003). The liquid syrups, food beverages, paints, dyes, inks and synthetic solvents having designated dissolution potency are all modulated viatuning the hydrophilicity or hydrophobicity. A very poised example fitting in this situation is how water is able to wet a larger area once a drop of it is put on a surface, while honey is not. If additives not altering functional activities are added to honey, possibilities of modulated interaction patterns via altered physical expression of variable compositional ingredients are significantly numerous. To track these changes, there is an immense need to efficiently measure those PCPs which are most affected by the chemical changes in the parent molecule: two of such parameters are surface tension, η, and viscosity, γ.

Before the last 10 to 15 years, these two PCPs were measured with distinct instruments, each putting the sample being examined under a different level of external stress. Furthermore, individual measurements required the sample to be transferred from one location to another entirely different one. This puts the sample being examined at the risk of contamination along with consuming more time and making the measurements more susceptible to errors. Thereby, time-dependent, robust tracking of PCPs is swiftly emerging as the need of the hour, owing to which this article sheds light on a many-in-one device (Survismeter) to track γ, η, interfacial tension (IFT), wetting coefficient, hydrodynamic diameter and several others.

15.2 SURVISMETER: CONCEPTUALIZATION AND EMERGENCE

The sole aim of dispersion is to minimize the solute-solute interactions and achieve its maximum transformation into solute-solvent interacting forces. Initially, a liquid mixture comprises only solvent molecules and all the interactions are amongst the solvent molecules. The forces are therefore only between one kind of molecules recognized as CF. The introduction of solute is the source of exploiting dissimilar chemical activities, expressing themselves distinctly through unique solute-solvent proportions. These combinative activities are retrieved through a CF to intermolecular forces (IMF) transformation. An interesting aspect is that all solvent molecules produce a similar extent of IMFs with variable stoichiometry of solutes. So, interacting efficacy of solute-solvent is highly sensitive with respect to the interacting sites in the solute. For example, it is reasonably well known that polyphenols having low bioavailability are endowed with multifold interactive domains such as delocalized pi-conjugations, lone pair electrons, substitution of electronegative or electropositive species and several other features. Similarly, all drugs are bioactive and express their structural activities only through maintenance of native structures. Thereby, it becomes imperative that interactions are not too strong and merely provides energetic chemical influence to generate sustained molecular oscillations in the native structure. These artefacts serve as genesis of multiple interactive abilities within a same molecule that are summed up as entropy within a same molecular structure (i.e. tentropy). In general, finer dispersion is attributed to higher η (IMF) and low γ(CF). Molecular structures and PCPs share an implicit relationship with each other via bridging the structural characteristics caused out of specific constitutional make-up, hybridizations and stress-strain distribution within a molecule (Ameta and Singh, 2015; Singh et al., 2017). The PCPs are outcomes of peculiar shape, geometry and dimension of the molecular structures forming distinctive bond lengths with specified electron densities of the localized sites in a molecule. The unequal distribution of electrons in constituent atoms causes greater opposite localizations in the prevailing molecular forces. These forces are the conceptualized factors of molecular forces that concurrently manifest density, boiling point, melting point, surface tension, viscosity, friccohesity, refractive index, optical density, osmotic pressure, vapour pressure and IMFs in the liquid mixtures.

Forages, liquid mixtures have been investigated for their performance via η and γ measurements, determined separately with different devices. The η and γ measurements are routinely made with viscometer and tensiometer respectively. These instruments are costly and environmentally threatening, therefore leading to wastage of time, chemicals and manpower. Apart from such issues, multiple transferring of analyzed sample and the energy input of external probes make the measurements time-consuming and error prone. The Survismeter is a device for simultaneous η and γ measurement, operates manually and does not need any external probe to track the molecular motions and their distributional impacts. The operation of this instrument is user-friendly and requires more physical carefulness and observation rather than sophisticated technical setup. It replaces the use of viscometers and stalagmometer, used for individual η and γ measurements. With its ecofriendly mode of operation, the Survismeter is a breakthrough in today's resource-crunch world. Details of its

Physicochemical Analysis of Liquid Solutions 231

working along with the data standardization could be traced in several eminent investigations, ranging from drug to nanoparticle dispersion in varying solvents, prediction of structure making or breaking activities of ionic liquids, chiral molecules and naturally active molecules in chemical environments of dissimilar polarities (Singh, 2006, 2007).

15.3 WORKING MECHANISM FOR η AND γ MEASUREMENT

Figure 15.2a depicts the front view of the Borosil Mansingh Survismeter (BMS), typically fractionated into surface tension, viscosity and pressure regulating limbs. Apart from these provisions, there is another projection to enter the sample within the device. Typically 25 to 30 mL of a liquid sample is needed, which is separately analyzed for η and γ. While measuring η, the sample is pushed up using the sucker pump operation after closing the γ and pressure limbs. Similarly, for γ measurement, the sample is pushed up in the γ limb while the η and pressure limbs remain closed. The γ and η measurements are respectively made using Pendant Drop Number (PDN) and Viscous Flow Time (VFT) methods and recorded using drop counter and stop-watch respectively. All measurements are temperature sensitive since the molecular forces are weakened with increasing temperatures, on account of higher kinetic energy (KE) of the constitutional molecules. The measurements require the PDN (n_o) and VFT (t_o) determination of reference sample first, which is usually the solvent of the sample being analyzed. The mathematical algorithms for η and γ calculations are given in equations 15.1 and 15.2.

$$\gamma = \left[\left(\frac{n_o}{n} \right) \left(\frac{\acute{A}}{\acute{A}_o} \right) \right] \gamma_o \tag{15.1}$$

$$\eta = \left[\left(\frac{t}{t_o} \right) \left(\frac{\acute{A}}{\acute{A}_o} \right) \right] \eta_o \tag{15.2}$$

Here, n, n_o, ρ, ρ_o (densities), t, t_o denote the PDN, ρ and VFT for sample and solvents, respectively. The measurements require the η and γ values of the solvent system to be known at the specific temperature being considered, which are usually taken from relevant literature sources. Figure 15.2b depicts the Survismeter enclosure within the water jacketed case, where apt provisions are made to mount the unit. The temperature of the measurements is maintained via connecting a temperature sensor and circulation of hot or cold water as per the need. For each analysis, generally four to five measurements are usually made after which values in close proximity are normally obtained. The sample holding unit is periodically cleaned using 1% aqueous chromic acid or standard aqua regia. Usually the PDN is comparatively less sensitive to temperature changes while VFT fluctuates in a sensitive manner. Higher PDN leads to a lower γ, due to which the PDN count increases with increasing temperature. Similarly, greater VFT accounts for a higher η which decreases with increasing temperature. Figure 15.3 describes the sample carrying

232 Zero Waste

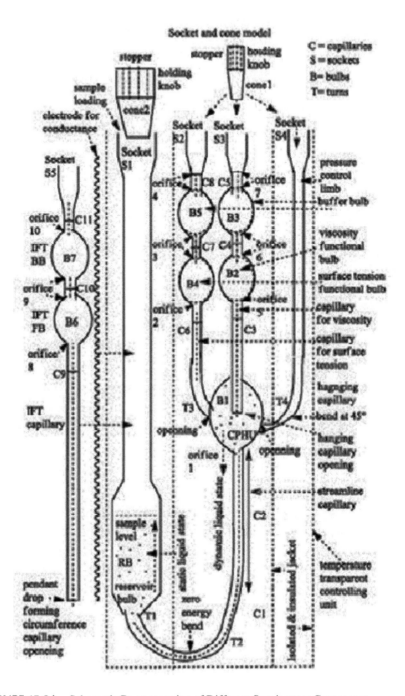

FIGURE 15.2A Schematic Representation of Different Survismeter Components

Physicochemical Analysis of Liquid Solutions 233

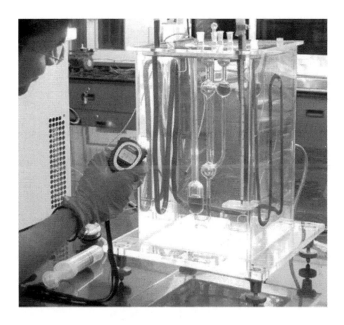

FIGURE 15.2B Typical View of Survismeter Unit in Jacketed Case for Temperature-Sensitive PCP Measurement

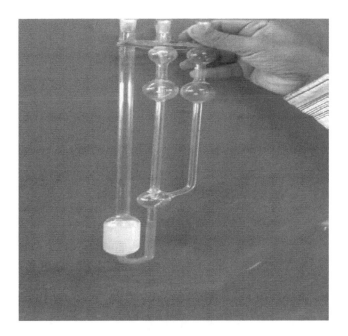

FIGURE 15.3A Working Snapshots of Survismeter

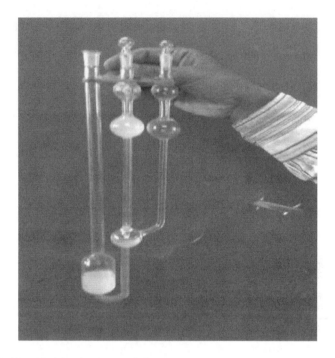

FIGURE 15.3B Working Snapshots of Survismeter

FIGURE 15.3C Working Snapshots of Survismeter

Physicochemical Analysis of Liquid Solutions 235

the Survismeter unit in the sample column, the connecting joint (central pressure unit, CPU) of all the limbs and the functional bulb. The positioning of γ limb is in a linear assembly while the one for η is slightly tapered, owing to which the fundamental mechanisms of droplet flow and shear distribution is significantly different, as indicated in Figure 15.4a and 15.4b.

During measurements, caution should be taken to prevent any external vibration from disturbing the measurement setup. The jerking of windows, doors and tabletops must be avoided for any vibrational interference. The γ predicts the forces between like molecules while η ascertains the value of adhesive forces, operational within two or more distinct species. Each PDN and VFT reading must be used as mean within standard deviation so that the results are obtained within the standard limits of errors. For increasing concentrations of similar solute-solvent mixtures, the unit need not be cleaned before changing the mixture components, but it is a must to clean the instrument for any residues. Except temperature control and water circulation (using a waterbath), the instrument does not require any energy from outside and is substantially energy savvy. For measurements, the PDN and VFT are measured from the top to bottom of graduated marks made on γ and η limbs, according to which calibrations and standardizations are established for deducing equations 15.1

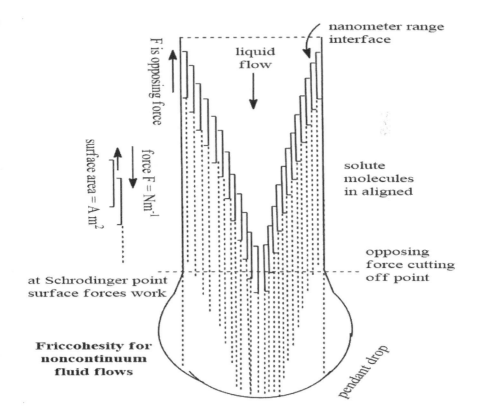

FIGURE 15.4A Typical Fluid Mechanics Models Operational in η

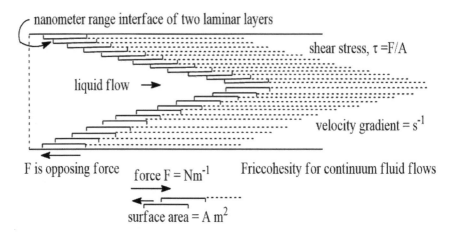

FIGURE 15.4B Typical Fluid Mechanics Models Operational in γ, Measurement Using Survismeter

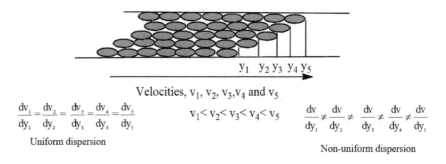

FIGURE 15.4C Typical Fluid Mechanics Models Operational in the Constancy of Velocity Gradients for a Uniform Dispersion

and 15.2. These equations are derived after an adequate consideration of contact angle and convex-concave dynamics of the liquid interfaces, whereby buoyancy corrections are also implemented. The details of these calculations can be looked for in some related sources, focused more on fundamental sciences of Survismeter operation (Singh, 2008a, 2017).

15.4 SIGNIFICANCE OF SURVISMETER AS CUTTING EDGE OF ROUTINES

Erstwhile of academia, Survismeter science has grown by leaps and bounds in the overlapping of domains of pharmaceutical sciences, quality control practices, analytical sciences of designer solvents and several others. Some practical studies conductible with Survismeter or where the Survismeter usage can affect the product quality and yield are included here.

Physicochemical Analysis of Liquid Solutions

15.4.1 PHARMACEUTICAL SCIENCES

The fastness of life with added stress levels has only added to the prevailing pressures, leading to the need of medications which act quickly, efficiently and deliver the corrective measures accurately. The most diligent factor affecting the performance of a drug molecule is its long-term stability once it is within the physiological boundaries. It is highly essential that the consumed drug is evenly distributed and reaches to its target location in an unaltered form in as much low time as possible. While reaching timely to the designated site might not be exactly dependent on PCPs, the distribution regime is no doubt, highly sensitively affected by its native state γ and η. For instance, recent studies on major polyphenolic constituent of turmeric, curcumin, have revealed its below par absorption from the gastrointestinal tract, owing to which it is a dire necessity that a drug needs to be delivered with a carrier medium enabling its spontaneous release once it is surrounded in a specific chemical environment. Similarly, many reports claim the drug structure is impaired after it reaches its targeted site, which is why nowadays numerous stabilizers are included to prolong the half-life as well as entire shelf-life maintained to a significantly high extent. Such possibilities make it mandatory for a rigorous physicochemical screening of a drug formulation at the pH and temperature of localized physiological environment (where the drug is intended to spend most of its time). Another area of concern is the thermodynamic stability of liquid medicines or syrups, owing to which it is important to identify the correct proportion of stabilizers to be added in a required amount of liquid medicinal content, without altering its native structural expression. The overall aim is to monitor the self-interactions, once the drug is inside the living environment. Survismeter, with its efficient temperature-based η and γ measurements, can authentically track the PCPs of a drug formulation at a specific temperature and in a specific chemical composition. Several pharmaceutical enterprises (such as Cadila) and health centres of leading academic institutes have already procured Survismeter. The screening is not only limited to pharmaceuticals since several other products, such as lotions, gels, beverages and testing of biological fluids (in case of disease diagnosis) are all measurable through a tracking of PCPs.

15.4.2 QUALITY CONTROL

With mounting needs and resource pressure, qualities of several consumables are being badly affected. The η and γ estimation offers an incentive in this domain since these properties are temperature and composition dependent. Any liquid material, be it a milk sample, water sample or any other, will have fixed η and γ values in its native unadulterated state. Even a slight adulteration would lead to a loss of native state of these liquid materials, which would be timidly analyzed through η and γ measurement. The stoichiometric dependence of PCPs can be of crucial significance in proper compositions of paints, varnishes, shaving solutions, lubricants, inks and wetting agents, since a low γ accounts for low spreadability through enhanced CF. Academia have encouraged this notion and Central University of Gujarat (Gandhinagar, India) even runs a six-month certificate course for the visually challenged in collaboration with the Government of Gujarat. This course utilizes a sister version of Survismeter, the Visionmeter, details of which can be found in its specified article (Singh et al., 2008b). The tracking of PCPs in this regard can help in maintaining desired quality of consumables, consumer products like paints, varnishes, lotions and several others. These capabilities of Survismeter make it beneficial to ensure an implementation of fair manufacturing practices and uplift the industrial standards and economic outputs.

15.4.3 Working Fascinations

The differences provided by Survismeter in terms of timid PCP tracking make it substantially different from the conventional instruments used till date. The biggest thrust factor is the Indian origin of invention alongside the commercialization by well-known Borosil Glassworks Pvt. Ltd. At present, many universities in different parts of India (at state and central level) have procured this device. More than 15 PhDs have been awarded through the felicitation of this device; regulatory bodies of Indian academia have introduced this device in the syllabi of master's courses in many universities across the world. The ease of operation, with limited occupying space, lower dependence on outside energy and less technical skill requirements are some of the distinctions which further augment this device as a safer physiochemical screener. The sample volume is recoverable and measurements are highly reproducible and could be evidenced in the publications in leading international journals (who have published the studies) (Table 15.1). Table 15.2 compares the Survismeter working distinctions with conventional techniques, from where it is well-justified why Survismeter should be chosen rather than the devices normally

---

TABLE 15.1

Salient Studies Conducted via Survismeter Usage. The Diversity Lasts from Determining Interaction Extents to Exploring Biological Activities of Metalsurfactant Complexes and Ascertaining the Dispersion Kinetics of Nanoemulsions.

S. No.	Broad Theme of the Study	Peculiar Aspect Analyzed	Ref.
1.	Optimizing antioxidant activity of curcumin	Physicochemical profile of edible oil nanoemulsions with variable surfactants	Malik et al., 2014
2.	Selection of optimum ZnO concentrations for its antioxidant potential	Physicochemical studies of Guava leaf extract comprising aqueous ZnO dispersions	Gupta et al., 2019
3.	Aqueous aggregation behaviour of HCl, H_2SO_4 and H_3PO_4 stabilized nanoemulsions	Physicochemical activities of mineral acid proticity attributes of oil-water dispersion pattern	Jangid et al., 2018
4.	Effect of alkali metal sizes on the aqueous interaction behaviour of their phosphates	Elucidation of structure making or breaking activities via γ, η and entropic activities	Ameta and Singh, 2005

(Continued)

Physicochemical Analysis of Liquid Solutions

TABLE 15.1 (Continued)

Salient Studies Conducted via Survismeter Usage. The Diversity Lasts from Determining Interaction Extents to Exploring Biological Activities of Metalsurfactant Complexes and Ascertaining the Dispersion Kinetics of Nanoemulsions.

S. No.	Broad Theme of the Study	Peculiar Aspect Analyzed	Ref.
5.	Effect of imidazolium ionic liquids on HSA-lanthanide chlorides-citric acid interactions at variable temperatures	Charge density driven enhanced interactions of ILs and HSA	Kumar et al., 2017
6.	Monitoring the dispersion kinetics of magnetic nanoparticles for methorexate binding and release activities	Prediction of stability and DNA binding activities of titanium dioxide nanoparticles in aqueous and cell-culture growth media	Patel et al., 2016
7.	Effect of increasing hydrophobicity on the micellization behaviour of platinum-surfactant complexes	Enhanced micellization on increasing alkyl chain, also ascertained via SEM, AFM and DLS techniques	Sharma et al., 2016
8.	Ascertaining the graphene oxide (GO) dispersion impact on curcumin nanoemulsions	Pro-surfactant acidic effect of GO and 5-aminoindazole functionalized GO on curcumin dispersion	Maktedar et al., 2017
9.	Effect of increasing hydrophobicity and temperature on the interaction behaviour of urea with different solvents	From urea to dimethyl and trimethyl urea, interactions gain more stability through hydrophobic capping	Singh et al., 2008
10.	Physicochemical study of fluorescence active compounds on lanthanide salt and dendrimer interactions	First study on lanthanide salt induced dendrimer aggregation, could be used in multidisciplinary dendrimer utility	Vashistha et al., 2018
11.	Physicochemical studies of anticancer platinum (IV) complexes in their interactions with calf-thymus DNA	In vitro drug-DNA interactions predicting DNA binding, disruption and intercalation activities of benzylamine supported platinum complexes	Sharma et al., 2016

TABLE 15.2

Comparison of Survismeter Working Distinctions with Normally Used Instruments, Indicating the Advantages of Using Survismeter

S. No.		With Survismeter	Without Survismeter
1.	Sample amount	25–30 mL	Higher
2.	Technical skill	Bare Minimal, due to manual operation	Higher due to automated operation
3.	User friendly	Yes	Much lesser

(Continued)

240 Zero Waste

TABLE 15.2 (Continued)

Comparison of Survismeter Working Distinctions with Normally Used Instruments, Indicating the Advantages of Using Survismeter

S. No.		With Survismeter	Without Survismeter
4.	Glassware Interference	Bare minimal, high quality Borosilicate glass is used	Much higher
5.	Multiple analysis	Yes	Not possible, that's why risk of sample contamination is more
6.	Time saving	Yes, owing to benefit of no sample transfer for another PCP measurement	Not much
7.	Eco-friendly	Yes	Not to same extent
8.	Sample recovery	Yes	Not possible, due to intervention of external probes
9.	Scale-up feasibility	Yes	Not to same extent
10.	Risk of toxic sample analysis	Not much, except minimal routine caution	Much higher

used for separate η and γ measurements. Although the device has circulation provision to adjust the working temperature, this could also be fixed manually through addition of hot or cold water. An exclusive inspiration in the science of Survismeter is the illustrated significance of frictional forces (FF) in the interactions, since all mixtures having entire CF at the beginning undergo mutual interlayer movements that are summed up as shear activities. Since all CFs are not always entirely converted into IMFs, a significant proportion always prevails as FF, which often remains unaccounted. The science of Survismeter unveils a quantification of this missing gap, via measurement of a new parameter, termed as friccohesity. The friccohesity is a direct index of CF to FF transformation, accounting for larger kinetic activities since FF does necessitate a mutual movement of intersecting layers. The genesis of friccohesity ascertains the interactive activities of dispersed phase, so the higher are the binding sites in a molecule, the greater will be its friccohesity in a given environment. The details of friccohesity can be looked for in more specific literature sources especially featuring it (Singh, 2006; Ameta and Kale, 2014; Chandra et al., 2013; Kumar et al., 2016; Singh, 2017).

15.5 ANALYTICAL TOOL FOR ROBUST PHYSICOCHEMICAL SENSING

Several situations create a need of determining the distribution kinetics sharply influenced by hydrophilic and hydrophobic force factors. The PCPs are highly accurate sensors of population-specific interacting activities, for example increase in surfactant concentration is likely to lower the γ to a higher extent. Similarly, if dispersed phase population or content is more, the η and γ values are likely to be significantly different. A good discussion in this reference opens the interface for γ reduction

Physicochemical Analysis of Liquid Solutions

through varying the surfactant nature, with multiple studies suggesting the use of nonionic surfactants for finer dispersion of drugs. This is so as nonionic surfactants bind the dispersed phase through weaker non-covalent forces while ionic surfactants induce vigorous hydration sphere formation of water molecules through intervening electrostatic interactions of their counter ions. Several studies report better dispersion of ionic surfactants via partial inclusion of nonionic surfactants. The tracking of surfactant along with estimation of varying hydrophilicity and hydrophobicity can be accurately made through PCP estimation since a peculiar hydrophilicity will always produce only a specific γ reduction. Similarly, η and γ measurement of pretreated waste streams could be a boon in sustainable waste management, providing authentic information about the discard of a desired threshold of chemically cumbersome and heterogeneous constituents. The most remarkable breakthrough is in the agriculture domain, where pesticide formulations could be designed to carry minimum pesticide content and attain its maximum distribution at a given temperature and pressure. This will not only save the crops from quality deterioration but also prevent harm to the environment and soil. Furthermore, the expenditure on buying the extra pesticide quantities could be mitigated to improve the economic condition of farmers and economy, on the whole. So, PCPs could offer exclusive selection of the best performing dispersion vehicle through tracking their operational sciences in terms of temperature-dependent PCPs, preventing the duplication of energy, money, resources and manpower.

15.6 OUTCOME CONCEPTS OR MECHANISMS

The science of Survismeter has already made its mark in the academic domain, with dozens of PhDs being successfully completed. This section focuses on some characteristic working models rationalized on the basis of Survismeter physicochemical studies for proposing the interaction regimes. The present day research significantly focuses on exploring the biological activities of new formulations or novel metal inorganic or organic complexes. Expression of adequate structural expressions is of paramount importance in the pursued interaction mechanisms, with all activities requiring a maximum CF to IMF interconversion. To meet such constraints, it is a must that binding forces (BF) in the designed formulations or prepared complexes are moderate so that the self-bonds can be broken and the new bonds with the formulation moieties or the fabricated complex domains and the damage causing target agent can be formed. In this reference, a fundamental interaction model is proposed through mutual η and γ modulations via (CF to FF) friccohesity maximization. The interaction model is described in a 2013 Royal Society publication as the terminology Drug-Friccohesity Interaction (DFI). The proposed DFI mechanism is used to justify the DNA binding activities of inorganic platinum complexes which require the phosphate backbone in DNA strands to be broken for intercalating activity of the administered complexes (Ameta et al., 2013). Similarly, there are several situations when the mixing of two or more different chemical substances causes both η and γ increments. In such cases, it is difficult to conclude whether CF or IMF are being enhanced. The element of interest is that if a particular species has OH groups and it is interacting with an OH group of another species, how would the first molecule recognize the OH groups of the second molecule as being non-native? In such cases, a new terminology works under the domain of CF operating between combined solute-solvent ensembles. These interactions are recognized as Secondary CF, the terminology finding its mention in a 2015 international publication as the Ameta-Singh Interaction Model (Ameta and Singh, 2015). Another pertinent situation relates to above-average functional expressions of the prepared complexes or formulations even when their dispersion is heterogeneous (Polydispersity Index, PDI >1). Such situations are caused by unequal distribution of constitutional molecular energies and amount to distinct levels of rotational, translational, vibrational and electronic distributions. Every single interacting molecule has several susceptible sites in its structure that are more likely to take part in the interactions, so the stress and strain experienced by different regions of a same chemical species are very different from each other. This causes the fragmentation of molecular domains with unequal kinetic energies, ultimately interacting with differential potency in an identical chemical environment. So, scenarios where the CF expression is still higher and still functional activities are considerable are due to moderate binding forces only.

15.7 CONCLUSION

The science of Survismeter is a pioneering breakthrough in modern science, with the device enabling robust investigation of examined interactions in a much shorter time with reasonable repeatability and reproducibility. With its multiple analysis approach and no specific technical skill requirement, the Survismeter revolutionizes the identification of molecular interactions through its easier operation with minimal dependence on external energy. The comparison of drug formulation physicochemical properties with the physiological pH of blood and other fluids can be established as the new conceptualization of determining biocompatibility of tested drug dispersions. Similarly, the physicochemical approach could be rationalized for effluent discard of different industries in varying chemical environments. The distribution efficacy of fuels, paints, inks, varnishes along with quality control standards of food samples (milk, curd, buttermilk, beverages, wines) are further incentives of the Survismeter scientific depth. With the ongoing efforts to make the instrument digitalized along with a provision of understanding the impact of UV light on the sample being analyzed, the multidimensionality of the device is poised to reach further heights.

15.8 ACKNOWLEDGEMENTS

Authors are highly thankful to Central University of Gujarat for providing infrastructure facilities. Rakesh Kumar Ameta* is also thankful to Council of Scientific and Industrial Research, India, for associating him as Senior Research Associate/ Pool Scientist. Authors are also thankful to Prof. Man Singh, inventor of Survismeter for administrative help and guidance.

REFERENCES

Ameta, R. K., Singh, M. and Kale, R. K., 2013. Synthesis and structure-activity relationship of benzylamine supported platinum (IV) complexes. *New Journal of Chemistry.* 37(5):1501–1508. doi:10.1039/C3NJ41141A

Ameta, R. K. and Singh, M., 2014. SAR and DFI studies of supramolecular tetraammonium platinate + DNA matrix with UV/Vis spectrophotometry and physicochemical analysis at 298.15 K. *Journal of Molecular Liquids.* 190:200–207. doi:10.1016/j.molliq.2013.11.009

Ameta, R. K. and Singh, M., 2015. Surface tension, viscosity, apparent molal volume, activation viscous flow energy and entropic changes of water + alkali metal phosphates at T = (298.15, 303.15, 308.15) K. *Journal of Molecular Liquids.* 203:29–38. doi:10.1016/j.molliq.2014.12.038

Avramov, M., Dimitrov, K. and Radoev, B., 1995. Brownian motion at liquid-gas interfaces. 5. Effect of insoluble surfactants-nonstationary diffusion. *Langmuir.* 11(5):1507–1510. doi:10.1021/la00005a017

Behrens, R. W., 1964. The physical and chemical properties of surfactants and their effects on formulated herbicides. *Weeds.* 12(4):255–258. doi:10.2307/4040747

Chandra, A., Patidar, V., Singh, M. and Kale, R. K., 2013. Physicochemical and friccohesity study of glycine, L-alanine and L-phenylalanine with aqueous methyltrioctylammonium and cetylpyridinium chloride from T = (293.15 to 308.15) K. *Journal of Chemical Thermodynamics.* 65:18–28. doi:10.1016/j.jct.2013.05.037

Chapter 12, Fluid Dynamics and its Biological and Medical Applications. 397–428. Pressbooks Open Publishing. Open Stax College, Creative commons attribution 4.0 international license, contributors: Open Stax College Physics and Rice University.

Conboy, J. C., Messmer, M. C. and Richmond, G. L., 1996. Investigation of surfactant conformation and order at the liquid-liquid interface by total internal reflection sum-frequency vibrational spectroscopy. *Journal of Physical Chemistry.* 100(18):7617–7622. doi:10.1021/jp953616x

Gómez-Díaz, D., Mejuto, J. C. and Navaza, J. M., 2001. Physicochemical properties of liquid mixtures. 1. viscosity, density, surface tension and refractive index of cyclohexane + 2,2,4-trimethylpentane binary liquid systems from 25°C to 50°C. *J. Chem. Eng. Data* 46(3):720–724. doi:10.1021/je000310x

Gupta, A., Eral, H. B., Hatton, T. A. and Doyle, P. S., 2016. Controlling and predicting droplet size of nanoemulsions: Scaling relations with experimental validation. *Soft Matter* 12(5):1452–1458. doi:10.1039/c5sm02051d

Gupta, R., Malik, P., Das, N. and Singh, M., 2019. Antioxidant and physicochemical study of *Psidium guajava* prepared zinc oxide nanoparticles. *Journal of Molecular Liquids* 275:749–767. doi:10.1016/j.molliq.2018.11.085

Jangid, A. K., Malik, P. and Singh, M., 2018. Mineral acid monitored physicochemical studies of oil-in-water nanoemulsions. *Journal of Molecular Liquids.* 259:439–452. doi:10.1016/j.molliq.2018.03.005

Jelali, M. and Kroll, A., 2003. *Physical Fundamentals of Hydraulics, Chapter-3 in Hydraulic Servo-Systems*, Springer, London, New York. ISBN: 978-1-85233-692-9.

Jones, R. G., Wilks, E. S., Metanomski, W. V., Kahovec, J., Hess, M., Stepto, R. and Kitayama, T., 2009. *Compendium of Polymer Terminology and Nomenclature (IUPAC Recommendations 2008)*, 2nd ed., Royal Society of Chemistry 464. ISBN: 978-0-85404-491-7.

Kumar, D., Chandra, A. and Singh, M., 2016. Effect of Pr $(NO_3)_3$, Sm $(NO_3)_3$ and Gd $(NO_3)_3$ on aqueous solution properties of urea: A volumetric, viscometric, surface tension and friccohesity study at 298.15 K and 0.1 MPa. *Journal of Solution Chemistry.* 45(5):750–771. doi:10.1007/s10953-016-0466-x

Kumar, D., Chandra, A. and Singh, M., 2017. Influence of imidazolium ionic liquids on the interactions of human hemoglobin with $DyCl_3$, $ErCl_3$ and $YbCl_3$ in aqueous citric acid at T = (298.15, 303.15, and 308.15) K and 0.1 MPa. *J. Chem. Eng. Data.* 62 (2):665–683. doi:10.1021/acs.jced.6b00695

Lee, S. H., 1996. Shear viscosity of model mixtures by non-equilibrium molecular dynamics. II. Effect of dipolar interactions. *Journal of Chemical Physics.* 105:2044. doi:10.1063/1.472073

Maktedar, S. S., Malik, P., Avashthi, G. and Singh, M., 2017. Dispersion enhancing effect of sonochemically functionalized graphene oxide for catalysing antioxidant efficacy of curcumin. *Ultrasonics Sonochemistry.* 39:208–217. doi:10.1016/j.ultsonch.2017.04.006

Malik, P., Ameta, R. K. and Singh, M., 2014. Preparation and characterization of bionanoemulsions for improving and modulating the antioxidant efficacy of natural phenolic antioxidant curcumin. *Chem. Biol. Interactions.* 222:77–86. doi:10.1016/j.cbi.2014.07.013

Malik, P., Ameta, R. K. and Singh, M., 2016. Physicochemical study of curcumin in oil driven nanoemulsions with surfactants. *Journal of Molecular Liquids.* 220:604–622. doi:10.1016/j.molliq.2016.04.126

Mandal, A. and Bera, A., 2012. Surfactant stabilized nanoemulsion: Characterization and application in enhanced oil recovery. *International Journal of Chemical, Molecular, Nuclear, Materials and Metallurgical Engineering.* 6(7):537–542.

Moreno, J. S., Jeremias, S., Moretti, A., Panero, S., Passerini, S., Scrosati, B. and Appetecchi, G. B., 2015. Ionic liquid mixtures with tunable physicochemical properties. *Electrochimica Acta.* 151:599–608. doi:10.1016/j.electacta.2014.11.056

Parthasarathy, S., Ying, T. S. and Manickam, S., 2013. Generation and optimization of palm oil-based oil-in-water (o/w) submicron-emulsions and encapsulation of curcumin using a liquid whistle hydrodynamic cavitation reactor (LWHCR). *Industrial and Engineering Chemistry Research.* 52(34):11829–11837. doi:10.1021/ie4008858

Patel, S., Patel, P., Undre, S. B., Pandya, S. R., Singh, M. and Bakshi, S., 2016. DNA binding and dispersion activities of titanium dioxide nanoparticles with UV/vis spectrophotometry, fluorescence spectroscopy and physicochemical analysis at physiological temperature. *Journal of Molecular Liquids.* 213(304–311). doi:10.1016/j.molliq.2015.11.002

Sando, G. M., Dahl, K. and Owrutsky, J. C., 2005. Surfactant charge effects on the location, vibrational spectra, and relaxation dynamics of cyanoferrates in reverse micelles. *Journal of Physical Chemistry B.* 109 (9):4084–4095. doi:10.1021/jp045287r

Schmitt, M. and Stark, H., 2016. Marangoni flow at droplet interfaces: Three-dimensional solution and applications. *Phys. Fluids* 28:012106. doi:10.1063/1.4939212

Sharma, N. K., Ameta, R. K. and Singh, M., 2016. Spectrophotometric and physicochemical studies of newly synthesized anticancer Pt (IV) complexes and their interactions with CT-DNA. *Journal of Molecular Liquids.* 222:752–761. doi:10.1016/j.molliq.2016.07.101

Singh, M., 2006. Survismeter-type I and II for surface tension, viscosity measurements of liquids for academic, and research and development studies. *J. Biochemical Biophysical Methods.* 67(2–3):151–161. doi: 10.1016/j.jbbm.2006.02.008

Singh, M., 2007. Survismeter, 2-in-1 for viscosity and surface tension measurement, an excellent invention for industrial proliferation of surface forces in liquids. *Surface Review Letters* 14:978–983. doi:10.1142/S0218625X0701055X

Singh, M., 2008a. Survismeter for simultaneous viscosity and surface tension study for molecular interactions. *Surface Interface Analysis.* 40(1):15–21. doi:10.1002/sia.2663

Singh, M., 2008b. Visionmeter: A novel instrument for teaching chemical sciences to the visually handicapped. *Developments, Applications and Tutorials in Experimental Mechanics Techniques.* 32(2):53–57. doi:10.1111/j.1747-1567.2007.00266

Singh, M., 2017. *Friccohesity and Tentropy: New Models of Molecular Sciences, Chapter 10 in Green Nanotechnology-Overview and Further Prospects*, InTECH Publishers.

Singh, M., Singh, S., Inamuddin and Asiri, A.M., 2017. IFT and friccohesity study of formulation, wetting, dewetting of liquid systems using Oscosurvismeter. *Journal of Molecular Liquids*. 244(7–18). https://doi.org/10.1016/j.molliq.2017.08.067.

Stefaniak, A.B., 2017. *Principal Metrics and Instrumentation for Characterization of Engineered Nanomaterials (Chapter-8) in Metrology and Standardization of Nanotechnology: Protocols and Industrial Innovations*, edited by Mansfield, E., Kaiser, D.L., Fujita, D. and Voorde, M.V., Wiley-VCH, Verlag, Weinheim. 151–174. doi:10.1002/9783527800308, ISBN:9783527800308

Vashistha, N., Chandra, A. and Singh, M., 2018. Influence of rhodamine B on interaction behaviour of lanthanide nitrates with 1st tier dendrimer in aqueous DMSO: A physicochemical, critical aggregation concentration and antioxidant activity study. *Journal of Molecular Liquids*. 244:7–18. doi:10.1016/j.molliq.2018.03.056

Walstra, P., 1983. *Formation of Emulsions (Chapter 2) in Encyclopedia of Emulsion Technology*, edited by Becher, P., Marcel Dekker, New York.

Yuan, C.L., Xu, Z.Z., Fan, M.X., Liu, H.Y., Xie, Y.H. and Zhu, T., 2014. Study on characteristics and harm of surfactants. *Journal of Chemical and Pharmaceutical Research*. 6(7):2233–2237.

16 Recent Advances in Bioremediation of Wastewater for Sustainable Energy Products

Dolly Kumari and Radhika Singh

CONTENTS

16.1 Introduction 248
16.2 What Is Wastewater? 249
    16.2.1 Types of Wastewater 249
        16.2.1.1 Grey Water 249
        16.2.1.2 Yellow Water..249
        16.2.1.3 Brown Water 250
        16.2.1.4 Black Water..250
    16.2.2 Sources of Wastewater..250
    16.2.3 Composition of Wastewater..251
    16.2.4 Characterization of Wastewater..251
16.3 Wastewater Treatment Methods 253
    16.3.1 Primary Treatment...253
    16.3.2 Secondary Treatment...253
    16.3.3 Tertiary Treatment...253
    16.3.4 Wastewater Treatment on the Basis of Their State of Matter..........254
        16.3.4.1 Physical Treatment...254
        16.3.4.2 Chemical Treatment...255
        16.3.4.3 Biological treatment...255
16.4 Bioremediation of Different Wastewater Sources 256
    16.4.1 Treatment Process of Some Wastewater Effluents 256
        16.4.1.1 Treatment Systems for Brown, Black, Yellow and Grey Water 256
        16.4.1.2 Phenolic Wastewater Treatment..256
        16.4.1.3 Treatment of Oil Mill Wastewater (OMW) 256
        16.4.1.4 Domestic Wastewater Treatment 257
        16.4.1.5 Other Industrial Wastewater Treatment............................257

248 Zero Waste

16.4.2 Microorganisms Used in Bioremediation 257
16.4.3 Mechanisms Involved in Bioremediation Process 259
16.5 Methods of Wastewater Digestion Leading to Formation of
Energy Products 261
16.5.1 Common Methods Used for Energy Generation from Wastewater 261
16.5.1.1 Anaerobic Digestion (AD) 261
16.5.1.2 Aerobic Digestion 262
16.5.1.3 Biosolids Incineration 262
16.5.2 Types of Useful Energy Products Recovered from Wastewater 263
16.5.2.1 Biohydrogen Production 263
16.5.2.2 Biomethane Production 263
16.5.2.3 Bioethanol Production 264
16.5.2.4 Biodiesel Production 264
16.5.2.5 Bioelectricity Generation 265
16.5.2.6 Microbial Fuel Cell (MFC) 265
16.6 Process and Mechanism of Energy Production from Wastewater 266
16.7 Future Perspectives for Energy Product Recovery from Effluents 267
16.7.1 Scaling Up MFCs to Practical Level 267
16.7.2 Monitoring and Control of Process 267
16.7.3 Mathematical Modelling 267
16.7.4 Hybrid Approach of Wastewater Treatment 267
16.7.5 Use of Controlled Microbiology 268
16.7.6 Addition of Catalytic Substances 268
16.8 Conclusion 268
References 268

16.1 INTRODUCTION

Water is a necessary commodity for all living beings, and life without water is beyond imagination. Scarcity of pure water is increasing day by day because of the addition of various harmful and toxic substances into open as well as groundwater sources. Basic necessity of present is to solve the problem of wastewater generation and to overcome the energy demand (Chaudhary et al., 2017). The energy prerequisite is increasing progressively due to rapid increase in population which is coupled with an increase in generation of wastewater (Brain et al., 2015). To accomplish both of these obligations, utilization of wastewater should be done in such a manner, such that the process used would treat the wastewater along with the production of some cherished products which can be reutilized further (Jadhav et al., 2017). The use of wastewater for energy generation is economic as this does not require expensive phenomenon. There is an urgent need for sustainable wastewater treatment which can be achieved by natural or biological treatment of the effluent (Uche et al., 2002). While many new technologies are developing for source recovery from wastewater, biological methods offer the strongest promise to efficiently recover valuable resources from streams. This chapter will focus broadly on different methods to recover resources from domestic and industrial wastewater and industrial wastes. The next generation of domestic wastewater treatment plants is targeting energy

Recent Advances in Wastewater Bioremediation

neutrality and complete utilization of wastewater to generate energy (Largus et al., 2004). There are also increasing drivers to recover valuable products from wastes and wastewaters of different nature, such as those from the industrial manufacturing and mining extraction. The fossil sources are limited and may deplete in upcoming future so alternative source of energy are being developed. Best alternative methods are anaerobic digestion with biogas production and bio-solid incineration with electricity generation which could satisfy at least some of the demand of energy (Tontti et al., 2016). These energy recovery policies could help counterbalance the electricity ingestion of the wastewater sector and represent possible areas for supportable energy strategy implementation. Our analysis considers energy consumption and potential savings only; the economics of energy recovery from wastewater treatment, while highly relevant, is reserved for a separate analysis (Singhal and Singh, 2015; Sivaramakrishna et al., 2014). Energy recovery at wastewater treatment plants denotes an important policy pedal for sustainability.

16.2 WHAT IS WASTEWATER?

The term wastewater is placed for discarded or previously used water in municipality or industry. When various contaminants are added in fresh water due to natural or human activities, the fresh water becomes wastewater (Rai et al., 2004; Agdag and Sponza, 2007). Wastewater can be classified on the basis of colour and the source from which it is generated.

16.2.1 TYPES OF WASTEWATER

Wastewater can be classified in various ways based upon its various characteristics. On the basis of colour, household wastewater types are grey water, yellow water, brown water and black water.

16.2.1.1 Grey Water

The wastewater generated from home (bathrooms, kitchen, washing machines etc.) is known as grey water (Brain et al., 2015). Grey water is generated from both residential and commercialized buildings in a huge amount. The treatment of grey water is easy through natural systems (soil, plants and microorganisms) as they absorb/degrade and filter its nutrients and returns clear water to water cycle (Lancaster, 2010). Grey water is not suitable for drinking purposes but can be utilized for irrigation of trees and shrubs whose woody stems perform as a filter for contaminants present in grey water.

16.2.1.2 Yellow Water

The term yellow water is stated for the water from separated toilets and urinals. In yellow water a large amount of nitrogen is present which is removed by costly nitrification and denitrification techniques. Yellow water is nutrient rich and can be used as fertilizer for plants if collected separately (Telkamp, 2006). Yellow water can be a good substitute of urea as it is nitrogen rich and contains lower amounts of heavy metals as compared to chemical fertilizers.

16.2.1.3 Brown Water

Brown water is the type of wastewater that contain along with faecal matter, chemicals and soil added due to various human activities. Soil degradation is well known worldwide and it is caused by over-cropping, soil erosion and use of chemical fertilizers (Balkema, 2003). Synthetic fertilizers are not capable to fulfil organic matter requirement of soil and also deplete various properties of soil and when this gets mixed with human wastes ultimately lead to the formation of brown water.

16.2.1.4 Black Water

This is wastewater that originates from toilet fixtures, dishwashers and food preparation sinks. It is made up of all the things that one can imagine going down the toilets, bath and sink drains. They include urine, faeces, toilet paper and wipes, body cleaning soapy liquids, anal cleansing water and so on. They are known to be highly contaminated with dissolved chemicals, particulate matter and its highly pathogenic (Selma and Ayaz, 2001). Thus, it can be said that black water is a mixture of yellow water, grey water and brown water. This is also called sewage water or sewage sludge.

16.2.2 SOURCES OF WASTEWATER

On the basis of the source from which the wastewater is being generated there are various types of wastewater, some of them are summarized in Table 16.1 as given below:

TABLE 16.1

Types and Sources of Different Wastewater Effluents

Types of Wastewater	Sources	References
Domestic wastewater (DW)	Toilets	Selma and Ayaz (2001)
Sewage sludge (SS)	Sewage	Sosnowski et al. (2003)
Dairy wastewater (DWW)	Dairy or milk industry	Luo et al. (2011); Venkata et al. (2007); Kumari et al. (2018)
Winery wastewater	Wine industry	Riano et al. (2011)
Tannery wastewater (TW)	Tannery and leather industry	Chaudhary et al. (2017)
Textile wastewater (TxW)	Textile industry	Pathak et al. (2014)
Food wastewater (FWW)	Petha sweet industry	Singhal and Singh (2015)
Phenolic wastewater (PW)	Petroleum refinery, paints and dyes, pharmaceutical and pesticides	Satsangee and Ghose (1990); Parag (2008); Rathinam et al. (2011)
Palm oil mill effluent (POME)	Palm oil mills	Deshon et al. (2017)

(Continued)

Recent Advances in Wastewater Bioremediation

TABLE 16.1 (Continued)
Types and Sources of Different Wastewater Effluents

Types of Wastewater	Sources	References
Olive mill wastewater (OMW)	Olive mills	Maragkaki et al. (2016)
Carpet mill effluent (CME)	Carpet mills	Chinnasamy et al. (2010)
Slaughter house effluent (SHE)	Slaughter house	Fuchs et al. (2003)
Medicine wastewater (MW)	Pharmaceutical industry	Sivaramakrishna et al. (2014)
Beverage wastewater (BW)	Beverage industry	Li et al. (2012)
Petha wastewater (PWW)	Petha sweet industry	Kumari et al. (2018)

16.2.3 COMPOSITION OF WASTEWATER

Wastewater contains chemical and biological matter which varies from source to source from which the wastewater is generated. Industrial wastewater contains a large amount of toxic chemicals and heavy metals, such as nickel, zinc, copper, cadmium, lead, arsenic, antimony and mercury (Ahmaruzzaman, 2011) with lower levels of biological matter, whereas household wastewater contains lower levels of chemicals as compared to industrial wastewater but high amount of organic matter. Agriculture wastewater contains high levels of chemicals (pesticides, weedicides, fertilizers etc.) with biological substances (algae, fungi, bacteria etc.) (Kong et al., 2010; Steinbusch et al., 2011). Chemical composition of wastewater comprises 70% organic and 30% inorganic compounds along with various gases. The organic compounds present in wastewater are mainly carbohydrates, proteins and fats. Inorganic mater consists of phosphorus, heavy metals, chloride, nitrogen, sulphur etc. Various gases commonly dissolved in wastewater are hydrogen sulphide, methane, ammonia, oxygen, nitrogen and carbon dioxide (Alissara et al., 2016). Biologically, wastewater includes various micro-organisms, such as Protista, plants and animals. Protista includes bacteria, algae, fungi and protozoans. Plants are mainly liverworts, seedy plants, ferns and mosses. Various pathogens are also found in wastewater which comes from the human beings suffering with various diseases (Kong et al., 2010; Li et al., 2013).

16.2.4 CHARACTERIZATION OF WASTEWATER

Wastewater is generally characterized by physical and chemical parameters. Physical parameters are colour, odour and turbidity, chemical parameters are pH, alkalinity, dissolved oxygen (DO), biochemical oxygen demand (BOD), chemical oxygen demand (COD), total organic carbon (TOC), total suspended solids (TSS), total dissolved solids (TDS), conductivity, nitrogen, phosphorus, heavy metals, volatile solids (VS), fats, oil and grease and gases. Different types of wastewater sources along with their typical properties are listed these properties discussed below in Table 16.2.

TABLE 16.2

Characteristics of Some Wastewater Effluents

Wastewater Types[a]	pH	COD (mg/L)	BOD$_5$ (mg/L)	DO (mg/L)	TS (g/L)	TDS (g/L)	TSS (g/L)	VS (g/L)	Alkalinity (mg/L)	References
PWW	12	4,705	1,800	6.3	43.62	5.3	38.32	39.84	1200	Kumari et al. (2018)
DW	N/A	740	350	N/A	N/A	N/A	450	320	1,850	Rawat et al. (2011)
FWW	11.9	5,882	580	3.8	5.44	5.22	0.22	1.64	2,400	Singhal and Singh (2015)
POME	4.2	51,000	25,000⇒	N/A	N/A	N/A	18,000	N/A	N/A	Deshon et al. (2017)
OMW	4.8	13,2300	N/A	N/A	41.8	N/A	N/A	36.8	N/A	Maragkaki et al. (2016)
SHE	5.3–6.8	58,000–20,150	2,200–9,800	N/A	N/A	N/A	2.4–4.7	N/A	N/A	Fuchs et al. (2003)

⇒BOD$_5$, [a] Abbreviations used and explained in Table 16.1, N/A: Not available

16.3 WASTEWATER TREATMENT METHODS

The process of removal of major contaminants from wastewater and sewage sludge is known as wastewater treatment. The site or place where this process is performed is called a wastewater treatment plant. Various treatment methods are used for wastewater effluent treatment as explained below, based upon the different steps involved in the treatment.

16.3.1 Primary Treatment

Primary treatment comprises basic processes to remove suspended solid waste and moderate its BOD which is the amount of oxygen microorganisms must consume to interruption of organic material present in the wastewater. This in turn increases DO, which is good for aquatic organisms and food webs. Primary treatment can reduce BOD by 20% to 30% and suspended solids by up to 60% (Ge et al., 2010). This step includes reduction of oil, grease, fats, sand and coarse solids. This step is mainly performed through machines.

16.3.2 Secondary Treatment

Secondary treatment can remove up to 85% of BOD and total suspended solids. This step includes degradation of dissolved contents of the sewage within a biological degradation of system as shown in Figure 16.1. Microorganisms are used in this process to degrade dissolved components in water. The last step of secondary treatment is the removal of biological matter from the treated water with very low levels of organic material and suspended solids (Hennebel et al., 2015). Secondary treatment uses biological processes to remove the dissolved organic matter missed in primary treatment. Microbes consume the organic matter present in wastewater as food, transfiguring it to carbon dioxide, water, and energy (Telkamp, 2006).

16.3.3 Tertiary Treatment

Tertiary treatment can remove up to 99% of all impurities from sewage and but it is a cost-intensive process. In some cases, treatment plant operators add chlorine as a

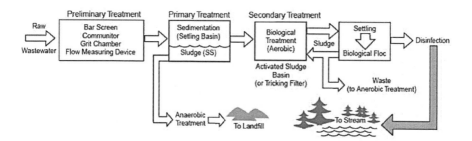

FIGURE 16.1 Schematic Representation of Waste Treatment Plant

disinfectant before discharging the treated water (Sharma and Sanghi, 2012). This step is useful to improve the quality of treated water to the standard required before discharging to water bodies. Sometimes nitrogen and phosphorus removal are done by tertiary treatment. The uppermost level of wastewater treatment is tertiary treatment, which is any process that goes elsewhere the previous steps and can include the use of sophisticated technology for further removal of contaminants or specific pollutants (Hernández, 2010).

16.3.4 Wastewater Treatment on the Basis of Their State of Matter

16.3.4.1 Physical Treatment

This includes influx (influent), removal of large particles and removal of sand and pre-precipitation. In this method no chemicals or biological processes are involved. Ultrafiltration, sedimentation, absorption, adsorption, sand filtration, microfiltration, reverse osmosis etc. are physical methods used for wastewater treatment as shown in Figure 16.2. There is a projecting method for physical water treatment known as sedimentation. Sedimentation is a procedure of hanging out the insoluble heavy particles from the wastewater. This method is best for purifying the water (Han et al., 2007). Pure water is removed when the insoluble materials settle down at the bottom of water. Aeration is another method which can be used for physical water treatment. In this process air is dispersed through the wastewater to deliver oxygen to it. It is also an operative method of cleaning the wastewater. Filtration can be used to filter out the contaminants. Special kinds of filters are used to pass sewage or water due to which insoluble particles or contaminants are separated. The most commonly used filters are the sand or pebbles. Some wastewaters contain grease, oil on their surface

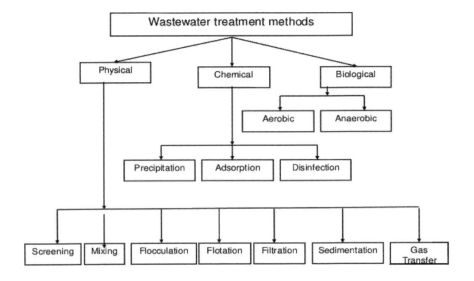

FIGURE 16.2 Different Methods of Wastewater Treatment

Recent Advances in Wastewater Bioremediation

they can also be removed from the surface easily through this method. Jiayi and Junqi (2001) have reported that small scale grey water treatment using a combination of sequence batch reactors and slow sand filter has achieved the same quality required for recharge of groundwater.

16.3.4.2 Chemical Treatment

As the name shows in this method chemicals are used to clean the water or sewage. Most commonly used chemical is chlorine and this method is called chlorination. Chlorine is the oxidizing agent that kill the bacteria which decompose the water by but add contaminant to it. Another oxidizing agent (disinfect) called ozone is used to purify the wastewater. These chemicals are very useful as they inhibit the bacteria to reproduce in the water thus making it pure (Mantzavinos and Kalogerakism, 2005). When the treatment of industrial wastewater is concerned then special methods are used for neutralization. A base or acid added to the water to neutralise its pH. Lime is the base which is most commonly used in the acidic wastewater. Most commonly used chemicals in the chemical wastewater treatment are the metals which have more than one valence shell. There are some methods of purifying sewage and wastewater which are used in the both physical and chemical wastewater treatment (Hernández, 2010). Carbon usage is the best example which has the capacity to adsorb harmful substances from the wastewater. This process is usually combined with settling of organic matter and other processes to remove solids followed by filtration. Various chemical processes such as foam flotation, precipitation, ion-exchange, adsorption, electrolysis, oxidation and disinfection also used in this method (Figure 16.2). Many agro-industrial wastes are used for treatment of wastewater, such as zeolites, silica gel–activated carbon and activated alumina (Bhatnagar and Sillanpaa, 2010).

16.3.4.3 Biological treatment

When water is used for drinking and cleaning purpose, it should be clean and free of any contaminants which can affect human health. For this purpose, biological treatment of water is used in which microorganism especially bacteria play an important role (Lovley, 2003). They decompose the wastewater biochemically or they break the organic materials and improve the quality of water so that it can be used for domestic purposes. Biological treatment of wastewater is a promising approach of treating wastewater due to environmentally friendly nature and cost effectiveness (Paraskeva and Diamadopoulos, 2006). Biological treatment systems can be divided into different categories such as aerobic and anaerobic systems or according to growth system of microorganism suspended growth and attached growth (Figure 16.2). Biological treatments methods are categorized according to the existence of the air in the treatment system. In aerobic systems, oxygen is present in biological systems as the electron acceptor (Mantzavinos and Kalogerakism, 2005). On the other hand, oxygen is inattentive in the anaerobic system and the electron acceptor is the organic material. The oxygen is used for oxidation of organic material to carbon dioxide, water, and new cells are produced as final products in aerobic system under anaerobic treatment absence of air (thus molecular free oxygen) is there and methane, carbon dioxide and new cells are produced at the end of the digestion process (Steinbusch et al. 2011). Biological treatment of wastewater involves participation of microbes (aerobic and anaerobic) and plants for generation of valuable products and removal of harmful components from the wastewater (Varia, 2014). Biological treatment removes organic substances as well as inorganic nutrients from wastewater. Microalgae are also used as an adsorbent of nutrients with biofuel production from wastewater (Kong et al., 2010).

16.4 BIOREMEDIATION OF DIFFERENT WASTEWATER SOURCES

Different types of wastewater are treated by various methods to make it reusable and for generation of useful products.

16.4.1 TREATMENT PROCESS OF SOME WASTEWATER EFFLUENTS

16.4.1.1 Treatment Systems for Brown, Black, Yellow and Grey Water

For the treatment of brown and black water two types of treatment; wet and dry treatments are used. No flush water is used in dry treatment whereas wet treatment utilizes flush water. The sanitization of less diluted or non-diluted faecal matter should be done before reusing it for agricultural practices (Mani and Kumar, 2014). This can be done by vermicomposting, composting, dehydration and biological processes like anaerobic digestion or anaerobic fermentation. Organic components/contaminants of wastewater are removed by using biological processes but treated effluent should be filtered through membrane filters to prevent the passage of solids into the effluent so that the effluent can be usable (Rayu, 2012). For this purpose, treatment with membrane bioreactors is the choice of future for wastewater treatment. Urine (yellow water) being relatively sterile can be used for agricultural practices without any pretreatment (Fuchs, 2003; Malaeb et al., 2013) although several methods of treatment of yellow water have been investigated.

16.4.1.2 Phenolic Wastewater Treatment

Chloro (p-chlorophenol, o-chlorophenol) and nitro (p-aminophenol, p-nitrophenol) derivatives present in wastewater effluents cause harmful effects on environment because of their higher levels of toxicity and unmanaged disposal (Pinto et al., 2003). For removal of phenolic wastewater cavitation is a useful technique which can be further modified by some other additives; Ultrasound/H_2O_2, Ultrasound/O_3, cavitation assisted by catalysts, sono-photocatalytic oxidation and cavitation coupled with biological oxidation are some of them (Parag, 2008). Cavitation generates hot spots and highly reactive free radicals due to which cavitation is found more effective in the field of wastewater treatment. Anaerobic biodegradation of phenol present in wastewater can be done using natural microflora (Rathinam et al., 2011; Satsangee and Ghose, 1990).

16.4.1.3 Treatment of Oil Mill Wastewater (OMW)

OMW is generated from a large number of oil mills seasonally in oil producing countries. OMW has a large amount of organic load and phyto-toxicity along with recalcitrant to nature (Deshon et al., 2017). A number of physical, physico-chemical,

Recent Advances in Wastewater Bioremediation

biological (composting, aerobic and anaerobic treatment and co-digestion) and combined processes have been used for treatment of OMW. Physical treatment of OMW (dilution, sedimentation, filtration, evaporation and centrifugation are not much effective to remove the pollutants to the permeable limits (Paraskeva and Diamadopoulos, 2006). Electro coagulation is an effective physico-chemical method used for OMW and other industrial wastewater treatment in which aluminium or iron electrodes generates coagulants in the solution resulting in in situ hydroxide formation of respective metals which removes the pollutants (Federico et al., 2016). Oxidation and advanced oxidation are effective in recalcitrant and toxic compound's breakdown and removal (Benitez et al., 1999). Biological and combined treatments resulted in high degree of COD removal as compared to other treatment methods.

16.4.1.4 Domestic Wastewater Treatment

Domestic wastewater is the water that has been used by a community and which contains all the substances added to the water during its use. It is thus composed of human body wastes (faeces and urine) together with the water used for flushing toilets, and silage, which is the wastewater resulting from personal washing, laundry, food preparation and the cleaning of kitchen utensils (Telkamp, 2006; Rawat et al., 2011). Since the domestic wastewater varies in its composition, a single method cannot be used for its treatment. Its treatment involves the use of all physical, chemical and biological methods to make it fit for use again for various purposes. Treatment of wastewater obtained from toilets has been done using Phragmites, Cyperus, Rush, Iris, Lolium, Canna, and Paspalum as natural filters. Iris was found best with 94% COD and 90% total nitrogen removal (Selma and Ayaz, 2001).

16.4.1.5 Other Industrial Wastewater Treatment

Castronovo et al. (2016) investigated biodegradation of artificial sweetener acesulfame in biological wastewater treatment. It is noticeable that potassium salt of acesulfame is widely used in food, pharmaceutical preparations and personal care products as artificial sweetener and excreted unchanged in urine without any metabolic activity after ingestion and absorption through gut (Von and Hanger, 2001). This results in higher concentration of acesulfame in wastewater and was degraded by technical process for drinking water through photolysis by UV light (Gan et al., 2014). Recent study presented activated sludge treatment and sand filtration of surface water by denitrifying-nitrifying process for removal of acesulfame from wastewater. Use of Ultrafiltration and reverse osmosis has been reported by De La et al. (2016) for salt removal from treated wastewater with secondary effluents. Wastewater with high ammonia concentration can be done by nitrification with nitrite accumulation of simulated industrial wastewater achieving 65% of loaded ammonia nitrogen as nitrite (Ruiz et al., 2003).

16.4.2 Microorganisms Used in Bioremediation

Although most organisms in biological wastewater treatment plants are microscopic in size, there are some organisms such as bristle worms and insect larvae that are macroscopic in size. There are five major groups of microorganisms generally found in the wastewater treatment process: (a) aerobic bacteria remove organic nutrients; (b) protozoa remove and digest dispersed bacteria and suspended particles; (c) metazoa dominate longer age systems including lagoons; (d) filamentous bacteria bulking sludge (poor settling and turbid effluent); and (e) algae and fungi are present with pH changes and older sludge (Kaewkannetra et al., 2009; Kong et al., 2010). The archae-bacteria and eubacteria are the most important microorganisms in biological treatment of wastewater. A complex consortium used as microbial source may contain hydrolytic, methanogenic and fermentative bacteria (Christy et al., 2014). Most of the bacteria are strictly anaerobes, some are facultative anaerobes and others are aerobes. A group of microbes called acidogenic bacteria convert the sugar/starchy substrates and proteins into carbon dioxide, hydrogen, ammonia and organic acids (Table 16.3). The maximum amount of VFA is converted into acetate and hydrogen by the activity of hydrogen-producing acetogenic bacteria. Only a few species of microbes are able to convert acetate into methane and carbon dioxide. Carbon dioxide and hydrogen is utilized by methanogenic bacteria to form methane (Schink, 1997). Along with bacteria some fungi like yeast, white rot fungi and algae are also used in bioremediation process of wastewater (Kong et al., 2010).

TABLE 16.3

Different Types of Aerobic and Anaerobic Microorganisms Used in Wastewater Treatment, Their Advantages and Disadvantages (Christy et al., 2014)

	Aerobic Treatment		
Microorganism Used	**Pollutants Degraded**	**Advantages**	**Disadvantages**
Pseuodomonas species	Mono and dichloro	High biomass	High energy required
Acinetobactor species	aromatic	concentration	
Alcaligenes species	compounds,	Short retention	Large surface area needed
Corynebacterium species	Alkanes, Cyanide,	time	for efficient treatment
Nocardia species	DDT, TNT,	Excellent	High treatment cost due to
Arthrobactor species	Oil spills, Household	process control	regular aeration
Mycobacterium species	waste, Xylene,	Handle a wide	
Clostridium species	Naphthalene,	range of flow	
Candida species	Octane, Camphor	Easily separate	
Gibeberella species	etc., Heavy metals	biomass from	
		waste stream	
Anaerobic Treatment			
Lactobacillus	All organic matter in	Low energy	Narrow temperature
Propionibacterium species	the wastewater	demand for	control range
Bacteroides species	Agricultural waste	aeration	
Fusobacterium species	Household waste	Low land	Accumulation of heavy
Porphyromonas species	Kitchen waste	requirement	metal and contaminants
Methanothermobacter species	Horticulture waste		in sludge

(Continued)

Recent Advances in Wastewater Bioremediation 259

TABLE 16.3 (Continued)

Different Types of Aerobic and Anaerobic Microorganisms Used in Wastewater Treatment, Their Advantages and Disadvantages (Christy et al., 2014)

Aerobic Treatment			
Microorganism Used	**Pollutants Degraded**	**Advantages**	**Disadvantages**
Methanobacterium species	Paper waste	Low sludge production	Installing and managing plants is difficult
Methanobrevibacter species	Fruit industry waste		
Methanococcus species		Less expensive than anaerobic system	High capital cost
Methanoculleus species,			
Methanogenium species			
Methanoflorens species		High organic removal efficiency	Low retention time required
Methanomicrobium species			
Methanopyrus species			
Methanoregula species		Production of energy rich methane	Generally required heating
Methanosaeta species			
Methanosphaera species			
Methanospirillium species		Low investment required	Corrosive compounds are produced during anaerobiosis
		Low nitrogen and phosphorus requirement	Hydrogen sulphide produced
		High loading capacity	Reactor may require extra alkalinity
		High treatment efficiency	Slow growth rate of anaerobes
		Effluent contain valuable fertilizer	Only used for pre-treatment of liquid waste
		Low operational cost	
		No foul smell	
		Low nutrient requirement	
		Rapid start-up possible after acclimation	

16.4.3 Mechanisms Involved in Bioremediation Process

Different mechanisms are involved for removal of different kinds of organic and inorganic wastes present in wastewater effluents; for biodegradable organics, bioconversion can be done by aerobic or anaerobic bacteria. Suspended solids can be removed by filtration and sedimentation process (Rayu et al., 2012). Nitrification/denitrification, plant uptake and volatilization can be used for removal of nitrogen

260 Zero Waste

and phosphorus. Heavy metal removal takes place through adsorption of plant roots and debris surfaces and by sedimentation. Trace organics can be removed by adsorption and biodegradation (Dubey and Shiwani, 2012). Remediation of pathogens can be achieved by natural decay, physical entrapment, filtration, predation, excretion of antibiotics from roots of the plants and sedimentation. The detailed mechanisms of bioremediation are summarized in Table 16.4.

TABLE 16.4
Various Mechanisms Involved in Bioremediation of Wastewater

Mechanism Involved	Pollutant Type	Sorbent/Microorganism Used	References
In situ bioremediation	Polluted soil and groundwater	Bacteria	Rayu et al. (2012)
Ex situ bioremediation	Ni, N_2, P, COD	*Amaranthus paniculatus*	Iory et al. (2013)
Biosparging	Petroleum contaminants of soil and groundwater	Naturally occurring bacteria	Machackova et al. (2012); Kumar and Mani (2012)
Molecular mechanisms	Hg reduction	Genetically engineered bacterium like *Deinococcus geothermalis, Cupriavidus metallidurans* strain MSR33	Rojas et al. (2011); Brim et al. (2003)
Physico-bio-chemical (Ion-exchange)	Zn (II), Cd(II)	*Saccharomyces cerevisiae*	Talos et al. (2009); Chen and Wang (2007)
Biosorption	Cu(II), Pb(II)	Seaweed, green macroalgae and lichen like *Portulacaoleracea, Cladoniarangiformis*	Dubey and Shiwani (2012); Jiang et al. (2012); Ekmekyapar et al. (2012)
Mycoremediation (Bioaccumulation)	Cr(VI), Ni(II), Cu(II), Pb(II)	Fungi like *Aspergillus versicolor, Aspergillus fumigates*	Tastan et al. (2010); Ramasamy et al. (2011)
Biotransformation (Detoxification)	Pollutants from soil and ecosystems	Aerobic and anaerobic microbes	Kumar et al. (2013)
Biomineralization	Pb ions in mine tailings	*Bacillus* species	Govarthanan et al. (2013)
Bioaugumentation	Contaminants of soil	Pre-grown microbial cultures	Tyagi et al. (2011)
Biodegradation (Reductive dechlorination)	Cu from soil, chlorine, U^{6+} to U^{4+}	*Kocuriaflava* CR1, *Geobacter* species	Achal et al. (2011); Lovley (2008)
Adsorption	Heavy metals	Microbes present in cellular structures	Fang et al. (2011)
Phytoremediation	Cd, Ni	Plants like *Arabidopsis halleri, Solanum nigrum*, and bacteria *Kluyveraascorbata* SUD 165	Wei et al. (2005); Wei et al. (2003); Burd et al. (1998)

(Continued)

Recent Advances in Wastewater Bioremediation

TABLE 16.4 (Continued)
Various Mechanisms Involved in Bioremediation of Wastewater

Mechanism Involved	Pollutant Type	Sorbent/Microorganism Used	References
Cyanoremediation	As, Pb^{2+}, Cu^{2+}	Bacteria and blue-green microalgae like *Synechocysis sp.* PCC6803, *Spirogyra* and *Cladophora*	Yin et al. (2012); Lee and Chang (2011); Kong et al., 2010
Phytomanagement	Metal removal	Alyssum biomass	Chaney et al. (2007)
Rhizosphere engineering	Heavy metal removal	Microbes or genetically modified plants	Lovley (2003)
Nano-biotechnology	Radioactive waste clean-up	*Deinococcus radiodurans*, a radioactive-resistant organism	Brim et al. (2000)
Bioventing	Heavy metals from soil and groundwater	Aerobic bacteria	Robinson et al. (2011)
Genomics (Gene targeted bioremediation)	Hg(II), Cd	Genetically modified metal tolerant plants like tobacco, yellow poplar, cottonwood and rice	Dhankher et al. (2011)

16.5 METHODS OF WASTEWATER DIGESTION LEADING TO FORMATION OF ENERGY PRODUCTS

In order to produce the various valuable energy products from the wastewater, it must be digested with the help of variety of microorganisms. Even after the digestion, the remaining sludge is again treated by different physical and chemical treated methods to get more energy from the wastewater and its leftover materials.

16.5.1 Common Methods Used for Energy Generation from Wastewater

Few common methods used for the energy generation from wastewater are as follows.

16.5.1.1 Anaerobic Digestion (AD)

AD is a biological process in which any organic waste can be converted into valuable products with the help of microbes. This process naturally occurs in ruminant stomach, marshes, sediment of lakes, municipal sewers and municipal landfills. AD is a complex process (Figure 16.3) involving a series of reactions (in absence of oxygen); hydrolysis, acidogenesis, acetogenesis, methanogenesis (Themelis and Ulloa, 2007; Agdag and Sponza, 2007). AD is very useful process which is applicable on a wide variety of waste effluents (sewage sludge, industrial wastewater, domestic wastewater etc.) for their conversion to useful products especially into various energy forms i.e. biohydrogen and methane (Jingura and Matengaifa, 2009). AD process is governed by various factors like temperature (25°–35°C), pH (~7), moisture, carbon source, nitrogen and C/N Ratio (Khalid et al., 2011). AD of sewage sludge used in treatment plants is very useful now days because of lower disposal costs and it is ecofriendly too (Appels et al., 2008).

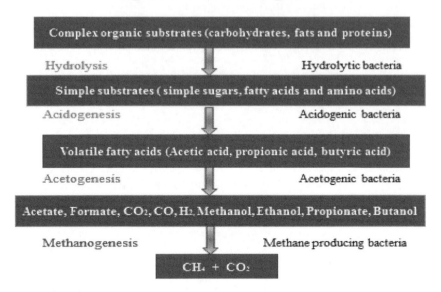

FIGURE 16.3 Showing Stepwise Anaerobic Digestion Process

16.5.1.2 Aerobic Digestion

Aerobic treatment is the process in which degradation of organic and inorganic components takes place in presence of molecular oxygen. Aerobic treatment of activated sludge results in 80%–85% COD removal (Rozzi and Malpei, 1996). A number of aerobic microbes such as *Aspergillus niger, Azotobacter chroococcum* and *Geotrichum candidum* were investigated for removal of phenol from wastewater. In a study for *Aspergillus niger*, 35%–64% of COD removal has been reported (Cereti et al., 2004). Two phenol resistant algae, *Ankistrodesmus braunii* and *Scenedesmus quadricauda* (Pinto et al., 2003) and a fungus *Phanerochaete chrysosporium* (Dhouib et al., 2006) were also investigated for 12% phenol removal and 20%–50% COD removal from OMW, respectively.

16.5.1.3 Biosolids Incineration

Prior to reuse or disposal, wastewater sludge must be treated to reduce odours and disease-causing agents such as pathogens and bacteria. Treated sludge is then referred to as biosolids. Biosolids have high water content and typically are dewatered prior to further treatment or disposal (Turovskiy, 2015). Some municipal wastewater treatment plants immolate dewatered biosolids as a means of disposal, which requires dewatering prior to immolation. Biosolids immolation with electricity generation is an effective biosolids disposal operation with potential for significant energy recovery. Two equipment options are commercially available for biosolids immolation: multiple hearth furnaces and fluidized bed furnaces (Lewis et al., 2002; Agdag and Sponza, 2007). Multiple hearth furnaces incinerate biosolids in multiple stages, allowing for hot air recycle to dry incoming biosolids and improve hearth generation

Recent Advances in Wastewater Bioremediation | 263

by reducing incoming moisture. Fluidized bed furnaces are newer technology that is more efficient, stable and easier to operate than multiple hearth furnaces, but are limited to continuous operation only (Kinney et al., 2006). Both incineration technologies require cleaning of exhaust gases to prevent emissions of odour, particulates, nitrogen oxides, acid gases, hydrocarbons, and heavy metals. Using either manifold hearth or fluidized bed furnaces, biosolids ignition can be used to command a steam cycle power plant, where high temperature from incineration is transferred to steam that turns a turbine connected to a generator, producing electricity (Roy et al., 2011).

16.5.2 Types of Useful Energy Products Recovered from Wastewater

A number of value-added products can be generated by controlled treatment of wastewater. Biological treatments of wastewater are compiled with extraction of useful products from wastewater effluents. These may be aerobic and anaerobic digestion or co-digestion of effluent. A large number of energy products in the form of biofuel are generated by the treatment of wastewater effluents which are as follows.

16.5.2.1 Biohydrogen Production

Biohydrogen is a carbon neutral fuel and would be the preferred clean product of microbial activity (Singh, 2013). Dark fermentation is the biological process involving hydrogen production from wastewater. As microbial consortium contains a wide variety of bacteria including few of them which retards hydrogen production (i.e. hydrogenophilic-methanogenesis) by its consumption (homo-acetogens and methanogens). For optimum hydrogen production the activity of inhibitory microbes can be suppressed either by heat pretreatment of inoculum or by high dilution rate (Chen et al., 2001; Singhal and Singh, 2014a) and by lowering the pH (Oh et al., 2003) of reactors. Another reliable process is catalytic methane to hydrogen conversion (Witt and Schmidt, 1996). Use of mixed microbial source is advantageous over pure microbial culture for increased biohydrogen production due to the presence of a wide variety of microbes (Pachapur et al., 2019). Different wastewaters have been used for hydrogen production, such as TxW (Jeihanipour and Bashiri, 2015); PWW (Singhal and Singh, 2015); rejected water and seed sludge (Cavinato et al., 2012); MW (Sivaramakrishna et al., 2014); OMW (Ghimire et al., 2016); and BW (Li et al., 2012).

16.5.2.2 Biomethane Production

Methane is another gas produced from anaerobic digestion of wastewater. Methane can be produced either by hydrogenophilic-methanogenesis (abiotically) or by fermentation (biologically) of organic matter present in wastewater (Villano et al., 2010). Anaerobic digestion is more efficient over aerobic process for methane production due to low energy constitutions, high energy construction in the form of methane, low sludge production with high organic ejection rates. Various wastewater effluents such as PWW (Singhal and Singh, 2015), brown water (Jun, 2011), rejected water and seed sludge (Cavinato et al., 2012) and acidic affluent discharged after the hydrogen fermentation of sugarcane juice (Alissara et al., 2016) have been used for methane production. Co-treatment of wastewater with food waste can result in enhanced methane yield and improved energy balance of wastewater treatment plants (Guven et al., 2019).

16.5.2.3 Bioethanol Production

Generally, bioethanol is produced from lignocellulosic biomass but now bioethanol can also be produced via wastewater treatment (Kumari and Singh, 2018). Bio-electrochemical conversion of waste biomass to ethanol is an innovative method in which biological acetate reduction occurs with hydrogen and bio-cathode works as an electron donor (Rosenbaum et al., 2011). A number of effluents (alone or with other wastes) have been utilized for ethanol production (OMW and olive pomace (Federico et al., 2016), apple pomace hydrolysate (Ezgi and Canan, 2015), rice straw with PWW and DWW (Kumari et al., 2018)). Recovery of carbon-rich compounds like ethanol and butanol in the cathodic chamber have been reported by Steinbusch et al. (2011).

16.5.2.4 Biodiesel Production

Algal growth in water bodies is an indication of water pollution. Blue-green algae plays dual role as in treating wastewater as well as in biomass production which in terns can be utilized for biofuel, especially biodiesel production as shown in Figure 16.4 (Pittman et al., 2011; Xiao and He, 2014). Algal biomass thus obtained can be also utilized as animal feed and also spread out as fertilizer (Lundquist et al.,

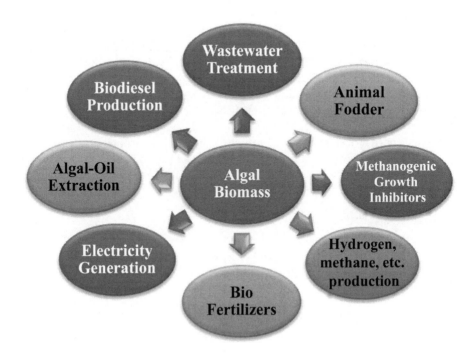

FIGURE 16.4 Showing Various Uses of Algal Biomass Grown on Wastewater Sources

Recent Advances in Wastewater Bioremediation **265**

2010). Certain species of algae, grown in wastewater is capable of produce up to 80% of oil as its storage product and produce about 23 times more oil than the best oil-seed plant (Paustian et al., 1998; Kong et al., 2010). Algal productivity outstrips palm oil by 10 bend (folds) and Jatropha, canola and sunflower crops by more than 30 bend in terms of oil produced per unit area (Barsanti and Gualtieri, 2006; Pittman et al., 2011). Oil trans-esterification to biodiesel, glycerol and algal biomass can be used to produce energy products and glycerol may be used for burning directly as fuel or can be converted to bioethanol and hydrogen by fermentation (Chisti, 2007; Asakuma, 2009). The practicability of algal biomass for biodiesel production can be enhanced by instantaneous wastewater treatment, using as animal feed and by secondary energy products production microalgae was also used in phycoremediation of wastewater sources (Kong et al., 2010; Rawat et al., 2011).

16.5.2.5 Bioelectricity Generation

Bio-electrochemical conversion process is involved in production of electricity from wastewater, in which catalytic community of microorganisms converts the chemical energy stored in organic matter into usable forms (Kassongo and Togo, 2011). Organic electron donors are catalyzed for oxidation in the anodic chamber by electrochemical bacteria and electrons are supplied to the anode, which can be arrested in the form of bioelectricity (Li et al., 2008). In this process electro-neutrality is maintained by the cation exchange of protons (H^+, generated during the catalytic conversion), from anodic chamber to cathodic chamber, where these protons are involved in the production of appreciated products (Jadhav et al., 2017). Microbial electrochemical cells (MECs) are also used for energy recovery from wastewater in tertiary treatment (Winkler and Straka, 2019). Algal biomass which is generated by photosynthesis can be a direct organic source of electricity generation in MFC and for producing biofuels (Xiao and He, 2014).

16.5.2.6 Microbial Fuel Cell (MFC)

MFCs have been recommended as a substitute for two-step biohydrogen production process. MFCs are similar to hydrogen fuel cells in which protons are exchanged from an anode compartment to cathode compartment through an electrolytic membrane (Logan, 2004; Han et al., 2005; Zhang et al., 2010). In a hydrogen fuel cell hydrogen is oxidized into electrons and protons on the anode and oxygen is reduced in water on the cathode. In MFC organic matter of wastewater is oxidized in the anode chamber and electrons are transferred to the cathode chamber. Anodophilic bacteria from various bacterial families (Desulfuromonaceae, Pasteurellaceae, Clostridiaceae, Aeromonadaceae, Comamonadaceae etc.) were able to transfer electrons to electrodes (Chaudhuri and Lovley, 2003; Pham et al., 2003). Electricity can be produced by urine and swine wastewater in MFC. Current density of 3.6 A/m^2 along with improved NH_4^+ ion recovery of about 61% was reported in MFC under electricity driven migration in a microbial electrochemical cell (Zhang et al., 2010; Haddadi et al., 2013). Singhal and Singh (2014b) applied single chamber microbial fuel cells (SCMFC) using graphite electrodes for bioelectricity generation from PWW.

16.6 PROCESS AND MECHANISM OF ENERGY PRODUCTION FROM WASTEWATER

The inter-relationship of energy products, water and the organic content of wastewater inspires to energy recovery operations. A wide range of energy products and value-added compounds can be recovered from wastewater; energy in the form of electricity can reduce the scarcity of electricity to some extent (Jingura and Matengaifa, 2009; Ashlynn et al., 2010; Jun, 2011). There are various techniques involved in the production of energy products from wastewater, namely, MFC, bio-electrochemical system, biochemical, chemical, biological (aerobic and anaerobic digestion of effluent) (Largus et al., 2004; Singhal and Singh, 2015). Table 16.5 shows various reactions involved in energy production from wastewater.

TABLE 16.5

Reactions Involved in Energy Production Processes from Organic Matter Present in Wastewater Largus et al. (2004)

Biotic and Abiotic Processes	Reaction Involved	Organic Matter Consumed
Hydrogen production process		
Glucose fermentation to acetic acid	$C_6H_{12}O_6 + 2H_2O \leftrightarrow 4H_2 + 2CH_3COOH + 2CO_2$	Glucose
Glucose fermentation to butyric acid	$C_6H_{12}O_6 \leftrightarrow 2H_2 + CH_3CH_2CH_2COOH + 2CO_2$	Glucose
Syntrophic propionic acid oxidation	$CH_3CH_2COOH + 2H_2O \leftrightarrow CH_3COOH + 3H_2 + CO_2$	Propionic acid
Syntrophic butyric acid oxidation	$CH_3CH_2CH_2COOH + 2H_2O \leftrightarrow 2CH_3COOH + 2H_2$	Butyric acid
Syntrophic acetic acid oxidation	$CH_3COOH + 2H_2O \leftrightarrow 4H_2 + 2CO_2$	Acetic acid
Catalytic methane conversion to syngas	$CH_4 + H_2O \leftrightarrow 3H_2 + CO$	Methane
Catalytic gas-shift fraction	$CO + H_2O \leftrightarrow H_2 + CO_2$	CO
Methane production process		
Hydrogenotrophic methanogenesis	$4H_2 + CO_2 \leftrightarrow CH_4 + 2H_2O$	Hydrogen
Acetoclastic methanogenesis	$CH_3COOH \leftrightarrow CH_4 + CO_2$	Acetic acid
Glucose to methane conversion	$C_6H_{12}O_6 \leftrightarrow 3CH_4 + 3CO_2$	Glucose
Ethanol production process		
Glucose fermentation to ethanol	$C_6H_{12}O_6 \leftrightarrow 2CH_3CH_2OH + 2CO_2$	Glucose
Reduction of acetic acid	$CH_3COOH + H_2 \leftrightarrow CH_3CH_2OH + H_2O$	Acetic acid
Cellulose bioconversion	$(—C_6H_{11}O_6—)n + aH_2O \leftrightarrow bCH_3COOH + cCH_3CH_2OH + dCO_2 + eH_2$	Cellulose
Electricity generation process		
Hydrogen fuel cell	$2H_2 + O_2 \leftrightarrow 2H_2O + electricity$	Hydrogen
Methane fuel cell	$CH_4 + 2O_2 \leftrightarrow CO_2 + 2H_2O + electricity$	Methane
Microbial fuel cell	$C_6H_{12}O_6 + 6O_2 \leftrightarrow 6CO_2 + 6H_2O + electricity$	Glucose
Formation of polyhydroxyalkanoates	$aC_6H_{12}O_6 + bO_2 \leftrightarrow (—COOH(CH_2)_3COO—)n + cH_2O$	Glucose
Formation of dyes	$(—C_6H_{11}O_6—)n + aNH_3^- + bH_2O \leftrightarrow cC_{23}H_{26}O_5 + dC_{22}H_{27}O_5\,N + eC_{21}H_{22}O_5$	Cellulose

16.7 FUTURE PERSPECTIVES FOR ENERGY PRODUCT RECOVERY FROM EFFLUENTS

Our future perspectives should be production of fertilizers and soil conditioners for healthy cultivation and not production of wastes for disposal (Otterpohl et al., 1999; Tang et al., 2019) in the form of pollutants. The aim should be to design an approach in such a way that wastes generated can be reconverted into pathogen free resources without any cost and polluting the environment. Some of the steps which may be taken are as follows.

16.7.1 SCALING UP MFCs TO PRACTICAL LEVEL

MFCs utilized now are designed for lab scale purposes, but to know the utility of MFC these should be scale up to practical level so the energy recovery should be enhanced and optimized (Eom et al., 2011). The restriction in scaling of MFCs are high scaling cost, pH buffers, high internal resistance, high material cost and low efficiency of mixed cultures on an electrode and these limitations may be overcome by reactor engineering, biological employment and material development (Logan, 2004).

16.7.2 MONITORING AND CONTROL OF PROCESS

The parameters and techniques used in a process should be monitored to an extent to achieve high yield of by-products and chemical compounds from the treatment of wastewater (Lefebvre et al., 2012). In case of electricity generation electro-chemical parameters (electric current, indicators and electrode potential) are some useful parameters which can be monitored for optimum electricity production (Peixoto et al., 2011). According to Tang et al. (2019) electrochemical, biological and bio-electrochemical technologies for wastewater treatment can generate increased amount of electricity required for the treatment of wastewater in a wastewater treatment plant and is also capable to reduce greenhouse gas emissions.

16.7.3 MATHEMATICAL MODELLING

More optimization of the process may be obtained by the use of mathematical models, especially two directional. Use of these models facilitate the complex process and make it easy to perform at practical level rather than lab scale (Deeke et al., 2012).

16.7.4 HYBRID APPROACH OF WASTEWATER TREATMENT

In the hybrid approach more than one useful product is produced in a single process for example integrated hydrogen and methane production, this leads to recover maximum energy from wastewater. The hybrid approach of MFC with anaerobic membrane bioreactor (Han et al., 2005; Malaeb et al., 2013) and integrated photobioreactor fuel cell (Eom et al., 2011) are hybrid processes and are more efficient than older ones. This technique is capable in treating high levels of wastewater along with polishing treatment for removal of specific type of pollutants from the effluent.

16.7.5 USE OF CONTROLLED MICROBIOLOGY

Energy generation by wastewater treatment can also be improved by selection of best microbial species responsible specific biofuel production and specific pollutant removal. This would lead to reduction in the operation cost and energy consumption for the process (Fischer et al., 2011).

16.7.6 ADDITION OF CATALYTIC SUBSTANCES

Addition of a very small amount of catalytic substance can increase the microbial process involved in the recovery of by-products from wastewater. Banu et al. (2019) achieved a higher biohydrogen yield of about 0.29 g COD/g COD in zeolite mediated microwave liquefaction (Ze-MWL) by adding 0.04 g/g SS of zeolite in waste activated sludge.

16.8 CONCLUSION

Wastewater is no longer a waste but it is a rich source of organic matter, nutrients and metal ions. These all components can be used for the production of a number of valuable energy products which can be converted to biofuels. Wastewater has a very complex nature and only one treatment process is not sufficient to effectively remove all pollutants. Biological, physical, and chemical processes are used separately or together in one process to support each other in a combined form to for wastewater treatment. Advanced technologies like membrane bioreactors can be used for the wastewater treatment but they have high investment and operational costs as compared to conventional processes. Moreover, continuous efforts should be on the development of suitable consortia that not only work on variable conditions but also help to recover the valuable products from the wastewater. Despite all these facts on one hand, there is a need of strict discharging standards to be implemented and at the same time on the other hand researchers and scientific community should develop the cutting-edge technologies to recover all the desired products in a sustainable manner.

REFERENCES

Achal, V., Pan, X. and Zhang, D., 2011. Remediation of copper-contaminated soil by Kocuria flava CR1, based on microbially induced calcite precipitation. *Ecological Engineering.* 37:1601–1605. https://doi.org/10.1016/j.ecoleng.2011.06.008

Agdag, O.N. and Sponza, D.T., 2007. Co-digestion of mixed industrial sludge with municipal solid wastes in anaerobic simulated landfilling bioreactors. *Journal of Hazardous Materials.* 140:75–85. https://doi.org/10.1016/j.jhazmat.2006.06.059

Ahmaruzzaman, M., 2011. Industrial wastes as low-cost potential adsorbents for the treatment of wastewater laden with heavy metals. *Advances in Colloid and Interface Science.* 166:36–59. https://doi.org/10.1016/j.cis.2011.04.005

Alissara, R., Sureewan, S. and Chakkrit, S., 2016. Methane production from acidic effluent discharged after the hydrogen fermentation of sugarcane juice using batch fermentation and UASB reactor. *Renewable Energy.* 86:1224–1231. https://doi.org/10.1016/j.renene.2015.09.051

Appels, L., Baeyens, J., Degréve, J. and Dewil, R., 2008. Principles and potential of the anaerobic digestion of waste-activated sludge. *Progress in Energy Combustion Science.* 34:755–781. https://doi.org/10.1016/j.pecs.2008.06.002

Asakuma, Y., Maeda, K., Kuramochi, H. and Fukui, K., 2009. Theoretical study of the transesterification of triglycerides to biodiesel fuel. *Fuel.* 88:786–791. https://doi.org/10.1016/j.fuel.2008.10.045.

Ashlynn, S. S., David, C. H. and Michael, E. W., 2010. Energy recovery from wastewater treatment plants in the United States: A case study of the energy-water nexus. *Sustainability.* 2:945–962. doi: 10.3390/su2040945

Banu, J. R., Eswari, A. P. and Kavitha, S., 2019. Energetically efficient microwave disintegration of waste activated sludge for biofuel production by zeolite: Quantification of energy and biodegradability modelling. *International Journal of Hydrogen Energy.* 44(4):2274–2288. https://doi.org/10.1016/j.ijhydene.2018.06.040

Barsanti, L. and Gualtieri, P., 2006. *Algae: Anatomy, Biochemistry and Biotechnology.* CRC Press, Taylor & Francis Group, Boca Raton, FL, 301. ISBN: 9781439867327-CAT# K13023

Benitez, F. J., Beltran-Heredia, J., Torregrosa, J. and Acero, J. L., 1999. Treatment of olive mill wastewaters by ozonation, aerobic degradation and the combination of both treatments. *Journal of Chemical Technology and Biotechnology.* 74:639–646. https://doi.org/10.1002/(SICI)1097-4660(199907) 74:7<639::AID-JCTB92>3.0.CO;2-I

Bhatnagar, A. and Sillanpaa, M., 2010. Utilization of agro-industrial and municipal waste materials as potential adsorbents for water treatment—a review. *Chemical Engineering Journal.* 157:277–296. https://doi.org/10.1016/j.cej.2010.01.007

Brain, R., Lynch, J. and Kopp, K., 2015. Greywater systems. *USU Extension Publication January 2015*: sustainability2015/01pr. Available online at: http://extension.usu.edu/files/publications/publication/Sustainability_2015-01pr.pdf

Brim, H., McFarlan, S. C. and Fredrickson, J. K., 2000. Engineering *Deinococcus radiodurans* for metal remediation in radioactive mixed waste environments. *Nature Biotechnology.* 18:85–90. doi: 10.1038/71986

Brim, H., Venkateshwaran, A., Kostandarithes, H. M., Fredrickson, J. K. and Daly, M. J., 2003. Engineering *Deinococcus geothermalis* for bioremediation of high temperature radioactive waste environments. *Applied and Environmental Microbiology.* 69:4575–4582. doi: 10.1128/AEM.69.8.4575-4582.2003

Burd, G. I., Dixon, D. G. and Glick, B. R., 1998. A plant growth-promoting bacterium that decreases nickel toxicity in seedlings. *Applied and Environmental Microbiology.* 64:3663–3668. PMCID: PMC106500

Castronovo, S., Wick, A., Scheurer, M., Nödler, K., Schulz, M. and Ternes, T. A., 2016. Biodegradation of the artificial sweetener acesulfame in biological wastewater treatment and sandfilters. *Water Research.* 110:342–353. doi: 10.1016/j.watres.2016.11.041

Cavinato, C., Bolzonella, D., Fatone, F., Cecchi, F. and Pavan, P., 2012. Optimization of two-phase thermophilic anaerobic digestion of biowaste for bio-hythane production through reject water recirculation. *Bioresource Technology.* 102(18):8605–8611. http://dx.doi.org/10.1016/j.biortech.2011.03.084

Cereti, C. F., Rossini, F. and Nassetti, F. 2004. Water supply reduction on warm season grasses in Mediterranean environment. *Acta Horticulturae.* 661(661):153–158. doi:10.17660/ActaHortic.2004.661.19

Chaney, R. L., Angle, J. S., Broadhurst, C. L., Peters, C. A., Tappero, R. V. and Sparks, D. L., 2007. Improved understanding of hyperaccumulation yields commercial phytoextraction and phytomining technologies. *Journal of Environmental Quality.* 36:1429–1443. doi: 10.2134/jeq2006.0514

Chaudhary, P., Chhokar, V., Kumar, A. and Beniwal, V., 2017. Bioremediation of tannery wastewater. *Advances in Environmental Biotechnology.* 125–144. doi: 10.1007/978-981-10-4041-2_7

Chaudhuri, S.K. and Lovley, D.R., 2003. Electricity generation by direct oxidation of glucose in mediatorless microbial fuel cells. *Nature Biotechnology.* 21:1229–1232. doi: 10.1038/nbt867

Chen, C.C., Lin, C-Y. and Chang, J-S., 2001. Kinetics of hydrogen production with continuous anaerobic cultures utilizing sucrose as the limiting substrate. *Applied Microbiology and Biotechnology.* 57:56–64. https://doi.org/10.1007/s002530100747

Chen, C.C. and Wang, J.L., 2007. Characteristics of Zn^{2+} biosorption by *Saccharomyces cerevisiae. Biomedical and Environmental Sciences.* 20:478–482.

Chinnasamy, S., Bhatnagar, A., Hunt, R.W. and Das, K.C., 2010. Microalgae cultivation in a wastewater dominated by carpet mill effluents for biofuel applications. *BioresourceTechnology.* 101:3097–3105. https://doi.org/10.1016/j.biortech.2009.12.026

Chisti, Y., 2007. Biodiesel from microalgae. *Biotechnology Advances.* 25:294–306. https://doi.org/10.1016/j.biotechadv.2007.02.001

Christy, P.M., Gopinath, L.R. and Divya, D., 2014. Microbial dynamics during anaerobic digestion of cow dung. *International Journal of Plant, Animal and Environmental Science.* 4(4):86–94.

Deeke, A., Sleutels, T.H.J.A., Hamelers, H.V.M. and Buisman, C.J.N., 2012. Capacitive bioanodes enable renewable energy storage in microbial fuel cells. *Environmental Science and Technology.* 46:3554–3560. doi: 10.1021/es204126r

De La, C.B.P., Gillerman, L., Gehr, R. and Oron, G., 2016. Nanotechnology for sustainable wastewater treatment and use for agricultural production: A comparative long-term study. *Water Research.* 110:66–73. doi: 10.1016/j.watres.2016.11.060

Deshon, S.P., Wei Tan Kaman, I.A. and Leonard, L.P.L., 2017. Palm oil mill effluent treatment using coconut shell—based activated carbon: Adsorption equilibrium and isotherm. *MATEC Web of Conferences.* 87:03009, ENCON 2016. doi: 10.1051/matecconf/20178703009

Dhankher, O.P., Doty, S.L., Meagher, R.B. and Doty, S., 2011. Biotechnological approaches for phytoremediation. In Altman, A., Hasegawa, P.M. (eds.), *Plant Biotechnology and Agriculture,* Academic Press, Oxford. 309–328. https://doi.org/10.1016/B978-0-12-381466-1.00020-1

Dhouib, A., Aloui, F., Hamad, N. and Sayadi, S., 2006. Pilot-plant treatment of olive mill wastewaters by *Phanerochaete chrysosporium* coupled to aerobic digestion and ultrafiltration. *Process Biochemistry.* 41:159–167. https://doi.org/10.1016/j.procbio.2005.06.008

Dubey, A. and Shiwani, S., 2012. Adsorption of lead using a new green material obtained from Portulaca plant. *International Journal of Environmental Science and Technology.* 9:15–20. doi: 10.1007/s13762-011-0012-8

Ekmekyapar, F., Aslan, A., Bayhan, Y.K. and Cakici, A., 2012. Biosorption of Pb(II) by Nonliving Lichen Biomass of Cladonia rangiformis Hoffm. *International Journal of Environmental Research.* 6(2):417–424. doi: 10.22059/IJER.2012.509

Eom, H., Chung, K., Kim, I. and Han, J-I., 2011. Development of a hybrid microbial fuel cell (MFC) and fuel cell (FC) system for improved cathodic efficiency and sustainability: The M2FC reactor. *Chemosphere.* 85(4):672–676. https://doi.org/10.1016/j.chemosphere.2011.06.072

Ezgi, E. and Canan, T., 2015. Production of bioethanol from apple pomace by using co-cultures: Conversion of agro-industrial waste to value added product. *Energy.* 88:775–782. doi: 10.1016/j.energy.2015.05.090

Fang, L., Wei, X. and Cai, P., 2011. Role of extracellular polymeric substances in Cu(II) adsorption on *Bacillus subtilis* and *Pseudomonas putida. Bioresource Technology.* 102:1137–1141. doi: 10.1016/j.biortech.2010.09.006

Federico, B., Giuseppe, M., Bernardo, R. and Fino, D., 2016. Selection of the best pretreatment for hydrogen and bioethanol production from olive oil waste products. *Renewable Energy.* 88:401–407. https://doi.org/10.1016/j.renene.2015.11.055

Fischer, F., Bastian, C., Happe, M., Mabillard, E. and Schmidt, N., 2011. Microbial fuel cell enables phosphate recovery from digested sewage sludge as struvite. *Bioresource Technology*. 102(10):5824–5830. https://doi.org/10.1016/j.biortech.2011.02.089

Fuchs, W., Binder, H., Mavrias, G. and Braun, R., 2003. Anaerobic treatment of wastewater with high organic content using a stirred tank reactor coupled with a membrane filtration unit. *Water Research*. 37:902–908.

Gan, Z., Sun, H., Wang, R., Hu, H., Zhang, P. and Ren, X., 2014. Transformation of acesulfame in water under natural sunlight: Joint effect of photolysis and biodegradation. *Water Research*. 64:113–122. doi: 10.1016/j.watres.2014.07.002

Ge, H., Jensen, P.D. and Batstone, D.J., 2010. Pre-treatment mechanisms during thermophilic-mesophilic temperature phased anaerobic digestion of primary sludge. *Water Research*. 44:123–130. https://doi.org/10.1016/j.watres.2009.09.005

Ghimire, A., Sposito, F. and Frunzo, L., 2016. Effects of operational parameters on dark fermentative hydrogen production from biodegradable complex waste biomass. *Waste Management*. 50:55–64. doi: 10.1016/j.wasman.2016.01.044

Govarthanan, M., Lee, K.J., Cho, M., Kim, J.S., Kamala-Kannan, S. and Oh, B.T., 2013. Significance of autochthonous Bacillus sp. KK1 on biomineralization of lead in mine tailings. *Chemosphere*. 8:2267–2272. doi: 10.1016/j.chemosphere.2012.10.038

Guven, H., Ersahin, M.E., Dereli, K.K., Ozgun, H., Sik, I. and Ozturk, I., 2019. Energy recovery potential of anaerobic digestion of excess sludge from high-rate activated sludge systems co-treating municipal wastewater and food waste. *Energy*. 172:1027–1036. https://doi.org/10.1016/j.energy.2019.01.150

Haddadi, S., Elbeshbishy, E. and Lee, H.S., 2013. Implication of diffusion and significance of anodic pH in nitrogen–recovering microbial electrochemical cells. *Bioresource Technology*. 142:562–569. doi: 10.1016/j.biortech.2013.05.075

Han, M., Kim, T. and Kim, J., 2007. Effects of floc and bubble size on the efficiency of the dissolved air flotation (DAF) process. *Water Science and Technology*. 56(10):109–115. doi: 10.2166/wst.2007.779

Han, S.K., Kim, S.H. and Shin, H.S., 2005. UASB treatment of wastewater with VFA and alcohol generated during hydrogen fermentation of food waste. *Process Biochemistry*. 40:2897–2905. https://doi.org/10.1016/j.procbio.2005.01.005

Hennebel, T., Boon, N., Maes, S. and Lenz, M., 2015. Biotechnologies for critical raw material recovery from primary and secondary sources: R&D priorities and future perspectives. *New Biotechnology*. 32:121–127. https://doi.org/10.1016/j.nbt.2013.08.004

Hernández, L.L., 2010. Removal of micropollutants from grey water—combining biological and physical/chemical processes. PhD thesis, Wageningen University.

Iory, V., Pietrini, F. and Cheremisina, A., 2013. Growth responses, metal accumulation and phytoremoval capability in Amaranthus plants exposed to nickel under hydroponics. *Water Air & Soil Pollution*. 224(2):1–10. doi: 10.1007/s11270-013-1450-3

Jadhav, D.A., Ray, S.G. and Ghangrekar, M.M., 2017. Third generation in bio-electrochemical system research—a systematic review on mechanisms for recovery of valuable by-products from wastewater. *Renewable and Sustainable Energy Review*. 76:1022–1031. https://doi.org/10.1016/j.rser.2017.03.096

Jeihanipour, A. and Bashiri, R., 2015. Perspectives of biofuels from wastes: A review. Chapter 2 in book *Biobutanol from Lignocellulosic Wastes*, edited by Karimi, K., Springer International Publishing, 18, Karlsruhe 76131, Germany. 37–83. doi: 10.1007/978-3-319-14033-9_2

Jiang, X., Wang, W., Wang, S., Zhang, B. and Hu, J., 2012. Initial identification of heavy metals contamination in Taihu Lake, a eutrophic lake in China. *Journal of Environmental Science*. 24(9):1539–1548. https://doi.org/10.1016/S1001-0742(11)60986-8

Jiayi, L. and Junqi, W., 2001. The practice, problem and strategy of ecological sanitation toilets with urine diversion in china. Paper presented in 1st international conference in ecological sanitation, Nanning, China, EcoSanRes, c/o Stockholm Environment Institute, Sweden, 5–8 November.

Jingura, R.M. and Matengaifa, R., 2009. Optimization of biogas production by anaerobic digestion for sustainable energy development in Zimbabwe. *Renewable Sustainable Energy Reviews.* 13:1116–1120. https://doi.org/10.1016/j.rser.2007.06.015

Jun, W.L., 2011. *Anaerobic Co-digestion of Brown Water and Food Waste for Energy Recovery. Daniel Thevenot, 11th Edition of the World Wide Workshop for Young Environmental Scientists (WWW-YES-2011)—Urban Waters: Resource or Risks?* Arcueil, France. 6–10 June.

Kaewkannetra, P., Imai, T., Garcia-Garcia, F.J. and Chiu, T.Y., 2009. Cyanide removal from cassava mill wastewater using Azotobactor vinelandii TISTR 1094 with mixed microorganisms in activated sludge treatment system. *Journal of Hazardous Material* 172:224–228. doi: 10.1016/j.jhazmat.2009.06.162

Kassongo, J. and Togo, C.A., 2011. Evaluation of full-strength paper mill effluent for electricity generation in mediator-less microbial fuel cells. *African Journal of Biotechnology* 10:15564–15570. http://dx.doi.org/10.5897/AJB11.146

Khalid, A., Arshad, M., Anjum, M., Mahmood, T. and Dawson, L., 2011. The anaerobic digestion of solid organic waste. *Waste Management.* 31:1737–1744. https://doi.org/10.1016/j.wasman.2011.03.021

Kinney, C.A., Furlong, E.T. and Zaugg, S.D., 2006. Survey of organic wastewater contaminants in biosolids destined for land application. *Environmental Science and Technology.* 40(23):7207–7215. doi: 10.1021/es0603406

Kong, Q.X., Li, L., Martinez, B. and Ruan, R., 2010. Culture of microalgae *Chlamydomonas reinhardtii* in wastewater for biomass feedstock production. *Applied Biochemistry and Biotechnology.* 160:9–18. doi: 10.1007/s12010-009-8670-4

Kumar, C. and Mani, D., 2012. *Advances in Bioremediation of Heavymetals: A Tool for Environmental Restoration.* LAP LAMBERT Academic Publishing AG & Co. KG, Saarbrücken.

Kumar, D., Shivay, Y.S., Dhar, S., Kumar, C. and Prasad, R., 2013. Rhizospheric flora and the influence of agronomic practices on them: A review. *Proceedings of National Academy of Sciences, India Sec B Biological Sciences.* 83(1):1–14. https://doi.org/10.1007/s40011-012-0059-4

Kumari, D., Chahar, P. and Singh, R., 2018. Effect of ultrasonication on biogas and ethanol production from rice straw pretreated with Petha waste water and dairy waste water. *International Journal of Current Engineering and Scientific Research (IJCESR).* 5(1):65–73. ISSN (PRINT):2393–8374 (ONLINE):2394–0697.

Kumari, D. and Singh, R., 2018. Pretreatment of lignocellulosic wastes for biofuel production: A critical review. *Renewable and Sustainable Energy Reviews.* 90:877–891. https://doi.org/10.1016/j.rser.2018.03.111

Lancaster, B., 2010. *Rainwater Harvesting for Dry Lands and Beyond: Vol 2 Water-Harvesting Earthworks,* Rainsource Press, Tucson, AZ.

Largus, T.A., Khursheed, K., Al-Dahhan, H.M., Wrenn, A. and Espinosa, R.D., 2004. Production of bioenergy and biochemicals from industrial and agricultural wastewater. *Trends Biotechnology.* 22(9). doi: 10.1016/j.tibtech.2004.07.001

Lee, Y.C. and Chang, S.P., 2011. The biosorption of heavy metals from aqueous solution by Spirogyra and Cladophora filamentous macroalgae. *Bioresource Technology.* 102(9):5297–5304. https://doi.org/10.1016/j.biortech.2010.12.103

Lefebvre, O., Neculita, C.M., Yue, X. and YongNg, H., 2012. Bio-electrochemical treatment of acid mine drainage dominated with iron. *Journal of Hazardous Materials.* 241–242:411–417. https://doi.org/10.1016/j.jhazmat.2012.09.062

Lewis, D.L., Gattie, D.K., Novak, M.E., Sanchez, S. and Pumphery, C., 2002. Interactions of pathogens and irritant chemicals in land-applied sewage sludges (biosolids). *BMC Public Health.* 12(4):409–423. https://doi.org/10.1186/1471-2458-2-11

Li, M., Cheng, X. and Guo, H., 2013. Heavy metal removal by biomineralization of urease producing bacteria isolated from soil. *International Biodeterioration & Biodegradation.* 76:81–85. https://doi.org/10.1016/j.ibiod.2012.06.016

Li, Y.C., Liu, Y.F. and Chu, C.Y., 2012. Techno-economic evaluation of biohydrogen production from wastewater and agricultural waste. *International Journal of Hydrogen Energy.* 37(20):15704–15710. https://doi.org/10.1016/j.ijhydene.2012.05.043

Li, Z., Zhang, X. and Lei, L., 2008. Electricity production during the treatment of real electroplating wastewater containing Cr^{6+} using microbial fuel cell. *Process Biochemistry.* 43:1352–1358. https://doi.org/10.1016/j.procbio.2008.08.005

Logan, B.E., 2004. Extracting hydrogen and electricity from renewable resources: A roadmap for establishing sustainable processes. *Environmental Science & Technology.* 38:160A–167A.

Lovley, D.R., 2003. Cleaning up with genomics: Applying molecular biology to bioremediation. *Nature Reviews Microbiology.* 1:35–44. doi: 10.1038/nrmicro731

Lovley, D.R., 2008. The microbe electric: Conversion of organic matter to electricity. *Current Opinion in Biotechnology.* 19(6):564–571. https://doi.org/10.1016/j.copbio.2008.10.005

Lundquist, T., Woertz, I., Quinn, N. and Benemann, J.R., 2010. *A Realistic Technology and Engineering Assessment of Algae Biofuel Production*, Energy Biosciences Institute, University of California. 1–178. Available online at http:// esd.lbl.gov/ files/ about/ staff/ nigelquinn/ EBI_Algae_Biofuel_Report_2010.10.25.1616.pdf

Luo, J., Ding, L., Benkun, Q., Jaffrin, M.Y. and Wan, Y., 2011. A two-stage ultrafiltration and nanofiltration process for recycling dairy wastewater. *Bioresource Technology.* 102:7437–7442. https://doi.org/10.1016/j.biortech.2011.05.012

Machackova, J., Wittlingerova, Z., Vlk, K. and Zima, J., 2012. Major factors affecting in situ biodegradation rates of jet-fuel during largescale biosparging project in sedimentary bedrock. *Journal of Environmental Science and Health A.* 47(8):1152–1165. doi: 10.1080/10934529.2012.668379

Malaeb, L., Katuri, K.P., Logan, B.E., Maab, H., Nunes, S.P. and Saikaly, P.E., 2013. A hybrid microbial fuel cell membrane bioreactor with a conductive ultra–filtration membrane biocathode for wastewater treatment. *Environmental Science & Technology.* 47(20):11821–11828. doi: 10.1021/es4030113

Mani, D. and Kumar, C., 2014. Biotechnological advances in bioremediation of heavy metals contaminated ecosystems: An overview with special reference to phytoremediation. *International Journal of Environmental Science & Technology.* 11:843–872. doi: 10.1007/s13762-013-0299-8

Mantzavinos, D. and Kalogerakism, N., 2005. Treatment of olive mill effluents. Part I. Organic matter degradation by chemical and biological processes-an overview. *Environment International.* 31:289–295. https://doi.org/10.1016/j.envint.2004.10.005

Maragkaki, A.E., Fountoulakis, M. and Gypakis, A., 2016. Pilot-scale anaerobic co-digestion of sewage sludge with agro-industrial by-products for increased biogas production of existing digesters at wastewater treatment plants. *Waste Management.* doi: 10.1016/j.wasman.2016.10.043

Oh, S.E., Ginkel, S.V. and Logan, B.E., 2003. The relative effectiveness of pH control and heat treatment for enhancing biohydrogen gas production. *Environmental Science & Technology.* 37:5186–5190. doi: 10.1021/es034291y

Otterpohl, R., Albold, A. and Olderburg, M., 1999. Source control in Urban sanitation and waste management: 10 Systems with reuse of resource. *Water Science and Technology.* 39(5):153–160. https://doi.org/10.1016/S0273-1223(99)00097-9

Pachapur, V.L., Kutty, P. and Pachapur, P., 2019. Seed pretreatment for increased hydrogen production using mixed-culture systems with advantages over pure-culture systems. *Energies.* 12(3):530. https://doi.org/10.3390/en12030530

Parag, R. G., 2008. Treatment of wastewater streams containing phenolic compounds using hybrid techniques based on cavitation: A review of the current status and the way forward. *Ultrasonics Sonochemistry.* 15:1–15. https://doi.org/10.1016/j.ultsonch.2007.04.007

Paraskeva, P. and Diamadopoulos, E., 2006. Technologies for olive mill wastewater (OMW) treatment: A review. *Journal of Chemical Technology and Biotechnology* 81:1475–1485. https://doi.org/10.1002/jctb.1553

Pathak, V. V., Singh, D. P., Kothari, R. and Chopra, A. K., 2014. Phycoremediation of textile wastewater by unicellular microalga *Chlorella pyrenoidosa*. *Cellular and Molecular Biology.* 60:35–40.

Paustian, K., Cole, C. V., Sauerbeck, D. and Sampson, N., 1998. CO_2 mitigation by agriculture: An overview. *Climatic Change.* 40:135–162. https://doi.org/10.1023/A:1005347017157

Peixoto, L., Min, B. and Martins, G., 2011. In situ microbial fuel cell-based biosensor for organic carbon. *Bioelectrochemistry.* 81:99–103. https://doi.org/10.1016/j.bioelechem.2011.02.002

Pham, C. A., Jung, S. J. and Phung, N. T., 2003. A novel electrochemically active and Fe(III)-reducing bacterium phylogenetically related to Aeromonas hydrophila, isolated from a microbial fuel cell. *FEMS Microbiology Letters.* 223:129–134. doi: 10.1016/S0378-1097(03)00354-9

Pinto, G., Pollio, A., Previtera, L., Stanzione, M. and Temussi, F., 2003. Removal of low molecular weight phenols from olive mill wastewater using microalgae. *Biotechnology Letters.* 25:1657–1659. https://doi.org/10.1023/A:1025667429222

Pittman, J. K., Dean, A. P. and Osundeko, O., 2011. The potential of sustainable algal biofuel production using wastewater resources. *Bioresource Technology.* 102:17–25. https://doi.org/10.1016/j.biortech.2010.06.035

Rai, Y., Price, M. and Alabaster, T., 2004. Proceedings of the water environment federation. *WEF/A&WMA Industrial Wastes.* 202–219.

Ramasamy, R. K., Congeevaram, S. and Thamaraiselvi, K., 2011. Evaluation of isolated fungal strain from e-waste recycling facility for effective sorption of toxic heavy metals Pb(II) ions and fungal protein molecular characterization-a Mycoremediation approach. *Asian Journal of Experimental Biological Sciences.* 2(2):342–347.

Rathinam, A., Rao, J. R. and Nair, B. U., 2011. Adsorption of phenol onto activated carbon from seaweed, determination of the optimal experimental parameters using factorial design. *Journal of the Taiwan Institute of Chemical Engineers.* 42:952–956. https://doi.org/10.1016/j.jtice.2011.04.003

Rawat, I., Ranjith, K. R., Mutanda, T. and Bux, F., 2011. Dual role of microalgae: Phycoremediation of domestic wastewater and biomass production for sustainable biofuels production. *Applied Energy.* 88:3411–3424. https://doi.org/10.1016/j.apenergy.2010.11.025

Rayu, S., Karpouzas, D. G. and Singh, B. K., 2012. Emerging technologies in bioremediation: Constraints and opportunities. *Biodegradation.* 23:917–926. doi: 10.1007/s10532-012-9576-3

Riano, B., Molinuevo, B. and Garcia-Gonzalez, M. C., 2011. Potential for methane production from anaerobic co-digestion of swine manure with winery wastewater. *Bioresource Technology.* 102:41–4136. https://doi.org/10.1016/j.biortech.2010.12.077

Robinson, C., Bromssen, M. V. and Bhattacharya, P., 2011. Dynamics of arsenic adsorption in the targeted arsenic-safe aquifers in Matlab, south-eastern Bangladesh: insight from experimental studies. *Applied Geochemistry.* 26:624–635. https://doi.org/10.1016/j.apgeochem.2011.01.019

Rojas, L. A., Yanez, C., Gonzalez, M., Lonos, S., Smalla, K. and Seeqer, M., 2011. Characterization of the metabolically modified heavy metal-resistant *Cupriavidus metallidurans* strain MSR33 generated for mercury bioremediation. *PLoS One.* 6(3):17555. doi: 10.1371/journal.pone.0017555

Rosenbaum, M., Aulenta, F., Villano, M. and Angenent, L. T., 2011. Cathodes as electron donors for microbial metabolism: Which extracellular electron transfer mechanisms are involved? *Bioresource Technology*. 102(1):324–333. https://doi.org/10.1016/j.biortech.2010.07.008

Roy, M. M., Dutta, A., Corscadden, K., Havard, P. and Dickie, L., 2011. Review of biosolids management options and co-incineration of a biosolid-derived fuel. *Waste Management*. 31(11):2228–2235. https://doi.org/10.1016/j.wasman.2011.06.008

Rozzi, A. and Malpei, F., 1996. Treatment and disposal of olive mill effluents. *International Biodeterioration Biodegradation*. 47:135–144. https://doi.org/10.1016/S0964-8305(96)00042-X

Ruiz, G., Jeison, D. and Chamy, R., 2003. Nitrification with high nitrite accumulation for the treatment of wastewater with high ammonia concentration. *Water Research*. 37:1371–1377. https://doi.org/10.1016/S0043-1354(02)00475-X

Satsangee, R. and Ghose, P., 1990. Anaerobic degradation of phenol using a cultivated mixed culture. *Applied Microbiology and Biotechnology*. 34(1):127–130. https://doi.org/10.1007/BF00170936

Schink, B., 1997. Energetics of syntrophic cooperation in methanogenic degradation. *Microbiology and Molecular Biology Reviews*. 61:262–280. PMCID: PMC232610

Selma, C. and Ayaz, L. A., 2001. Treatment of wastewater by natural systems. *Environment International*. 26:189–195. https://doi.org/10.1016/S0160-4120(00)00099-4

Sharma, S. K. and Sanghi, R., 2012. *Advances in Water Treatment and Pollution Prevention*. Springer (Accessed on 7 February 2013). ISBN: 9789400742048

Singh, R., 2013. Fermentative biohydrogen production using microbial consortia. In Gupta, V. and Tuohy, M. (eds.), *Biofuel Technologies*, Springer, Berlin, Heidelberg. 273–300. https://doi.org/10.1007/978-3-642-34519-7_11

Singhal, Y. and Singh, R., 2014a. Effect of microwave pretreatment of mixed culture on biohydrogen production from waste of sweet produced from Benincasa hispida. *International Journal of Hydrogen Energy*. 39(14):7534–7540. http://dx.doi.org/10.1016/j.ijhydene.2014.01.198

Singhal, Y. and Singh, R., 2014b. Comparison of graphite electrode types on generation of bioelectricity from Petha industry wastewater using Single Chamber Microbial Fuel Cells (SCMFC). *Advanced Science Letters*. 20(7–8):1578–1581. doi: 10.1166/asl.2014.5598

Singhal, Y. and Singh, R., 2015. Energy recovery from Petha industrial wastewater by anaerobic digestion. *International Journal of Science and Engineering*. 3(2):146–151. ISSN:2347-2200/V3N2/pp-146–151 /©IJSE

Sivaramakrishna, D., Sreekanth, D., Sivaramakrishnan, M., Kumar, B. S., Himabindu, V. and Narasu, M. L., 2014. Effect of system optimizing conditions on biohydrogen production from herbal wastewater by slaughterhouse sludge. *International Journal of Hydrogen Energy*. 39(14):7526–7533. https://doi.org/10.1016/j.ijhydene.2014.02.026

Sosnowski, P., Wieczorek, A. and Ledakowicz, S., 2003. Anaerobic co-digestion of sewage sludge and organic fraction of municipal solid wastes. *Advances in Environmental Research*. 7:609–616. https://doi.org/10.1016/S1093-0191(02)00049-7

Steinbusch, K. J., Hamelers, H. V., Plugge, C. M. and Buisman, C.J.N., 2011. Biological formation of caproate and caprylate from acetate: Fuel and chemical production from low grade biomass. *Energy & Environmental Science*. 4(1):216–224. doi: 10.1039/C0EE00282H

Talos, K., Pager, C. and Tonk, S., 2009. Cadmium biosorption on native *saccharomyces cerevisiae* cells in aqueous suspension. *Acta Univ Sapientiae Agric Environ*. 1:20–30.

Tang, J., Zhang, C., Shi, X., Sun, J. and Cunningham, J. A., 2019. Municipal wastewater treatment plants coupled with electrochemical, biological and bio-electrochemical technologies: Opportunities and challenge toward energy self-sufficiency. *Journal of Environmental Management*. 234:396–403. https://doi.org/10.1016/j.jenvman.2018.12.097

Tastan, B.E., Ertugrul, S. and Dönmez, G., 2010. Effective bioremoval of reactive dye and heavy metals by Aspergillus versicolor. *Bioresource Technology.* 101(3):870–876. doi: 10.1016/j.biortech.2009.08.099

Telkamp, P., 2006. *Separate Collection and Treatment of Domestic Wastewater in Norway.* Wageningen University, Wageningen.

Themelis, N.J. and Ulloa, P.A., 2007. Methane generation in landfills. *Renewable Energy.* 32:1243–1257. https://doi.org/10.1016/j.renene.2006.04.020

Tontti, T., Poutiainen, H. and Heinonen-Tanski, H., 2016. Efficiently treated sewage sludge supplemented with nitrogen and potassium is a good fertilizer for cereals. *Land Degradation & Development.* 28(2). doi: 10.1002/ldr.2528

Turovskiy, I.S., 2015. Biosolids or Sludge? The Semantics of Terminology. *Water and Wastes Digest* (Accessed on 24 April 2015).

Tyagi, M., Fonseca, M.M.D. and Carvalho, C.C.D., 2011. Bioaugmentation and biostimulation strategies to improve the effectiveness of bioremediation processes. *Biodegradation.* 22:231–241. doi: 10.1007/s10532-010-9394-4

Uche, J., Valero, A. and Serra, L., 2002. La desalación y reutilización como recursos alternativos. *Documentación administrativa*, Gobierno de Aragón, Universidad de Zaragoza.

Van Der Vlueten-Balkema, A.J., 2003. Sustainable wastewater treatment: Developing a methodology and selecting promising systems. PhD thesis, Eindhoven University, Eindhoven.

Varia, J.C., Martinez, S.S., Velasquez-Orta, S. and Bull, S., 2014. Microbiological influence of metal ion electrodeposition: Studies using graphite electrodes $(AuCl_4)^-$ and Shewanella putrefaciens. *Electrochimica Acta.* 115:344–351. https://doi.org/10.1016/j.electacta.2013.10.166

Venkata, M.S., Lalit, B.V. and Sarma, P.N., 2007. Anaerobic biohydrogen production from dairy wastewater treatment in sequencing batch reactor (AnSBR): Effect of organic loading rate. *Enzyme and Microbial Technology.* 41:506–515. https://doi.org/10.1016/j.enzmictec.2007.04.007

Villano, M., Aulenta, F., Ciucci, C., Ferri, T., Giuliano, A. and Majone, M., 2010. Bio-electrochemical reduction of CO_2 to CH_4 via direct and indirect extracellular electron transfer by a hydrogenophilic methanogenic culture. *Bioresource Technology.* 101(9):3085–3090. https://doi.org/10.1016/j.biortech.2009.12.077

Von, R.L.G.W. and Hanger, L.Y., 2001. Acesulfame-K. In *Alternative sweetener.* ed L. O'Brein Nabors. New York, Basel: Marcel Dekker.

Wei, S.H., Zhou, Q.X. and Wang, X., 2005. Cadmium-hyperaccumulator *Solanum nigrum* L. and its accumulating characteristics. *Environmental Science.* 26:167–171.

Wei, S.H., Zhou, Q.X., Zhang, K. and Liang, J., 2003. Roles of rhizosphere in remediation of contaminated soils and its mechanisms. *Ying Yong Sheng Tai Xue Bao.* 14:143–147.

Winkler, M.K.H. and Straka, L., 2019. New directions in biological nitrogen removal and recovery from wastewater. *Current Opinion in Biotechnology.* 57:50–55. https://doi.org/10.1016/j.copbio.2018.12.007

Witt, P.M. and Schmidt, L.D., 1996. Effect of flow rate on the partial oxidation of methane and ethane. *Journal of Catalysis.* 163:465–475. https://doi.org/10.1006/jcat.1996.0348

Xiao, L. and He, Z., 2014. Applications and perspectives of phototrophic microorganisms for electricity generation from organic compounds in microbial fuel cells. *Renewable and Sustainable Energy Reviews.* 37:550–559. https://doi.org/10.1016/j.rser.2014.05.066

Yin, X.X., Wang, L.H., Bai, R., Huang, H. and Sun, G.X., 2012. Accumulation and transformation of arsenic in the blue-green alga Synechocysis sp. PCC6803. *Water, Air & Soil Pollution.* 223(3):1183–1190. https://doi.org/10.1007/s11270-011-0936-0

Zhang, B.G., Zhou, S.G. and Zhao, H., 2010. Factors affecting the performance of microbial fuel cells for sulphide and vanadium (V) treatment. *Bioprocess and Biosystems Engineering.* 33:187–194. https://doi.org/10.1007/s00449-009-0312-2

17 Effect of Manure on the Metal Efficiency of *Coriandrum sativum*

Deepshekha Punetha, Geeta Tewari, Chitra Pande, Lata Rana and Sonal Tripathi

CONTENTS

17.1 Introduction 278
17.2 Materials and Methods 279
    17.2.1 Study Area and Plant Material 279
    17.2.2 Pot Experiment 279
    17.2.3 Sampling and Analysis 279
    17.2.4 Data Analysis 281
    17.2.5 Isolation of Essential Oil 281
    17.2.6 Analysis of the Essential Oil 281
    17.2.7 Statistical Analysis 281
17.3 Results and Discussion 281
    17.3.1 Effect of Manure Amendments on the Physicochemical Properties of Soil 281
    17.3.2 Effect of Manure Amendments on Growth Parameters of *Coriandrum sativum* 282
    17.3.3 Metal Accumulation by *Coriandrum sativum* 282
        17.3.3.1 Zinc (Zn) 282
        17.3.3.2 Nickel (Ni) 283
        17.3.3.3 Copper (Cu) 285
        17.3.3.4 Chromium (Cr) 287
    17.3.4 Metal Content in Extracted Water and Essential Oil 290
    17.3.5 Effect of Manure-Amended Contaminated Soil on Essential Oil Yield..290
    17.3.6 Effect of Manure-Amended Metal Contaminated Soil on Essential Oil Composition 290
    17.3.7 Cluster Analysis...296
17.4 Conclusion 297
References 297

17.1 INTRODUCTION

Accumulation of toxic heavy metals in the environment and their adverse effect on both public health and the natural environment is a major environmental problem that can influence both plant productivity as well as safety as a food and feed crops (McGrath et al., 2002). Geological and anthropogenic activities are the main causes of heavy metal contamination (Dembitsky, 2003). Sources of anthropogenic metal contamination consist of industrial activities such as electroplating, gas exhaust, mining and smelting of metalliferous ores, energy and fuel production, fertilizer and pesticide application and municipal waste disposal (Kabata-Pendias and Pendias, 1989). When present in excess amount, the metal contaminants (Zn, Cu, Ni, Pb, Cr, Hg etc.) are first accumulated by the plant roots and then transferred to the different tissues within the plants. Plants have been mainly used for the last two decades for the remediation of soil and water contaminated with heavy metals as it is a cost-effective and non-invasive method (Mojiri, 2012). Under this, phytoextraction is a rising method for cleaning up of contaminated sites (Ok and Kim, 2007).

Coriandrum sativum L. is an aromatic plant belonging to the family Apiaceae (Umbelliferace) and widely grown as a spice plant for its medicinal properties (aromatic, stimulant and carminative) or for the commercial production of essential oils for pharmaceutical, perfume and cosmetic industries. Various research reports have shown that some aromatic and medicinal plants could be safely grown on metal polluted soils (Zheljazkov and Nielson, 1996a,b; Zheljazkov and Warman, 2003; Zheljazkov et al., 2008; Prasad et al., 2011; Amir et al., 2012; Kunwar et al., 2015; Kunwar et al., 2018; Bisht et al., 2019)) because the final product (essential oil) of these crops was found to contain very small concentrations of metals (sometimes below detection limit) (Scora and Chang, 1997). As coriander (*Coriandrum sativum* L.), dill (*Anethum graveolens* L.), chamomile (*Chamomilla recutita* L.), fennel (*Foeniculum vulgare* Mill.), peppermint (*Mentha x piperita* L.), basil (*Ocimum basilicum* L.), and sage (*Salvia officinalis* L.) are the most commercially important essential oil producing aromatic plants, the effect of metals on the metal accumulation, growth and essential oil production are yet mostly unknown (Topalov, 1962).

Heavy metals can be immobilized by treating the contaminated soils with inorganic or organic amendment. Rock phosphate, apatite, liming agents, hydroxyapatite, iron and manganese oxide are the inorganic materials which can reduce cadmium availability to plants (Keller et al., 2005). Organic amendments such as cow/pig/chick manure, farmyard manure and composts are considered as low-cost adsorbents which improve water-holding capacity and reduce bioavailability of heavy metals in soil and food crops. They can also reduce phytotoxicity and increase plant growth (Li et al., 2001; Walker et al., 2003; Pichtel and Bradway, 2008).

Compost is applied in soil to maintain and recover soil structure (Lillenberg et al., 2010), and its organic matter content can neutralize the natural decline in intensively cultivated soils. It is also found that monitoring and assessment of heavy metal content in the different crops including vegetables and aromatic plants as well as change in heavy metal concentration with the application of manure has limited published data. Therefore, the present study was conducted to study the effect of manure addition on the heavy metal (Cu, Zn, Ni and Cr) accumulation in coriander (*Coriandrum sativum* L.).

Effect of Manure on Metal Efficiency

17.2 MATERIALS AND METHODS

17.2.1 STUDY AREA AND PLANT MATERIAL

The composite surface soil sample (0–20 cm) was collected from Moradabad (Karula nala), which was highly contaminated with different metals (Zn, Ni, Cu and Cr) due to the presence of electroplating industries near this area. The seeds were purchased from Vegetable Research Centre (VRC), Pantnagar, and the plants were identified at Botanical Survey of India, Dehradun.

17.2.2 POT EXPERIMENT

The soil and cow manure were air-dried and grounded well before use and analysis. The processed and analyzed soil was treated with four levels of cow manure (0 ton/hectare, 26 t/h, 52 t/h and 104t/h) in triplicate and then incubated for four weeks. The total weight of soil in each treatment was 3 kilograms.

After addition of cow manure, the amended soil treatments were again incubated for another four weeks, under three cycles of saturation water and air-drying. The samples were transferred into free-draining 5 kg earthen pots (10 inches in diameter). To prevent the leaching of metals, each pot was placed in a separate dish. Thirty seeds of *C. sativum* were sown in each pot. After 30 days (first cutting for metal accumulation analysis), thinning of plants was done and the plant number was reduced to ten per pot. The plants were watered regularly and were harvested after 90 days of growth for metal accumulation and essential oil analysis.

17.2.3 SAMPLING AND ANALYSIS

The pH and electrical conductivity (EC) of cow manure and amended soils were measured in soil-water suspension (1:2) (Jackson, 1958) using an electric shaker for half an hour, and organic carbon was measured by the Walkley and Black method (Jackson, 1958) (Table 17.1 and Table 17.2).

Well-mixed manure-amended soil samples were digested with 8 mL of aqua regia on a sand bath for two hours (Lokeshwari and Chandrappa, 2006). Soil samples were also extracted for DTPA (Diethylene triamine penta-acetic acid) extractable metals following the procedure developed by Lindsay and Norvell (1969). Dry plant tissue was finely ground and wet ashes using HNO_3:H_2SO_4: $HClO_4$ (10:1:4 v/v) (Jackson, 1958) (Table 17.3).

TABLE 17.1

Physicochemical Properties of Manure (Cow Dung)

Properties	Manure (Cow Dung)
pH	7.620 ± 0.500
EC (dS m^{-1})	1.2 ± 0.2
Organic carbon (%)	7.33 ± 0.20
Organic matter (%)	12.64 ± 0.50

TABLE 17.2
Effect of Manure Amendment on the Physicochemical Properties of Amended Soils and Growth Parameters of *C. sativum*

Amendments (t/h)	Soil pH (1:2) (soil: water)	EC (dS m^{-1}) (1:2) (soil: water)	Plant Height (cm) (first month)	Plant Height (cm) (third month)	% Oil Yield (w/w)	Plant Weight (g)	Seed Weight (g)
0	7.75^a±0.60	0.3^a±0.1	9^a±1	38^a±4	0.1^a±0.0	78.79^a±6.40	27.64^a±3.80
26	7.87^a±0.10	0.3^a±0.1	10^a±1	37^a±4	0.2^b±0.0	73.57^a±5.80	16.04^b±3.70
52	7.99^a±0.10	0.4^a±0.1	11^a±1	39^a±4	0.6^c±0.0	236.81^b±4.50	60.00^c±1.70
104	8.14^a±0.60	0.3^a±0.1	11^a±3	39^a±5	0.4^d±0.0	280.31^c±2.6	105.00^d±2.30

The mean followed by same alphabet (a–d) at superscript within the column of same soil are not significantly different ($p<0.05$) according to Duncan's test.

TABLE 17.3
Effect of Cow Dung Amendments on Zn Content in Karula Nala Soil and *C. sativum* (Permissible Limit: 50 mg kg^{-1})

Treatments	Soil (mg kg^{-1})		Aerial Parts (mg kg^{-1})		Roots (mg kg^{-1})	DTPA (mg kg^{-1})		Bioconcentration Factor (BCF) (Plant/Soil)		Translocation Factor (TLF) (Shoot/Root)
	I Cutting	II Cutting	I Cutting	II Cutting		I Cutting	II Cutting	I Cutting	II Cutting	
A$_1$	344.80^a±5.7	216.16^a±8.4	49.25^a±9.1	45.00^a±4.1	55.29^a±4.1	31.03^a±9.9	19.45^a±5.8	0.14^a±0.03	0.21^a±0.05	0.81^a±0.18
A$_2$	345.26^a±5.1	219.26^a±9.3	46.87^a±7.8	44.23^a±8.7	52.18^a±6.6	30.12^a±8.2	18.21^a±7.6	0.14^a±0.03	0.20^a±0.07	0.85^a±0.05
A$_3$	342.10^a±8.6	220.37^a±3.6	42.09^a±6.9	41.54^a±7.3	48.12^a±4.1	28.12^a±5.4	16.28^a±6.2	0.12^a±0.03	0.19^a±0.13	0.86^a±0.05
A$_4$	346.58^a±5.8	220.37^a±3.3	41.78^a±10.0	40.95^a±6.2	47.23^a±4.1	25.27^a±3.9	15.12^a±6.8	0.12^a±0.07	0.18^a±0.14	0.87^a±0.04

A$_1$, A$_2$, A$_3$ and A$_4$ are the 0, 26, 52 and 104 t/h amended soil samples; the mean followed by same alphabet at superscript within the column of same soil are not significantly different ($p<0.05$) according to Duncan's test.

Effect of Manure on Metal Efficiency

17.2.4 Data Analysis

Metal accumulation by *C. sativum* was performed by different data analysis tools, namely, transfer factor (TF), bioaccumulation factor (BCF) and translocation factor (TLF) (Ma et al., 2001).

Bioconcentration factor (BCF) =Metal concentration in roots/Metal concentration in soil

Translocation factor (TLF) =Metal concentration in aerial parts/Metal concentration in roots

Transfer factor (TF) =Metal concentration in plant/Metal concentration in soil

17.2.5 Isolation of Essential Oil

The fresh plant material after harvesting (90 days of growth) was hydrodistilled for 3 h using a Clevenger apparatus. The hydrodistilled oil was dried over anhydrous sodium sulphate and stored at 4°C in a BOD incubator before use.

17.2.6 Analysis of the Essential Oil

Gas chromatographic (GC) data of the different essential oil samples were obtained by using GLC instrument (Shimadzu, 2010) and Gas chromatographic/mass spectrometric (GC/MS) analysis of the different essential oil samples was done on Shimadzu 2010 GC, coupled with Shimadzu QP 2010 plus MS. The working conditions of GC and GC/MS were similar to Punetha et al. (2014).

17.2.7 Statistical Analysis

The mean values and standard deviations (SD) were calculated for all the treatments using Microsoft Excel. The recorded data were subjected to one-way analysis of variance (ANOVA) to determine the significant difference between groups, considering a level of significance of less than 5% by using SPSS 16.0 software. Correlation coefficients were checked at probability levels of $p<0.01$ and $p<0.05$ using SPSS 16.0.

17.3 RESULTS AND DISCUSSION

17.3.1 Effect of Manure Amendments on the Physicochemical Properties of Soil

Physicochemical properties of contaminated soil and manure amended contaminated soil as well as manure samples are shown in Table 17.1 and Table 17.2. Manure (cow dung) taken for this study had the pH 7.62 and electrical conductivity 1.2 dS m^{-1}. It had 7.33% organic carbon and 12.64% organic matter.

In the treated soils, pH varied from 7.75 to 8.14, and it was found that Karula nala soil had the highest pH value at 104t/h amendment and the lowest pH value was observed for the control soil. Electrical conductivity varied from 0.3–0.4 dS m^{-1}. The highest EC value was noted at 52 t/h cow dung amendment and remaining amendments had the same EC value (0.3 dS m^{-1}). Application of cow dung caused an increase in both pH and EC of soil samples. These results were similar to that of Courtney and Mullen (2008), who observed the highest pH value after addition of

282   Zero Waste

compost I (mainly sewage sludge with pH 8.0), whereas soil fertilized with mineral
forms of nutrients had the lowest pH.

17.3.2 EFFECT OF MANURE AMENDMENTS ON GROWTH PARAMETERS OF *CORIANDRUM SATIVUM*

When the manure was added at different levels to the contaminated soil, a significant
change was seen in the growth parameters in terms of plant height, plant weight,
seed weight and oil yield as shown in Table 17.2. Plant height was measured in two
different months and it was found that the plant height was increased regularly as the
manure concentration was enhanced and the highest plant height was found at 52t/h
along with the highest oil yield. Increased essential oil yield with compost treatments
was also reported by Swaminathan et al. (2008) and Kumar et al. (2009) in *Artemisia
pallens*. Plant and seed weight also increased significantly and showed the highest
value at 104t/h (280.31 g and 105.00 g for plant and seed weight, respectively).

17.3.3 METAL ACCUMULATION BY *CORIANDRUM SATIVUM*

The availability of metals to the plants depends upon soil pH, organic carbon content,
redox potential, cation exchange capacity, calcium carbonate content etc. (Alloway
et al., 1988). There was a significant difference seen when coriander grown in contaminated soil with different manure amended contaminated soil samples. The results
of this study on the accumulation of different metals can be explained as follows.

17.3.3.1 Zinc (Zn)

Zinc content in *Coriandrum sativum* when grown in contaminated soil (high concentration of Cu, Zn and Cr) as well as manure-amended contaminated soil ranged
from 41.87 to 49.25 mg kg^{-1} and 40.95 to 45.00 mg kg^{-1} respectively for first and
second cuttings (Figure 17.1; Table 17.3). Zinc accumulation was found higher in the
second cutting in the control treatment, while after manure addition Zn accumulation was higher for the first cutting.

It was found that Zn concentration tended to reduce in the aerial parts as well as
in roots (for both the cuttings) as manure concentration increased in the soil as shown
by BCF and TLF having the following trend:

0.14 (0t/h) = 0.14 (26t/h) > 0.12 (52t/h) = 0.12 (104t/h) (BCF; First cutting)
0.21 (0t/h) > 0.20 (26t/h) > 0.19 (52t/h) > 0.18 (104t/h) (BCF; Second cutting)
0.87 (104t/h) >0.86 (52t/h)> 0.85 (26t/h) >0.81 (0t/h) (Translocation factor)

Aerial parts of the plant had higher accumulation of Zn than the roots as shown
above, and this sequence was similar to Herrero et al.(2003), who showed that leaf
was the main organ for Zn accumulation, while Pb was found in roots in higher concentration in the sunflower plant.

There was an increase in soil pH as the cow dung concentration increased, which
may be the cause of lower Zn content in plant tissue as well as roots and was also
observed by Subrahmanyam et al. (1992), who found that at higher pH or higher content of lime (calcium carbonates), the availability of most metals decreases to such

Effect of Manure on Metal Efficiency

extent that, in the case of essential microelements, it may lead to acute deficiency symptoms and retard development of certain susceptibility plants (iron chlorosis in *Arnica montana* and *Mentha arvensis* L.).

Zinc content in all the soil and manure amended samples was found to be lower than the permissible limit given by Awasthi (2000) (50 mg kg^{-1}). The results were similar to Tlustoš et al. (1997), who also showed a reduction in cadmium concentration in oat biomass after manure application into the experimental soil. Correlations between Zn, growth parameters of coriander and physicochemical properties of soil have been given in Table 17.4.

17.3.3.2 Nickel (Ni)

Nickel content showed a variation from 7.14 to 8.15 mg kg^{-1} and 8.48 to 10.21 mg kg^{-1} in the plant for both the cuttings respectively as shown in Figure 17.1 and Table 17.5. Metal accumulation was higher in the second cutting than the first cutting both with and without manure applications. Heavy metals present in trace amounts could have a serious effect in the environment through bioaccumulation, toxicity and food chain (Ogri et al., 1998). However, it was seen in the present study that the phytoextraction of Ni by *Coriandrum sativum* decreased with the addition of manure. This decrease can be understood better with the help of bioconcentration and translocation factor.

0.41 (0t/h) > 0.37 (26t/h) > 0.35 (52t/h) > 0.32 (104t/h) (BCF; First cutting)
1.02 (0t/h) > 0.88 (26t/h) > 0.82 (52t/h) > 0.70 (104t/h) (BCF; Second cutting)
0.93 (104t/h) >0.90 (52t/h)> 0.83 (26t/h) > 0.67 (0t/h) (Translocation factor)

TABLE 17.4
Correlation between Zn, Growth Parameters of Coriander and Physicochemical Properties of Karula Nala Soil

	Zn Content in Soil	Zn Content in Plant	Zn Content (DTPA)	BCF	pH	EC	Plant Height	Plant Weight	Seed Weight
Zn content in soil	1	.370	.940**	−.954**	.014	.010	−.998**	−.011	−.012
Zn content in plant		1	.631	−.096	−.841**	−.836**	−.423	.842**	.760*
Zn content (DTPA)			1	−.806*	−.319	−.265	−.959**	−.311	−.303
BCF				1	−.292	−.271	.935**	−.295	−.275
pH					1	.850**	.047	.943**	.910**
EC						1	.047	.791*	.641
Plant height							1	.052	.047
Plant weight								1	.973**
Seed weight									1

** Correlations are significant at *p*<0.01. * Correlations are significant at *p*<0.05.

TABLE 17.5
Effect of Cow Dung Amendments on Ni Content in Karula Nala Soil and *Coriandrum sativum* (Permissible Limit: 1.5 mg kg⁻¹)

Metal	Sample	Soil (mg kg⁻¹)		Aerial Parts (mg kg⁻¹)		Roots (mg kg⁻¹)	DTPA (mg kg⁻¹)		Bioconcentration Factor (BCF) (Plant/Soil)		Translocation Factor (TLF) (Shoot/Root)
		I Cutting	II Cutting	I Cutting	II Cutting		I Cutting	II Cutting	I Cutting	II Cutting	
Ni	B_1	$20.00^a\pm1.5$	$10.00^a\pm1.9$	$8.15^a\pm2.1$	$10.21^a\pm3.6$	$15.24^a\pm4.2$	$1.80^a\pm0.7$	$0.90^a\pm0.1$	$0.41^a\pm0.0$	$1.02^a\pm0.1$	$0.67^a\pm0.1$
	B_2	$21.34^a\pm5.6$	$11.34^a\pm6.2$	$8.01^a\pm3.5$	$9.99^a\pm2.3$	$12.09^a\pm6.7$	$1.42^{a,b}\pm0.3$	$0.87^a\pm0.1$	$0.37^a\pm0.1$	$0.88^{a,b}\pm0.0$	$0.83^{a,b}\pm0.0$
	B_3	$21.97^a\pm1.8$	$11.09^a\pm2.4$	$7.68^a\pm4.0$	$9.08^a\pm1.3$	$10.11^a\pm1.5$	$1.02^{a,b}\pm0.1$	$0.78^a\pm0.0$	$0.35^a\pm0.0$	$0.82^{a,b}\pm0.2$	$0.90^b\pm0.1$
	B_4	$22.18^a\pm7.0$	$12.16^a\pm6.3$	$7.14^a\pm2.5$	$8.48^a\pm3.4$	$9.12^a\pm3.4$	$0.89^b\pm0.2$	$0.75^a0.1\pm$	$0.32^a\pm0.1$	$0.70^b\pm0.1$	$0.93^b\pm0.1$

Note: B1, B2, B3 and B4 are the 0, 26, 52 and 104 t/h amended soil samples; the mean followed by same alphabets at superscript within the column of same soil are not significantly different ($p<0.05$) according to Duncan's test.

Effect of Manure on Metal Efficiency

TABLE 17.6

Correlation between Ni, Growth Parameters of Coriander and Physicochemical Properties of Karula Nala Soil

	Ni Content in Soil	Ni Content in Plant	Ni Content (DTPA)	BCF	pH	EC	Plant Height	Plant Weight	Seed Weight
Ni content in soil	1	−.889**	.564	−.975**	.144	.135	−.980**	.120	.109
Ni content in plant		1	−.314	.952**	−.522	−.425	.805*	−.522	−.513
Ni content (DTPA)			1	−.529	−.590	−.577	−.700	−.562	−.509
BCF				1	−.281	−.246	.933**	−.252	−.239
pH					1	.850**	.047	.943**	.910**
EC						1	.047	.791*	.641
Plant height							1	−.052	−.047
Plant weight								1	.973**
Seed weight									1

** Correlations are significant at $p<0.01$. * Correlations are significant at $p<0.05$.

Bioconcentration and translocation factor showed the ability of plant to accumulate metal in the roots and transfer to the aerial parts of the plant. TLF was lower than the BCF for control as well as the 26t/h manure-amended sample, which means that there was higher accumulation of metal in roots rather than transfer to the aerial part, while TLF was higher in the other two treatments (52 and 104t/h) than the BCF, which showed that there was transfer of metal to the aerial parts rather than to accumulate in roots (Table 17.5). Correlations between Ni, growth parameters of coriander and physicochemical properties of soil have been given in Table 17.6.

17.3.3.3 Copper (Cu)

Effect of manure addition to the contaminated soil and their relation with the metal accumulation is given in Figure 17.1 and Table 17.7. Copper accumulation in *Coriandrum sativum* was observed to be more in the second cutting as compared to the first cutting in case of contaminated soil and manure amended contaminated soil.

Copper content in soil varied from 245.05 to 404.59 mg kg^{-1} and the BCF of *C. sativum* showed that there was a reduction in metal accumulation as the manure concentration increased.

The order of metal in roots of *C. sativum* showed that as the pH value increased there was a higher accumulation of Cu in roots and the TLF decreased from 0.78 to 0.53, indicating that there was reduction of metal in aerial parts rather than to accumulate in roots. These results showed similarity with other studies such as Gigliotti et al. (1996), who reported that there was a very low transfer of heavy metals to plant tissues for maize after addition of 90 t ha^{-1}y^{-1} of MSW compost (at 70.5% DS) to calcareous clay loam soil (pH 8.3) for 6 years (total application=540 t ha^{-1}). The order of bioconcentration factor and translocation factor in *C. sativum* for Cu is as follows:

TABLE 17.7
Effect of Cow Dung Amendments on Cu Content in Karula Nala Soil and *C. sativum* (Permissible Limit: 30 mg kg^{-1})

Metal	Sample	Soil (mg kg^{-1})		Aerial Parts (mg kg^{-1})		Roots (mg kg^{-1})	DTPA (mg kg^{-1})		Bioconcentration Factor (BCF) (Plant/Soil)		Translocation Factor (TLF) (Shoot/Root)
		I Cutting	II Cutting	I Cutting	II Cutting		I Cutting	II Cutting	I Cutting	II Cutting	
Cu	C$_1$	400.97[a]±9.5	245.83[a]±9.4	16.22[a]±3.3	14.76[a]±1.9	20.46[a]±2.1	36.09[a]±1.2	22.12[a]±3.5	0.04[a]±0.0	0.06[a]±0.0	0.72[a,b]±0.1
	C$_2$	400.19[a]±2.5	246.39[a]±5.2	15.18[a]±4.6	14.02[a]±2.6	18.39[a]±2.9	35.33[a]±1.3	22.02[a]±4.0	0.04[a]±0.0	0.06[a]±0.0	0.76[a]±0.1
	C$_3$	401.28[a]±9.9	245.05[a]±8.1	13.98[a]±4.0	12.12[a]±6.5	15.59[a]±3.0	33.69[a,b]±1.7	21.29[a]±2.7	0.03[a]±0.0	0.05[a]±0.0	0.78[a]±0.2
	C$_4$	404.59[a]±2.9	248.88[a]±2.9	10.48[a]±1.1	10.74[a]±5.9	20.27[a]±3.5	31.35[b]±2.9	20.14[a]±2.8	0.02[a]±0.0	0.04[a]±0.0	0.53[b]±0.0

Note: C1 , C2 , C3 and C4 are the 0, 26, 52 and 104 t/h amended soil samples; the mean followed by same alphabets at superscript within the column of same soil are not significantly different ($p<0.05$) according to Duncan's test.

TABLE 17.8
Correlation between Cu, Growth Parameters of Coriander and Physicochemical Properties of Karula Nala Soil

	Cu Content in Soil	Cu Content in Plant	Cu Content (DTPA)	BCF	pH	EC	Plant Height	Plant Weight	Seed Weight
Cu content in soil	1	.253	.973**	−.780*	.015	.006	−.998**	.014	.017
Cu content in plant		1	.461	.385	−.930**	−.720*	−.312	−.902**	−.902**
Cu content (DTPA)			1	−.625	−.196	−.148	−.985**	−.193	−.194
BCF				1	−.609	−.462	.738*	−.629	−.633
pH					1	.850**	.047	.943**	.910**
EC						1	.047	.791*	.641
Plant height							1	.052	.047
Plant weight								1	.973**
Seed weight									1

** Correlations are significant at $p<0.01$. * Correlations are significant at $p<0.05$.

0.04 (0t/h) = 0.04 (26t/h) > 0.03 (52t/h) > 0.02 (104t/h) (BCF; First cutting)
0.06 (0t/h) = 0.06 (26t/h) > 0.05 (52t/h) > 0.04 (104t/h) (BCF; Second cutting)
0.81 (104t/h) > 0.78 (52t/h) > 0.76 (26t/h) > 0.72 (0t/h) (Translocation factor)

Results of this study also showed that a high pH value was observed to be responsible for the reduction in availability of Cu. Vaca-Paulín et al. (2006), also found an increased Cu sorption capacity and lower content of available copper in soil amended with composts compared to unfertilized soil. Correlations between Cu, growth parameters of coriander and physicochemical properties of soil have been given in Table 17.8.

17.3.3.4 Chromium (Cr)

Chromium concentration varied from 66.12 to 70.00 mg kg^{-1} and the accumulation was found to be higher in the second cutting as compared to the first cutting for all the soil samples (Figure 17.1; Table 17.9). Addition of manure (cow dung) decreased Cr accumulation in the aerial parts (in both the cuttings) as well as in roots of *Coriandrum sativum*. Manure addition may cause the increase in the pH value which can be responsible for the bioavailability of Cr as suggested by Subrahmanyam et al. (1992).

Bioconcentration and translocation factor for Cr metal showed that the Cr content accumulated more in roots rather than transfer to the aerial parts as the BCF was higher than that of TLF. The order of bioconcentration factor and translocation factor in *C. Sativum* for Cr is as follows:

0.86 (0t/h) > 0.84 (26t/h) > 0.82 (52t/h) > 0.79 (104t/h) (BCF; First cutting)
1.43 (0t/h) > 1.42 (26t/h) > 1.39 (52t/h) > 1.31 (104t/h) (BCF; Second cutting)
0.94 (104t/h) > 0.93 (52t/h) > 0.89 (26t/h) = 0.89 (0t/h) (Translocation factor)

Correlations between Cr, growth parameters of coriander and physicochemical properties of soil have been given in Table 17.8. Various significant correlations among different metals (Zn, Ni, Cu and Cr), soil properties and growth parameters of *C. sativum* at significant levels $p<0.1$ and $p<0.5$ are shown in Table 17.11.

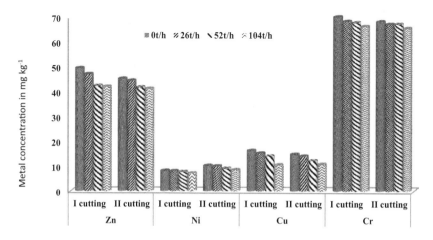

FIGURE 17.1 Metal Content in the Aerial Parts of Coriander Grown in Karula Nala Soil

TABLE 17.9

Effect of Cow Dung Amendments on Cr Content in Karula Nala Soil and *C. sativum* (Permissible Limit: 20 mg kg⁻¹)

Metal	Sample	Soil (mg kg⁻¹)		Aerial Parts (mg kg⁻¹)		Roots (mg kg⁻¹)	DTPA (mg kg⁻¹)		Bioconcentration Factor (BCF) (Plant/Soil)		Translocation Factor (TLF) (Shoot/Root)
		I Cutting	II Cutting	I Cutting	II Cutting		I Cutting	II Cutting	I Cutting	II Cutting	
Cr	D_1	81.66^a±6.5	47.63^a±7.5	70.00^a±9.1	68.05^a±5.3	76.39^a±3.2	7.35^a±1.0	4.29^a±0.8	0.86^a±0.1	1.43^a±0.5	0.89^a±0.2
	D_2	81.24^a±9.9	47.15^a±3.9	68.16^a±2.3	67.04^a±3.9	74.85^a±9.3	6.87^a±3.7	4.17^a±0.8	0.84^a±0.2	1.42^a±0.1	0.89^a±0.1
	D_3	82.72^a±7.6	48.25^a±4.2	67.63^a±4.0	67.03^a±5.9	72.11^a±6.6	6.18^a±0.8	3.79^a±1.8	0.82^a±0.2	1.39^a±0.3	0.93^a±0.0
	D_4	83.79^a±8.6	49.83^a±7.4	66.12^a±2.4	65.38^a±4.6	69.84^a±7.4	5.69^a±2.0	3.35^a±1.9	0.79^a±0.1	1.31^a±0.1	0.94^a±0.1

D_1, D_2, D_3 and D_4 are the 0, 26, 52 and 104t/h amended soil samples; the mean followed by same alphabet at superscript within the column of same soil are not significantly different $p<0.05$, according to Duncan's test.

TABLE 17.10

Correlation between Cr, Growth Parameters of Coriander and Physicochemical Properties of Karula Nala Soil

	Cr Content in Soil	Cr Content in Plant	Cr Content (DTPA)	BCF	pH	EC	Plant Height	Plant Weight	Seed Weight
Cr content in soil	1	.376	.909**	−.996**	.052	.036	−.994**	.056	.058
Cr content in plant		1	.708*	−.311	−.865**	−.701	−.458	−.757*	−.744*
Cr content (DTPA)			1	−.880**	−.352	−.301	−.947**	−.343	−.331
BCF				1	−.124	−.091	.983**	−.120	−.123
pH					1	.850**	.047	.943**	.910**
EC						1	.047	.791*	.641
Plant height							1	.052	.047
Plant weight								1	.973**
Seed weight									1

** Correlations are significant at $p<0.01$. * Correlations are significant at $p<0.05$.

TABLE 17.11
Correlation between Metals (Zn, Ni, Cu and Cr) for C. *sativum* Grown in Karula Nala Soil

	1	2	3	4	5	6	7	8	9	10	11	12	13	14	15	16
1	1	.370	.940**	−.954**	.989**	−.841**	.664	−.951**	1.00**	.253	.974**	−.777*	.998**	.408	.926**	−.992**
2		1	.631	−.096	.245	.092	.898**	−.160	.364	.878**	.546	.224	.330	.894**	.678	−.278
3			1	−.806*	.890**	−.622	.830*	−.812*	.940**	.565	.990**	−.527	.927**	.671	.995**	−.899**
4				1	−.981**	.949**	−.465	.986**	−.955**	.004	−.874**	.915**	−.965**	−.165	−.782*	.979**
5					1	−.889**	.564	−.975**	.989**	.142	.938**	−.838**	.992**	.280	.867**	−.996**
6						1	−.314	.952**	−.840**	.260	−.719*	.977**	−.860**	.061	−.602	.896**
7							1	−.529	.658	.750*	.794*	−.165	.635	.860**	.872**	−.601
8								1	−.949**	−.019	−.882**	.891**	−.958**	−.176	−.797*	.976**
9									1	.253	.973**	−.780*	.999**	.406	.924**	−.993**
10										1	.461	.385	.217	.923**	.587	−.153
11											1	−.625	.964**	.590	.985**	−.945**
12												1	−.805*	.182	−.496	.843**
13													1	.376	.909**	−.996**
14														1	.708*	−.311
15															1	−.880**
16																1

1= Zn concentration in soil, 2= Zn concentration in plant, 3= Zn (DTPA), 4=Zn (BCF), 5= Ni concentration in soil, 6= Ni concentration in plant, 7= Ni(DTPA), 8=Ni (BCF), 9= Cu concentration in soil, 10= Cu concentration in plant, 11= Cu (DTPA), 12=Cu (BCF), 13= Cr concentration in soil, 14= Cr concentration in plant, 15= Cr (DTPA), 16=Cr (BCF) (Karula soil).** Correlations are significant at $p<0.01$.* Correlations are significant at $p<0.05$.

Coriandrum sativum, when grown in contaminated and cow dung amended soil samples, showed variation in the accumulation and transfer of all the metals.

Cr >Ni>Zn> Cu (On the basis of BCF for both cuttings)
Cr > Zn > Ni>Cu (On the basis of TLF)

Chromium was the highest accumulated metal in roots as well as the least transferred metal to the aerial parts respectively. This can also be explained that the effects of organic matter amendments on heavy metal bioavailability depend on the nature of the organic matter, and on the particular soil type and metals concerned (Clemente et al., 2005).

17.3.4 METAL CONTENT IN EXTRACTED WATER AND ESSENTIAL OIL

The results of oil and extracted water samples showed no or a very low concentration of metals (below detectable limit). These results were also supported by Zheljazkov and Nielsen (1996a, 1996b). As the essential oil extracted from the leaves and stems of *C. sativum* shows potential natural immunotoxicity effects (Chung et al., 2012), the essential oil extracted from the present study can also be used for such purpose.

17.3.5 EFFECT OF MANURE-AMENDED CONTAMINATED SOIL ON ESSENTIAL OIL YIELD

Effect of manure amended metal contaminated soils on the essential oil yield of *C. sativum* was significant ($p< 0.05$) (Table 17.2). Improvement in the oil yield was observed with increasing the manure concentration. This positive correlation may be due to the ability of manure to increase the physical, chemical and biological of soil, which is related to a good balance of water and nutrients in the root zone (Gharib et al., 2008). The highest oil yield was noted for 52t/h (0.6% w/w). Similar reports on different plant species were also found (Abdelaziz et al., 2007; Kumar et al., 2009).

17.3.6 EFFECT OF MANURE-AMENDED METAL CONTAMINATED SOIL ON ESSENTIAL OIL COMPOSITION

The GC (Figure 17.2 and Figure 17.3) and GC-MS analysis of *Coriandrum sativum* essential oil showed the presence of 35 (control), 29 (26t/h), 29 (52t/h) and 38 (104t/h) compounds representing 91.87%, 88.37%, 91.68% and 94.0% respectively of the total oil (Table 17.12). Linalool (10.01%–61.73%), (*E*)-3-pentadecen-2-ol (1.23%–23.10%), caryophyllene oxide (1.51%–10.27%), (2*E*)-decanal (2.45%–19.24%) and n-octacosane (0.00–17.68%) were the major volatile constituents (Figure 17.4). Some earlier reports showed linalool (60%–70%) as the main active constituent (Guenther, 1950; Nadeem et al., 2013; Punetha et al., 2018).

A study by Kumar et al. (2015) revealed that nutrient supplement through organic sources influenced the growth, yield and quality of coriander. Similarly, essential oil composition of coriander in the present study showed that the content of linalool and n-octacosane significantly increased with manure application and had the maximum percentage at 52t/h for linalool (61.73%) and 104t/h for n-octacosane (17.68%).

TABLE 17.12
Effect of Cow Dung Amendments on the Essential Oil Composition of C. *sativum*

Name of Compounds	RI (Calculated)	RI (Adams, 2007)	% of Oil (0 t/h)	% of Oil (26 t/h)	% of Oil (52 t/h)	% of Oil (104 t/h)	Mode of Identification
α–Pinene	925	932	0.02	–	–	0.04	a, b
Camphene	947	946	0.01	–	–	0.01	a, b
Myrcene	987	988	0.31	0.26	0.25	0.18	a, b
Limonene	1022	1024	0.02	0.23	0.31	0.02	a, b
n–Octanol	1060	1063	0.02	0.31	0.41	0.02	a, b
Linalool oxide (*cis*) furanoid	1067	1067	1.51	0.69	0.69	1.68	a, b
Fenchone	1082	1083	1.51	0.40	0.30	1.00	a, b
Linalool	1090	1095	15.81	46.19	61.73	10.01	a, b
n–Nonanal	1093	1100	–	–	–	0.01	a, b
Linalool oxide (*cis*) pyranoid	1169	1170	0.26	–	–	0.43	a, b
Linalool oxide (*trans*) pyranoid	1172	1173	0.02	1.00	0.61	0.24	a, b
n–Decanal	1199	1201	3.23	1.85	0.80	1.86	a, b
(2E)–Decenal	1258	1260	19.24	10.0	2.45	18.88	a, b
(2E)–Decen–1–ol	1265	1268	0.34	0.16	0.24	0.27	a, b
Carvacrol	1297	1298	2.90	1.09	0.26	2.75	a, b
(8Z)–Undecenal	1323	–	–	0.66	0.13	0.56	a, b
Geranyl acetate	1375	1379	–	1.32	0.43	1.29	a, b
n–Dodecanal	1405	1408	1.29	0.12	0.17	t	a, b
(E)–3–Pentadecen–2–ol	1428	–	23.10	9.52	1.23	17.89	a, b

(Continued)

TABLE 17.12 (Continued)

Effect of Cow Dung Amendments on the Essential Oil Composition of *C. sativum*

Name of Compounds	RI (Calculated)	RI (Adams, 2007)	% of Oil (0 t/h)	% of Oil (26 t/h)	% of Oil (52 t/h)	% of Oil (104 t/h)	Mode of Identification
(2E)–Dodecenal	1463	1464	0.31	–	0.39	0.25	a, b
Tridecen–1–al <2E>	1562	1567	0.78	0.39	–	0.01	a, b
Caryophyllene oxide	1575	1582	10.27	4.68	1.51	7.88	a, b
Myristaldehyde	1623	–	0.04	–	0.29	0.02	a, b
Trans–tetradec–2–enal	1654	–	0.01	0.22	0.33	0.01	a, b
Hexadec–(8Z)–enal<14–methyl>	1890	–	0.08	0.16	0.59	0.03	a, b
n–Hexadecanol	1880	1874	0.71	–	–	0.12	a, b
Diidobutyl phthalate	1910	–	0.29	0.44	2.91	0.20	a, b
Phytol	1950	1942	0.03	–	–	0.11	a, b
Isopropyl palmitate	2032	–	0.15	0.35	1.69	0.03	a, b
n–Nonadecanol	2160	–	0.64	0.16	–	0.06	a, b
Heptyldecanoate	2224	–	0.07	–	–	0.08	a, b
Octyllaurate	2325	–		0.26	–	0.22	a, b
Octadecanoic acid, methyl ester	2778	–	0.53	0.41	0.35	1.44	a, b
Octacosane	2801	2800	4.61	–	2.08	17.68	a, b
Octyl–Octanoate	2857	–	0.28	1.94	2.62	0.01	a, b
Myristic acid tetradecyl ester	2999	–	0.74	2.79	4.87	1.50	a, b
Myristylpalmitate	3187	–	2.25	1.21	1.77	6.66	a, b
n–Pentatriacontane	3523	3500	0.31	0.58	0.66	0.53	a, b
n.d	3535	–	0.18	0.98	1.61	0.05	a, b
Total			91.87%	88.37%	91.68%	94.03%	

TABLE 17.13
Correlation between the Growth Parameters and Major Constituents of *C. sativum* in Karula Nala Soil

	Height	Oil Yield	Fresh Plant Weight	Fresh Seeds Weight	Fenchone	Linalool	n–Decanal	2(E)–Decenal	(E)–3–pentadiene	Caryophyllene oxide	n–Octacosane	Myristyl Palmitate	pH	EC
Height	1	.751	.900	.846	.060	–.111	–.393	–.057	–.121	–.126	.598	.593	.673	.592
Oil yield		1	.797	.559	–.600	.544	–.884	–.682	–.746	–.750	.102	.132	.654	.898
Fresh plant weight			1	.946	–.208	–.040	–.608	–.132	–.271	–.290	.680	.703	.925	.826
Fresh seeds weight				1	.055	–.355	–.349	.191	.039	.016	.881	.895	.910	.641
Fenchone					1	–.874	.893	.890	.950	.958*	.424	.363	–.326	–.723
Linalool						1	–.737	–.985*	–.944	–.933	–.742	–.706	–.066	.465
n–Decanal							1	.828	.919	.930	.096	.041	–.625	–.941
(2E)–Decenal								1	.979*	.971*	.624	.585	–.082	–.595
(E)–3– Pentadiene									1	.999**	.481	.432	–.268	–.731
Caryophyllene oxide										1	.456	.405	–.297	–.750
n–Octacosane											1	.997**	.710	.247
Myristylpalmitate												1	.751	.300
pH													1	.850
EC														1

* Correlations are significant at $p < 0.05$.
** Correlations are significant at $p < 0.01$.

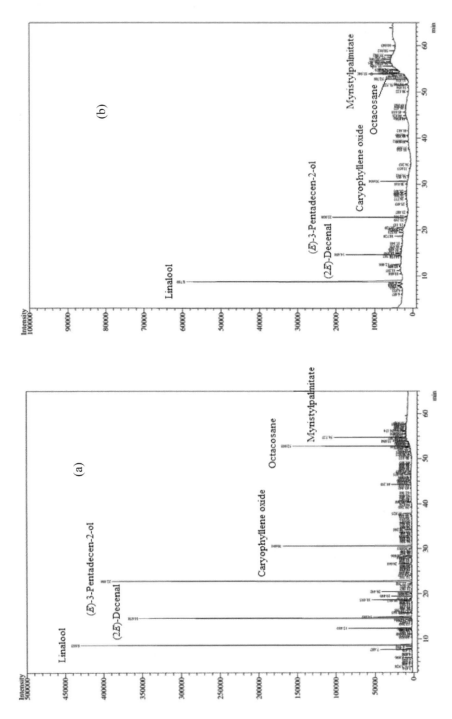

FIGURE 17.2 Gas Chromatogram of Essential Oil of *C. Sativum* Cultivated in Cow Dung Amended Karula Nala Soil (a) 0 t/h (b) 26 t/h

Effect of Manure on Metal Efficiency 295

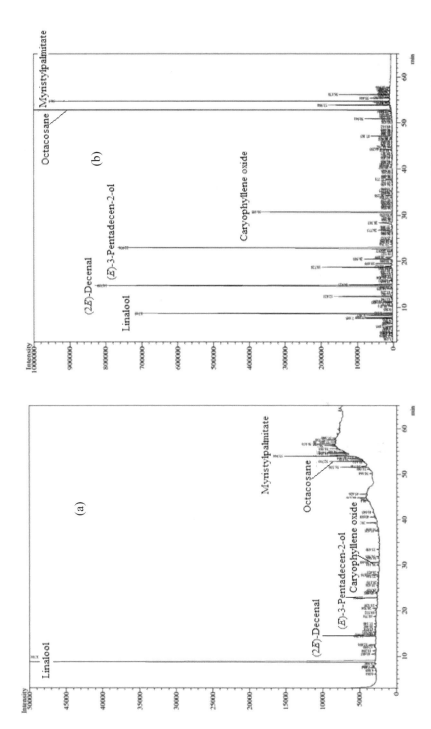

FIGURE 17.3 Gas Chromatogram of Essential Oil of *C. sativum* Cultivated in Cow Dung Amended Karula Nala Soil:(a) 52 t/h; (b) 104t/h

Furthermore, the oil yield was also observed to be positively related with the manure application as it also showed significant increase at 52t/h. Caryophyllene oxide had the highest percentage (10.27%) in control treatment and it decreased significantly in various manure amended metal contaminated soil samples. Similarly, the percentage of (2E)-Decenal (19.24% in control) and (E)-3-pentadecen-2-ol (23.10% in control) significantly decreased after the addition of manure as shown in Figure 17.4. According to Younesian et al. (2013), positive effect on one constituent simultaneously showed negative effect on the percentage of other constituent after the application of vermicompost. Previous reports suggested the effect of different manure applications on the essential oil composition of various medicinal plants (Omidbaigi, 2000; Kaplan et al., 2009). Correlations among major constituents of the essential oil of *C. sativum* are shown in Table 17.13.

17.3.7 Cluster Analysis

The essential oils of *C. sativum* from the different cow dung amendments were analyzed by cluster analysis as shown in Figure 17.5, which grouped these essential oils

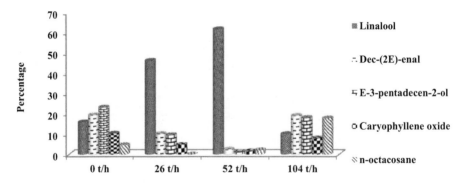

FIGURE 17.4 Variations in the Major Constituents of *C. sativum* Grown in Cow Dung Amended Karula Nala Soil

```
* * * * * * * * * * * * * * * * * * H I E R A R C H I C A L   C L U
S T E R     A N A L Y S I S * * * * * * * * * * * * * * * * * *

 Dendrogram using Ward Method

                        Rescaled Distance Cluster Combine

       C A S E      0         5        10        15        20        25
     Label     Num  +---------+---------+---------+---------+---------+

     0t/h         1  ┐
     104t/h       4  ┘
     26t/h        2
     52t/h        3
```

FIGURE 17.5 Agglomerative Hierarchical Clustering Analysis (Dendrogram) of Essential Oil Constitutes of *C. sativum* Grown in Manure Amended Karula Nala Soil

Effect of Manure on Metal Efficiency 297

into two clusters depending on the differences in their chemical constituents and allowing them to be characterized into two different groups.

Group I Essential oil from 0t/h and 104t/h (Linalool, (2E)-decanal and (E)-3- pentadecen-2-ol)
Group II Essential oil from 26t/h and 52t/h (Linalool)

17.4 CONCLUSION

From the above study, it may be concluded that when seeds of coriander were grown in Karula nala soil (metal contaminated) amended with manure, there was reduction in the metal accumulation in aerial parts as well as in roots. This reduction was best in 104t/h amendment for Zn (15.17% reduction from control contaminated soil in aerial parts), 104t/h amendment for Cu (35.39% reduction from control contaminated soil in aerial parts), 104t/h amendment for Cr (5.54% reduction from control contaminated soil in aerial parts) after the first cutting for all the three metals and 104t/h amendment for Ni (16.94% reduction from control contaminated soil in aerial parts), after the second cutting. On this basis *C. sativum* may be used in phytoremediation process.

In the present study linalool (a biologically active substance used in pharmaceutical industry, in eco-friendly pest management, in the synthesis of vitamin A and vitamin E) significantly increased with the manure addition and showed best results at 52t/h. Apart from this, the other major aliphatic compound (2E)-decanal (a biological active compound due to bactericidal activity against *Salmonella choleraesuis* ssp.), maintained its percentage with manure application along phytoremedial potential.

REFERENCES

Abdelaziz, M., Pokluda, R. and Abdelwahab, M., 2007. Influence of compost, microorganisms and NPK fertilizer upon growth, chemical composition and essential oil production of *Rosmarinus officinalis* L. *Not. Bot. Horti. Agrobo.* 35:86–90.

Adams, R. P., 2007. *Identification of Essential Oil Components by Gas Chromatography/ Mass Spectrometry*. 4th ed., Allured Publishing, Carol Stream, IL.

Alloway, B. J., Thornton, I., Smart, G. A., Sherlock, J. C. and Quinn, M. J., 1988. Metal availability. *Sci. Total Environ.* 75:41–69.

Amir, M. S., Moghaddam, P. R., Kocheki, A., Danesh, S. and Fotovat, A., 2012. Effect of cadmium and lead on quantitative and essential oil traits of peppermint (*Mentha* piperita L.). *Notul ae Scientia Biologicae.* 4:101–109.

Awasthi, S. K., 2000. *Prevention of Food Adulteration Act no 37 of 1954. Central and State Rules as Amended for 1999*, 3rd ed., Ashoka Law House, New Delhi.

Bisht, M., Pande, C., Tewari, G., Bhatt, S. and Tripathi, S., 2019. Effect of zinc on the growth and essential oil composition of Ocimum gratissimum L. *J. Essent. Oil. Bear Pl.* 22(2):441–454.

Chung, I. M., Ahmad, A., Kim, S. J., Naik, P. M. and Nagella, P., 2012. Composition of the essential oil constituents from leaves and stems of Korean *Coriandrum sativum* and their immunotoxicity activity on the *Aedes aegypti* L. *Immunopharm. Immunot.* 34(1):152–156.

Clemente, R., Walker, D. J. and Bernal, M. P., 2005. Uptake of heavy metals and As by *Brassica juncea* grown in a contaminated soil in Aznacóllar (Spain). The effect of soil amendments. *Environ. Pollut.* 138:46–58.

Courtney, R. G. and Mullen, G. J., 2008. Soil quality and barley growth as influenced by the land application of two compost types. *Bioresour. Technol.* 99(8):2913–2918.

Dembitsky, V., 2003. Natural occurrence of Arseno compounds in plants, lichens, fungi, algal species, and microorganisms. *Plant Sci.* 165:1177–1192.

Gharib, F. A., Moussa, L. A. and Massoud, O. N., 2008. Effect of compost and bio-fertilizers on growth, yield and essential oil of sweet marjoram (*Majorana hortensis*) plant. *Int. J. Agr. Biol.* 10:381–387.

Gigliotti, G., Businelli, D. and Giusquiani, P. L., 1996. Trace metals uptake and distribution in corn plants grown on a 6-year urban waste compost amended soil. *Agric. Ecosyst. Environ.* 58:199–206.

Grath, M., Zhao, S.P.F.J. and Lombi, E., 2002. Phytoremediation of metals, metalloids and radionuclides. *Adv. Agron.* 75:1–56.

Guenther, E., 1950. *The Essential Oil*, REK Publishing Company, Florida, Vol. IV. 602–615.

Herrero, E. M., Lopez-Gonzalvez, A., Ruiz, M. A., Lucas-Garcia, J. A. and Barbas, C., 2003. Uptake and distribution of zinc, cadmium, lead and copper in *Brassica napus* var. *oleifera* and *Helianthus annuus* grown in contaminated soils. *Int. J. Phytoremediation.* 5:153–167.

Jackson, M. L., 1958. *Soil Chemical Analysis*, Prentice Hall, Englewood Cliffs, NJ. 498.

Kabata-Pendias, A. and Pendias, H., 1989. The use of plants for the removal of toxic metals from contaminated soil. In Mitch, M. L.(ed.), *American Association for the Advancement of Science Environmental Science and Engineering Fellow, Trace Elements in the Soil and Plants*, CRC Press.

Kaplan, M., Kocabas, I., Sonmez, I. and Kalkan, H., 2009. The effects of different organic manure applications on the dry weight and the essential oil quantity of sage (*Salvia fruticosa* Mill). *Acta Hortic.* 826:147–152.

Keller, C., Marchetti, M., Rossi, L. and Lugan-Moulin, N., 2005. Reduction of cadmium availability to tobacco (*Nicotiana tabacum*) plants using soil amendments in low cadmium-contaminated agricultural soils: A pot experiment. *Plant Soil.* 276:69–84.

Kumar, R., Singh, M. K., Kumar, V., Verma, R. K., Kushwah, J. K. and Pal, M., 2015. Effect of nutrient supplementation through organic sources on growth, yield and quality of coriander (*Coriandrum sativum* L.). *Indian J. Agric. Res.* 49(3):278–281.

Kumar, T. S., Swaminathan, V. and Kumar, S., 2009. Influence of nitrogen, phosphorus and biofertilizers on growth, yield and essential oil constituents in ratoon crop of davana (*Artemisia pallens* Wall.). *Electronic J. Environ. Agric. Food Chem.* 8(2):86–95.

Kunwar, G., Pande, C., Tewari, G., Bhatt, S. and Tripathi, S., 2018. Phytoremedial potential of a new chemotype of *Ocimum kilimandscharicum* Guerke from Kumaun Himalaya. *J. Essent. Oil Bear. Pl.* 21(3):623–639.

Kunwar, G., Pande, C., Tewari, G., Singh, C. and Kharkwal, G., 2015. Effect of heavy metals on terpenoid composition of *Ocimum basilicum* L. and *Mentha spicata* L. *J. Essent. Oil Bear. Pl.* 18(4):818–825.

Li, Z., Ryan, J. A., Chen, J.L. and Al-Abed, S. R., 2001. Adsorption of cadmium on biosolids amended soil. *J. Environ. Qual.* 30:903–911.

Lillenberg, M., Yurchenko, S., Kipper, K., Herodes, K., Pihl, V., Lõhmus, R., Ivask, M., Kuu, A., Kutti, S., Litvin, S. V. and Nei, L., 2010. Presence of fluoroquinolones and sulfonamides in urban sewage sludge and their degradation as a result of composting. *Int. J. Environ. Sci. Technol.* 7(2):307–312.

Lindsay, W. L. and Norvell, W. A., 1969. Development of a DTPA micronutrient soil test. *Agro. Abstr.* 84.

Effect of Manure on Metal Efficiency 299

Lokeshwari, H. and Chandrappa, G.T., 2006. Impact of heavy metal contamination of Bellandur Lake on soil and cultivated vegetation. *Curr. Sci.* 91(5):622–627.

Ma, L.Q., Komar, K.M., Tu, C. and Zhang, W.A., 2001. A fern that hyperaccumulates arsenic. *Nature.* 409:579.

McGrath, S. P., Zhao, F. J. and Lombi, E., 2002. Phytoremediation of metals, metalloids and radionuclides. *Advances in Agronomy.* 75:1–56.

Mojiri, A., 2012. Phytoremediation of heavy metals from municipal wastewater by *Typha domingensis. Afr. J. Microbiol. Res.* 6(3):643–647.

Nadeem, M., Anjum, F.M., Khan, M.I., Tehseen, S., El-Ghorab, A. and Sultan, J.I., 2013. Nutritional and medicinal aspects of coriander (*Coriandrum sativum* L.): A review. *J.British Food.* 115(5):743–755.

Ogri, R.O., Ibok, U.J. and Ogbuagu, I., 1998. A study on trace metals level in soil around the calabar cement company (cacemco) Nigeria during the rainy season. *Int. J. Environ. Educ. Inform.* 17(1):91–98.

Ok, Y.S. and Kim, J.G., 2007. Enhancement of cadmium phytoextraction from contaminated soils with *Artemisia princeps* var. *orientalis. J. Appl. Sci.* 7(2):263–268.

Omidbaigi, R., 2000. *Production and Processing of Medicinal Plants,* Astan Ghods Razavi Press, Tehran, Iran. 397.

Pichtel, J. and Bradway, D.J., 2008. Conventional crops and organic amendments for Pb, Cd and Zn treatment at a severely contaminated site. *Bioresour. Technol.* 99:1242–1251.

Prasad, A., Kumar, S., Khaliq, A. and Pandey, A., 2011. Heavy metals and arbuscular mycorrhizal (AM) fungi can alter the yield and chemical composition of volatile oil of sweet basil (*Ocimum basilicum* L.). *Biol. Fert. Soils.* 47(8):853–861.

Punetha, D., Pande, C., Tewari, G. and Tewari, K., 2014. Qualitative variation in *Coriandrum sativum* L. seed oil with manure applications. *Indian Perfumer.* 58(2):27–31.

Punetha, D., Tewari, G. and Pande, C., 2018. Compositional variability in inflorescence essential oil of *Coriandrum sativum* from North India. *J. Essent. Oil Res.* 30:113–119.

Scora, R.W. and Chang, A.C., 1997. Essential oil quality and heavy metal concentrations of peppermint grown on a municipal sludge amended soils. *J. Environ. Qual.* 26:975–979.

Subrahmanyam, K., Chattopadhyay, A.K. and Singh, D.V., 1992. Yield response and iron status of Japanese mint as influenced by soil and foliar applied iron. *Fert. Res.* 31:1–4.

Swaminathan, V., Kumar, T.S., Sadasakthi, A. and Balasubramanian, R., 2008. Effect of nitrogen and phosphorus along with biofertilizers on growth, yield and physiological characteristics of Davana (*Artemisia pallens* Wall.). *Adv. Plant Sci.* 21(2):693–695.

Tlustoš, P., Balík, J., Pavlíková, D. and Száková, J., 1997. Uptake of cadmium, zinc, arsenic and lead by chosen plants. *Rost. Výroba.* 43:487–494.

Topalov, V.D., 1962. *Essential Oil and Medicinal Plants,* Hr. G. Danov Press, Plovdiv, Bulgaria.

Vaca-Paulín, R., Esteller-Alberich, M. V., Lugo-dela, F. J. and Zavaleta-Mancera, H. A., 2006. Effect of sewage sludge or compost on the sorption and distribution of copper and cadmium in soil. *Waste Manag.* 26:73–81.

Walker, D.J., Clemente, R., Roig, A. and Bernal, M.P., 2003. The effect of soil amendments on heavy metal bioavailability in two contaminated Mediterranean soils. *Environ. Pollut.* 122(2):303–312.

Younesian, A., Taheri, S. and Moghaddam, P.R., 2013. The effect of organic and biological fertilizers on essential oil content of *Foeniculum vulgare* Mill (Sweet fennel). *Int. J. Agric Crop Sci.* 5(18):2141–2146.

Zheljazkov, V.D.and Nielsen, N.E., 1996a. Studies on the effect of heavy metals (Cd, Pb, Cu, Mn, Zn and Fe) upon the growth, productivity and quality of lavender (*Lavandula angustifolia* Mill) production. *J. Essent. Oil Res.* 8:259–274.

Zheljazkov, V. D. and Nielsen, N. E., 1996b. Effect of heavy metals on peppermint and cornmint. *Plant Soil*. 178:59–66.

Zheljazkov, V. D., Craker, L. E., Baoshan, X., Nielsen, N. E. and Wilcox, A., 2008. Aromatic plant production on metal contaminated soils. *Sci. Total Environ*. 395:51–62.

Zheljazkov, V. D. and Warman, P. R., 2003. Application of high Cu compost to Swiss chard and basil. *Sci. Total Environ*. 302:13–12.

18 Fly Ash
A Potential By-Product Waste

Bhavana Sethi and Saurabh Ahalawat

CONTENTS

18.1 Introduction 301
18.2 Characterization of Fly Ash 302
    18.2.1 Physical Characteristics..303
    18.2.2 Chemical Composition 303
    18.2.3 Mineralogical Characteristics of Fly Ash..304
18.3 Utilization of Fly Ash...305
    18.3.1 Utilization of Fly Ash in Construction Materials.............................305
        18.3.1.1 Fly Ash in Portland Cement Concrete...............................306
        18.3.1.2 Fly Ash in Construction of Roads......................................306
        18.3.1.3 Fly Ash Bricks 307
        18.3.1.4 In Lightweight Aggregates..308
    18.3.2 Utilization of Fly Ash in Agriculture 308
        18.3.2.1 In Improving Soil Properties 308
        18.3.2.2 Fertilizer Value of Fly Ash–Based Composts 309
    18.3.3 Fly Ash in Recovery of Valuable Minerals 309
        18.3.3.1 Alumina Recovery from Fly Ash 309
        18.3.3.2 Silica Recovery from Fly Ash..310
    18.3.4 Fly Ash as Low-Cost Adsorbent...310
    18.3.5 Fly Ash–Based Polymer Composites as Wood Substitute...............310
    18.3.6 Utilization of Fly Ash in Development of Protective Coatings........310
    18.3.7 Fly Ash Utilization in Making Ceramic Tiles..................................311
    18.3.8 Fly Ash as a Reinforcement Filler..311
    18.3.9 Fly Ash–Based Geopolymer...311
    18.3.10 Fly Ash in Synthesis of Zeolites...312
18.4 Conclusion 312
References..313

18.1 INTRODUCTION

One of the biggest challenges of the present century is to tackle the problem of rapid industrialization and urbanization. Though these are the need of the present society and are mostly inevitable, these have led to increasing environmental degradation and have negative impacts on both environment as well as social life. The major

301

negative effect of these global processes is the production of enormous quantities of industrial wastes and the problems related with their safe management and healthy disposal. This problem has become one of the primary urban environmental issues. The process of disposal and safe management of the industrial waste is difficult and complex.

One of such major industrial waste around the world is fly ash, which is predominantly a by-product of coal combustion in thermal power plants. During combustion of coal, the volatile matter and carbon is burnt off and the coal impurities such as clays, shale, quartz and feldspar mostly fuse and remain in suspension (Mehta, 1993). These fused particles are passed along with flue gas. As the flue gas approaches the low temperature zone, the fused substances coagulate to form mainly spherical particles which are called fly ash. The remaining matter which agglomerate and settle down at the bottom of furnace are called bottom ash. The distribution of bottom ash and fly ash is 20% and 80%, respectively. Fly ash is captured by mechanical separators, electrostatic precipitators or bag filters. It is a mixture of particles varying in shape, size and composition. ASTM C-618 categorizes coal combustion fly ash into two classes: Class F fly ash and Class C fly ash (ASTM, 2012). The major difference between these classes is the amount of calcium, silica, alumina and iron content in the ash. The Class F fly ash is normally generated due to combustion of anthracite or bituminous coal and Class C is obtained by the burning of lignite or sub-bituminous coal combustion. Class C fly ash possesses CaO in excess (10%–40%) while Class F contains CaO less than 10%. Due to higher CaO content, Class C fly ash participate both in cementitious and pozzolanic reaction, whereas Class F fly ashes mostly contribute in pozzolanic reaction during hydration process (Surabhi, 2017).

Fly ash particles generally range in size from 1 to 150 μm (Berry et al., 1989). These are very fine particles and readily airborne even in mild wind and thus pollute the environment. Long term inhalation of fly ash causes bronchitis, silicosis, fibrosis of lungs, pneumonitis etc. in human beings. If not managed properly, fly ash disposal in water bodies can cause damage to aquatic life also. It can also contaminate the underground water resources with traces of toxic metals present in it. Flying fine particles of ash poses problems for people living near power stations, corrodes structural surfaces and affects horticulture. Eventual settlement of fly ash particles over many hectares of land in the vicinity of a power station brings about observable degeneration in soil characteristics (Upadhyay and Kamal, 2007). In addition to the serious environmental impacts, huge quantities of fly ash generation can invite substantial waste management costs and can also be a long-term liability for energy companies. Thus, alternative methods of management of fly ash waste (disposal, reuse and/or recovery) have been a growing area of research in past few decades.

This chapter emphasizes the different scope of utilization of industrial waste fly ash. Several key aspects governing the performance of fly ash in utilization, including physical, chemical and mineralogical characterization, have also been included.

18.2 CHARACTERIZATION OF FLY ASH

Characterization of fly ash is important because it assists in understanding the utilization and disposal of fly ash and problems associated with atmospheric dispersal of the particulate matter (Ramsden and Shiboaka, 1982).

18.2.1 Physical Characteristics

1. *Size and shape of fly ash*: The size of particles varies depending on the source of fly ash. Fly ash consists of silt-sized particles which are generally spherical, typically ranging in size between 10 to 100 microns. These small glass spheres help in improving the fluidity and workability of fresh concrete.
2. *Colour*: The colour of fly ash varies from tan to dark grey, depending on its chemical and mineral constituents. Tan and light colours indicate high lime content. A brownish colour is typically associated with the iron content. A dark gray to black colour typically represents the presence of higher quantity of unburned carbon content.
3. *Fineness*: The fineness of fly ash is most closely related to the operating condition of the coal crushers and the grindability of the coal itself. For fly ash use in concrete applications, fineness is defined as the percent by weight of the material retained on the 0.044 mm (No. 325) sieve. A coarser gradation can result in a less reactive ash and could contain higher carbon contents. Limits on fineness are addressed by ASTM and state transportation department specifications. Fly ash can be processed by screening or air classification to improve its fineness and reactivity. Some non-concrete applications, such as structural fills, are not affected by fly ash fineness. However, other applications such as asphalt filler are greatly dependent on the fly ash fineness and its particle size distribution (FHWA, 2003).
4. *Grain size distribution*: It indicates if a material is well graded, poorly graded, fine or coarse etc. and also helps in classifying the coal ash. Leonard and Bailey have reported the range of gradation for fly and bottom ashes which can be classified as silty sands or sandy silts (Leonard and Bailey, 1982). Figure 18.1 represents a SEM micrograph of coarse, medium and fine fly ash sample (Sethi, 2007).
5. *Specific gravity*: It is one of the important physical properties needed for the use of coal ashes for geotechnical and other applications. In general, the specific gravity of coal ashes is around 2.0 but can vary to a large extent (1.6 to 3.1). Because of the generally low value for the specific gravity of coal ashes compared to soils, ash fills tend to result in low dry densities. The reduction in unit weight is an advantage in the case of its use as a backfill material for retaining walls, since the pressure exerted on the retaining structure as well as the foundation structure will be less. The other application areas include embankments especially on weak foundation soils, reclamation of low-lying areas etc.

18.2.2 Chemical Composition

Chemically, fly ash is considered a mixture of ferro-aluminosilicate minerals. However, composition of fly ash depends mainly upon the geological factors related to coal deposits, combustion conditions and the type of devices used to collect ash. The major constituents of fly ash are primarily oxides of Si, Al, Fe, Ca, and Mg which constitute about 95%–99% of the total. Minor constituents of fly ash are Ti,

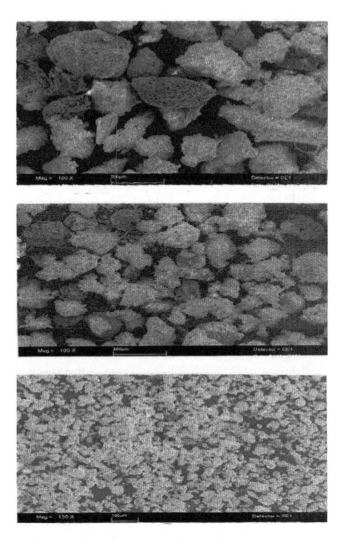

FIGURE 18.1 SEM Micrograph of (a) Coarse Fly Ash Sample; (b) Medium Fly Ash Sample; (c) Fine Fly Ash Sample

Na, K, and S, at 0.5%–3.5% of the total (Nawaz, 2013). The chemical properties of the coal ashes greatly influence the environmental impacts that may arise out of their use/disposal as well as their engineering properties. The adverse impacts include contamination of surface and subsurface water with toxic heavy metals present in the coal ashes, loss of soil fertility around the plant sites etc.

18.2.3 Mineralogical Characteristics of Fly Ash

The mineralogical composition of any fly ash sample is influenced by type and source. Due to the rapid cooling of burned coal in the power plant, fly ashes consist

TABLE 18.1
Qualitative Mineralogical Analysis of a Typical Fly Ash Sample (Sethi, 2007)
Fly Ash from Parricha Thermal Power Station, Parricha, Jhansi (U.P.)

i) *Microcrystalline*	
Quartz	Predominant
Cristobalite	Moderate
Felspar	Low
Augite	Traces
ii) *Amorphous*	Traces

of non-crystalline particles ($\leq 90\%$), or glass and a small amount of crystalline material. Depending on the system of burning, some unburned coal may be collected with ash particles. In addition to a substantial amount of glassy material, each fly ash may contain one or more of the major crystalline phases: quartz, cristobalite, mullite, magnetite and hematite. In sub-bituminous fly ash, the crystalline phases may include C3A, C4A3S, calcium sulphate and alkali sulphates. The reactivity of fly ash is related to the non crystalline phase or glass. The reasons for the high reactivity of high-calcium fly ash may partially lie in the chemical composition of the glass. The composition of glass in low-calcium fly ash are different from that in high-calcium fly ash. X-ray diffraction studies are carried out primarily to identify the mineral phases. A typical XRD profile mineralogical composition is shown in Table 18.1. The data in the table indicates the presence of quartz and cristobalite, as major micro-crystalline phases.

18.3 UTILIZATION OF FLY ASH

In view of the plethora of environmental problems caused by industrial waste fly ash, substantial research has been undertaken on the subject globally to ensure its potential utilization. Fly ash has already established itself in the construction industry. It is now successfully used in manufacture of cement as a filler material, concrete, ceramics, construction fills, road base and a mineral filler in asphalt mix. Another promising area of disposal of this waste is the filling of empty mines and land filling. It could possibly be used in agriculture land as a soil stabilizer. It has also been explored for the recovery of alumina and silica. Other miscellaneous applications include its use as an adsorbent, as a wood substitute, in reinforcement fillers, in synthesis of protective coatings, in the tile industry, in geopolymer technology and in synthesis of zeolites. Various possibilities for its use are under research and some of them are discussed in the following section.

18.3.1 UTILIZATION OF FLY ASH IN CONSTRUCTION MATERIALS

The reuse of fly ash as a building material primarily comes from its pozzolanic behaviour. When mixed with calcium hydroxide, it has many of the same properties as cement. Replacing a portion of the cement with fly ash creates a cementitious material that, when used as an input with aggregates, water and other compounds, produces a concrete mix that is well-suited to road, airport runway and bridge construction (Motz and Geiseler, 2001). Some potential areas of fly ash utilization in building construction are given below.

18.3.1.1 Fly Ash in Portland Cement Concrete

Fly ash is used in concrete admixtures to enhance the performance of concrete. Portland cement contains about 65% lime. Some of this lime becomes free and available during the hydration process. This liberated lime forms the necessary ingredient for reaction with fly ash silicates to form strong and durable cementing compounds, thus improving many of the properties of the concrete.

Merits to fresh concrete: Generally, fly ash benefits fresh concrete by reducing the mixing water requirement and improving the paste flow behaviour. The resulting benefits are as follows (Upadhyay and Kamal, 2007):

- Improved workability
- Decreased water demand
- Reduced heat of hydration.

Merits to hardened concrete: One of the primary benefits of fly ash is its reaction with available lime and alkali in concrete, producing additional cementitious compounds.

- Increased ultimate strength
- Reduced permeability
- Improved durability.

18.3.1.2 Fly Ash in Construction of Roads

The demand for road construction is increasing with increasing population and lack of mineral resources. Hence road construction is becoming very difficult and expensive. Judicious use of low-cost materials helps to utilize the available natural resources more efficiently. Mineral wastes or by-products such as fly ash produced from thermal power plants provide a great potential as a low-cost mineral resource for construction materials (Baykal et al., 2004). In the past few decades, fly ash has established itself as a useful material in road construction.

18.3.1.2.1 Fly Ash in Structural Fills/Embankments

Fly ash can be used as a material to construct fills and embankments. When fly ash is compacted in lifts, a structural fill is constructed that is capable of supporting highway buildings or other structures. Fly ash has been used in the construction of structural fills/embankments that range from small fills for road shoulders to large fills for interstate highway embankments. One of the most significant characteristics of fly ash in its use as a fill material is its strength. Well-compacted fly ash has strength comparable to or greater than soils normally used in earth fill operations. In addition, fly ash possesses self-hardening properties which can result in the development of shear strengths. The addition of cement can induce hardening in bituminous fly ash which may not self-harden alone. Significant increases in shear strength can

Fly Ash 307

be realized in relatively short periods of time and it can be very useful in the design of embankments.

18.3.1.2.2 Fly Ash in Stabilized Base Course

Course stabilization is a method of improving the road structure in which the subbase or the base course, or the upper part of the base course is bound using bitumen, cement or blast furnace slag or a combination of these. Stabilization is carried out both when constructing new roads and renovating old roads. Typically, base course stabilization is used for improving the structure of an old road, and the method utilizes existing material from the old road. Fly ash when combined with lime and aggregate can be used to produce a quality stabilized base course. These road bases are referred to as pozzolanic-stabilized mixtures (PSMs). Typical fly ash contents may vary from 12% to 14% with corresponding lime contents of 3% to 5%. Portland cement may also be used in lieu of lime to increase early age strengths. The resulting material is produced, placed and looks like cement-stabilized aggregate base (Upadhyay and Kamal, 2007).

18.3.1.3 Fly Ash Bricks

Bricks are the foundation in construction industry and their use in this industry is inevitable. These have been mainly produced from clay and shale for decades. Brick manufacturing requires continuous extraction of clay and the removal of the topsoil, which is giving rise to substantial depletion of virgin resources (Naganathan et al., 2015). In the manufacture of bricks, fly ash can be an alternative to clay. Fly ash can be used either with clay as part replacement or in combination with other materials like sand, lime, gypsum etc. to produce a substitute to conventional clay bricks. The technologies for the manufacture of fly ash bricks can broadly be classified into the following: clay–fly ash bricks, red mud–fly ash bricks, sand fly ash bricks, fly ash–lime bricks, fly ash–lime/gypsum bricks. A general comparison of properties of clay bricks and fly ash bricks is given in Table 18.2.

TABLE 18.2
Properties: Clay Bricks vs. Fly Ash Bricks

Properties	Clay Bricks	Fly Ash Bricks
Colour	Differs from burnt red to light brown depending on the type of clay used for manufacture of the bricks	Cement colour
Size and shape	Less uniform	Highly uniform in size and shape
Plastering requirement	Required to create a smooth surface	Have a very smooth finish hence plaster is not required
Porosity	More Porous	Less porous
Weight	Heavier	Lighter
Resistance to salinity and water seepage	Less resistant	More resistant
Requirement of water in construction	More	Less

18.3.1.4 In Lightweight Aggregates

Lightweight aggregates are used as a filler material in the production of concrete products such as blocks and panels. They provide volume and hardness. These aggregates provide the reduction in weight, which results in a variety of implications. The concrete made with these aggregates offers the same strength at a lesser weight, the use of such concrete makes longer and slimmer spans of structural concrete possible. The overall engineering design of the structure can thus improve and can even yield structures that would otherwise not be possible. In addition to some potential benefits to the concrete itself, the use of fly ash as a lightweight aggregate in place of natural aggregate offers a wealth of benefits both economically and environmentally. The use of fly ash as a lightweight aggregate reduces the need for companies to mine virgin aggregate materials for use in concrete production. The use of concrete made with these aggregates also helps to reduce structural costs; lighter weight structural elements such as beams and footings can be used, cutting down on the cost of materials. The beneficial reuse of fly ash reduces disposal costs for producers also (Kayali, 2008).

18.3.2 Utilization of Fly Ash in Agriculture

Consumption of fly ash in agriculture provides a possible alternative for its safe disposal without serious ill effects. Fly ash improves the physical, chemical and biological properties of soils. Also, when it is used along with biosolids, it can serve as a source of readily available plant macro- and micronutrients. Fly ash varies widely in its physical and chemical composition, therefore, the mode of use in agriculture depends on the characteristics of soil or soil type. The major areas of application of fly ash in agriculture include the following.

18.3.2.1 In Improving Soil Properties

Fly ash can be used as a potential nutrient supplement for degraded or marginal soils. However, the bioaccumulation of toxic heavy metals and their critical levels for human health in plant parts and soil needs to be investigated before its application. The final objective would thus be to utilize fly ash in degraded/marginal soils to such an extent as to achieve enhanced fertility without affecting the soil quality and minimizing the accumulation of toxic metals in plants below critical levels for human health. There are several potential beneficial effects of fly ash application in soil.

Beneficial Effects of Fly Ash Application on Soil (Kishor et al., 2010):

- Improvement in soil texture
- Reduction in bulk density of soil
- Improvement in water holding capacity
- Optimization of pH value
- Increase in soil buffering capacity
- Improvement in soil aeration, percolation and water retention in the treated zone (due to dominance of silt-size particles in fly ash)
- Reduction in crust formation
- Source of micronutrients like Fe, Zn, Cu, Mo, B etc.

Fly Ash 309

- Source of macronutrients like K, P, Ca etc.
- Reduction in the consumption of soil ameliorants (fertilizers, lime).

In spite of several beneficial effects of fly ash on soil, investigations should be done before its use, as it might result in a few negative effects:

- Reduction in bioavailability of some nutrients due to high pH (generally from 8 to 12)
- High content of phytotoxic elements, especially boron.

18.3.2.2 Fertilizer Value of Fly Ash–Based Composts

Composting is a widely acceptable alternative for converting waste into a more useful eco-friendly fertilizer known to improve soil fertility. Sewage sludge, cow dung, paper and food waste are the organic substrates that are most commonly tested as sources of C and N in fly ash composting. In this case, decomposition rate, and hence nutrient release, is strongly influenced by the fly ash: organic waste mixing ratio, as well as the C:N ratio of the organic waste. The fly ash composts show great potential to supply the major elements, especially P, in crop production. A major drawback to biological decomposition of fly ash appears to be the reduction of microbial activity, population and diversity. Earthworms and special microbial cocktails are a potential solution to this problem. Research is required to identify the microbes that tolerate high concentrations of fly ash modification during composting. Fly ash composting appears viable mostly at low incorporation rates ranging from 5% to 25%; and at these low application rates, the heavy metals emanating from fly ash composting may not be a serious challenge. However, repeated applications of fly ash composts to the soil over time may increase the heavy metal load to toxic levels (Mupambwa et al., 2015).

18.3.3 Fly Ash in Recovery of Valuable Minerals

A wide range of opportunity exists for the beneficial reuse and/or recovery of materials from fly ash.

18.3.3.1 Alumina Recovery from Fly Ash

Despite of the high demand for alumina, bauxite is currently the only commercially viable source of alumina. Researchers have been working to find an alternative, as the Bayer process yields staggering amounts of red mud—another toxic industrial waste that poses a number of problems(Shemi et al., 2015). Fly ash is considered as a potential source of aluminium and other strategic metals. Most fly ash contains between 25% and 30% aluminium oxide (Al_2O_3). Different extraction processes of alumina from fly ash are studied by researchers; mainly being limestone sintering process (Zhang and Zhou, 2007; Matjie et al., 2005), soda-lime sintering process (Tang and Chen, 2008; Liu et al., 2006) and acid leaching process (Nayak and Panda, 2010; Thamilselvi and Balamurugan, 2018). However, challenges to commercial viability of these processes still exist, and research is still ongoing in this direction.

18.3.3.2 Silica Recovery from Fly Ash

Fly ash has also been used for extraction of silica. Out of the numerous research work done on recovery of silica from fly ash, alkaline leaching seems to be the best method to extract the largest amounts of Si from fly ash. A group of researchers have determined that Si dissolution was higher when NaOH in used as the leaching reagent (Panagiotopoulou et al., 2007). Extraction of Si from fly ash in the form of sodium silicates is also advantageous since sodium silicate itself has numerous uses.

In addition to alumina and silica, fly ash is also being studied as a potential alternative source to recover materials like iron, titanium, silicon, vanadium, gallium, germanium, selenium, lithium, molybdenum, uranium, gold, silver, platinum, rare earth elements and magnetic materials (Shemi et al., 2015).

18.3.4 Fly Ash as Low-Cost Adsorbent

Adsorbents are widely used in applications to remove targeted components (often contaminants) from liquids and gases. The rate of adsorption depends upon surface area, pore size and composition of adsorbent. Studies have revealed that fly ash possesses adsorption capacity for removal of gaseous pollutants in air, inorganic ions and organic compounds in water. Comparison with other adsorbents, fly ash could be an effective adsorbent depending on the compositions and treatment. To improve removal efficiencies and adsorption capacities, chemical modifications of fly ash needs to be conducted. The unburned carbon in fly ash also plays an important role for adsorption. The unburned carbon can be converted to activated carbon, which will enhance the adsorption capacity (Wang and Wu, 2006). Recently, fly ash composites are also employed efficiently as adsorbents. Swarnima et al. have worked on the tannic acid adsorption efficiencies for fly ash (FA), NaOH treated fly ash (NaFA) and chitosan-modified NaOH treated fly ash (chitosan/NaOH/fly ash composite) (Agarwal et al., 2018).

18.3.5 Fly Ash–Based Polymer Composites as Wood Substitute

Fly ash–based polymer composites can be utilized as a wood substitute. For manufacturing wood substitute composites, processed fly ash is mixed with polymer and catalyst and synthesized using natural fibre reinforcement in moulds of required dimensions. The fly ash–based polymer composite products exhibited better tensile, flexural and impact strength. Fly ash composites are weather and corrosion resistant, termite, fungus, rot and rodent resistant and fire retardant. This durable and abrasive resistant fly ash composite is stronger than wood and could be used as a substitute for timber. The timber substitute composite is cost-effective and maintenance-free and has wider applications in construction industry for use as doors, windows, ceilings, flooring, partitions, furniture etc. (Asokan et al., 2012).

18.3.6 Utilization of Fly Ash in Development of Protective Coatings

Fly ash is rich in metal oxides; it has tremendous potential as a coating material on structural and engineering components. It has been gainfully used as a potential

Fly Ash 311

cost-effective material for deposition of plasma spray coatings on metallic substrates. Premixing of aluminium powder with fly ash can produce metal-ceramic composite coatings of improved interfacial adhesion (Satapathy et al., 2009). Recently, the development of protective and decorative industrial coatings as solvent borne, cold curing, two-pack epoxy systems, using fly ash as an extender have been reported. The properties of fly ash that contribute to its usefulness as an extender are its abrasion resistance, chemical inertness, low oil absorption, and low specific gravity (Tiwari and Saxena, 1999).

18.3.7 Fly Ash Utilization in Making Ceramic Tiles

Researchers have developed a process that allows utilization of fly ash with moderately high carbon content in manufacturing ceramic tiles. Owing to its chemical composition and physical characteristics, fly ash can be used as a partial replacement for clay in the ceramic tile industry. The composition of fly ash is such that it can be used as a major raw ingredient for making wall and floor tiles. In fly ash, the rounded particles with a wide distribution in size are predominantly glassy. The glass content in fly ash is generally higher than 70%(Helmuth, 1987). The vitreous (glassy) phase plays an important role in sintering process (Kingery et al., 1976). The fine powder characteristic of fly ash makes them usable in same form as received without any further size reduction. Therefore, utilization of fly ash for making ceramic tiles is very beneficial (Mishulovich and Evanko, 2003).

18.3.8 Fly Ash as a Reinforcement Filler

Fly ash can be used as filler in rubber products. The main composition of the fly ash is silica. Similar to carbon black and precipitated silica, fly ash can also act as a reinforcing filler to improve the mechanical properties of rubber compounds. Palm-based fly ash can be used to improve the mechanical properties of thermoplastic materials (Ahmad et al., 2012). As the content of filler is relatively small in the rubber compounds, the reinforcing effect of fly ash without modification was similar to the reinforcing effect of silica (Sombatsompop et al., 2004). Silane coupling agents have also been used to modify the surface of fly ash. Their results showed that silane-treated fly ash would improve cure characteristics and mechanical properties of rubber compounds (Thongsang and Sombatsompop, 2006). Fly ash was also studied as an alternative to carbon black. Nano-sized fly ash (148 nm) exhibits superior reinforcing effect in comparison to carbon black, precipitated silica or unmilled fly ash in SBR composites as revealed by higher modulus, tensile strength and hardness (Paul et al., 2009).

18.3.9 Fly Ash–Based Geopolymer

The geopolymer technology offers an alternative solution to the utilization of fly ash without impacting the environment. The production of fly ash–based geopolymer is mainly based on alkali activated geopolymerization, which can occur under mild conditions and is considered as a cleaner process due to much lower CO_2 emissions than that from the production of cement. The geopolymer so produced can trap and fix the trace toxic metal elements from fly ash or external sources. The critical factors in a geopolymerization process are Si/Al ratios, the type and the amount of the alkali solution, the temperature, the curing conditions and the additives used. The properties of fly ash–based geopolymer including compressive strength, flexural and splitting tensile strength, and durability such as the resistance to chloride, sulphate, acid, thermal, freeze-thaw and efflorescence are inherently dependent upon the chemical composition, chemical bonding and the porosity. The mechanical properties and durability can be improved by fine tuning Si/Al ratios, alkali solutions, curing conditions, and adding slag, fibre, rice husk-bark ash and red mud. Fly ash–based geopolymer is expected to be used as a kind of novel green cement. It can also be used as a class of materials to adsorb and immobilize toxic or radioactive metals (Zhuang et al., 2016).

18.3.10 FLY ASH IN SYNTHESIS OF ZEOLITES

The potential of fly ash as a raw material for synthesis of microporous aluminosiliceous minerals, zeolites is studied by different researchers. The fly ash–based zeolites provide an opportunity for effective as well as value added management of the fly ash. Due to their excellent ion exchange capacity, high surface area and unique pore characteristics, zeolites have been used for removal of heavy metals (As, Cd, Cr, Cs, Cu, Fe, Hg, Mn, Ni, Pb, Sr, W and Zn) and ionic species (ammonium, chloride, fluoride, nitrate, phosphate and sulphate) from industrial sludges, acid mine drainage and wastewater from domestic and industrial sources. In addition, fly ash zeolites find their application as sorbent medium in permeable reactive barriers and contaminant barrier liners for immobilizing the contaminant plume in soil (Koshy and Singh, 2016). The processes of synthesizing zeolites from fly ash include different methodologies, which primarily indicate modifications or alterations in pretreatment steps, alkali addition etc. It results in the formation of different structures that have potential for application in many fields (ion exchange capacity). The synthesis of zeolites from coal fly ash with specific focus such as applications in water treatment acts as a more value-added high-technology utilization of fly ash (Shaila et al., 2015).

18.4 CONCLUSION

In view of the plethora of environmental problems caused by the industrial waste fly ash, substantial research has been undertaken on the subject globally to ensure its potential utilization. Fly ash has already well established itself in the construction industry. It is now successfully used in manufacture of cement as a filler material, concrete, ceramics, construction fills, road base and mineral filler in asphalt mix. It also finds application in the filling of empty mines and land filling. It could possibly be used in agriculture land as a soil stabilizer. However, its utilization in agriculture is a growing research area and needs more exploration because of the presence of heavy metals and radioactive elements. It has also been explored for the recovery of alumina and silica. Other miscellaneous applications include its use as an adsorbent,

Fly Ash 313

as a wood substitute, in reinforcement fillers, in synthesis of protective coatings, in the tile industry, in geopolymer technology and in synthesis of zeolites. The utilization of fly ash is beneficial not only in terms of energy savings but also for the environment. Thus, with a wide variety of reuse and recovery applications of fly ash, research is still ongoing to improve sustainability and reduce the risks associated with utilization of fly ash.

REFERENCES

Agarwal, S., Rajoria, P. and Rani, A., 2018. Adsorption of tannic acid from aqueous solution onto chitosan/NaOH/fly ash composites: Equilibrium, kinetics, thermodynamics and modeling. *Journal of Environmental Chemical Engineering.* 6(1):1486–1499. https://doi.org/10.1016/j.jece.2017.11.075.

Asokan, P., Saxena, M. and Morchhale, R. K., 2012. Fly ash—a promising engineering material—the technologies developed by CSIR AMPRI Bhopal. In NML (ed.), *Proceeding: Industry Meet on Fly Ash Based Products/Technologies*, CSIR-National Metallurgical Laboratory and Centre for Fly Ash Research and Management, New Delhi. pp. 35–49.

ASTM International, ASTM C618–12a, 2012. *Standard Specification for Coal Fly Ash and Raw or Calcined Natural Pozzolan for Use in Concrete*, ASTM International, West Conshohocken, PA. https://doi.org/10.1520/c0618-12.

Bahruddin Ahmad, A., Prayitno, A. and Satoto, R., 2012. Morphology and mechanical properties of palm-based fly ash reinforced dynamically vulcanized natural rubber/polypropylene blends. *Procedia Chemistry.* 4:146–153. https://doi.org/10.1016/j.proche.2012.06.021.

Baykal, G., Edinciler, A. and Saygili, A., 2004. Highway embankment construction using fly ash in cold regions. *Resources, Conservation and Recycling.* 42:209–222. https://doi.org/10.1016/j.resconrec.2004.04.002.

Berry, E. E., Hemmings, R. T., Langley, W. S. and Carette, G. G., 1989. Beneficiated fly ash: Hydration, microstructure, and strength development in Portland cement systems. In Malhotra, V. M. (ed.), *Proceedings of the Third International Conference on the Use of Fly Ash, Silica Fume, Slag, and Natural Pozzolans in Concrete, ACI Special Publication* SP-114, Trondheim, Norway. 241–273.

Federal Highway Administration (FHWA), June 2003. *Fly Ash Facts for Highway Engineers,* 4th ed., FHWA-IF-03-019.

Helmuth, R., 1987. Fly ash in cement and concrete, Portland cement association. *Journal of Cement and Concrete Research.* 30:201–204.

Kayali, O., 2008. Fly ash lightweight aggregates in high performance concrete. *Construction and Building Materials.* 22(12):2393–2399. https://doi.org/10.1016/j.conbuildmat.2007.09.001.

Kingery, W. D, Uhlmann, D. R. and Kent, B. H., 1976. *Introduction to Ceramics*, 2nd ed., Wiley, New York.

Kishor, P., Ghosh, A. K. and Kumar, D., 2010. Use of flyash in agriculture: A way to improve soil fertility and its productivity. *Asian Journal of Agricultural Research.* 4:1–14. https://doi.org/10.3923/ajar.2010.1.14.

Koshy, N. and Singh, D. N., 2016. Fly ash zeolites for water treatment applications. *Journal of Environmental Chemical Engineering.* 4(2):1460–1472. https://doi.org/10.1016/j.jece.2016.02.002.

Leonard, G. A. and Bailey, B., 1982. Pulverized coal ash as structural fill. *Journal of Soil Mechanics and Foundation Engg. Division, ASCE.* 108(GT4):517–531.

Liu, Y. Y., Li, L. S. and Wu, Y., 2006. Further utilization of fly ash-extracting alumina. *Journal of Light Metals (in Chinese).* 5:20–23.

Matjie, R.H., Bunt, J.R. and Van, H., 2005. Extraction of alumina from coal fly ash generated from a selected low rank bituminous South African coal. *Minerals Engineering.* 18(3):299–310. https://doi.org/10.1016/j.mineng.2004.06.013.

Mehta, P.K., 1993. Pozzolanic and cementitious by-products as mineral admixtures for concrete—a critical review. In *Proceedings in First International Conference on the Use of Fly Ash, Silica Fume, Slag and Other Mineral by-products in Concrete*, ACI SP-79, American Concrete Institute, Detroit. 1–48.

Mishulovich, A. and Evanko, J.L., 2003. Ceramic tiles from high-carbon fly ash. International Ash utilization symposium center for applied energy research, University of Kentucky, paper 18.

Motz, H. and Geiseler, J., 2001. Products of steel slags an opportunity to save natural resources. *Waste Management.* 21:285–293. https://doi.org/10.1016/s0956-053x(00)00102-1.

Mupambwa, H.A., Dube, E. and Mnkeni, P.N.S., 2015. Fly ash composting to improve fertiliser value—a review. *South African Journal of Science.* 111(7/8):1–6. https://doi.org/10.17159/sajs.2015/20140103.

Naganathan, S., Mohamed, A.Y.O. and Mustapha, K.N., 2015. Performance of bricks made using fly ash and bottom ash. *Construction and Building Materials.* 96:576–580. https://doi.org/10.1016/j.conbuildmat.2015.08.068.

Nawaz, I., 2013. Disposal and utilization of fly ash to protect the environment. *International Journal of Innovative Research in Science, Engineering and Technology.* 10(12):5259–5266.

Nayak, N. and Panda, C.R., 2010. Aluminium extraction and leaching characteristics of Talcher thermal power station fly ash with sulphuric acid. *Fuel.* 89(1):53–58. https://doi.org/10.1016/j.fuel.2009.07.019.

Panagiotopoulou, C.H., Kontori, E., Perraki, T.H. and Kakali, G., 2007. Dissolution of aluminosilicate minerals and by-Products in alkaline media. *Journal of Materials Sciences.* 42:2967–2973. https://doi.org/10.1007/s10853-006-0531-8

Paul, K.T., Pabi, S.K., Chakraborty, K.K. and Nando, G.B., 2009. Nanostructured fly ash-styrene butadiene rubber hybrid nanocomposites. *Polymer Composites.* 30:1647–1656. https://doi.org/10.1002/pc.20738

Ramsden, A.R. and Shiboaka, M., 1982. Characterization and analysis of individual fly ash particles from coal fired power stations by a combination of microscopy, quantitative electron microprobe analysis. *Atmospheric Environment.* 16(9):2191–2206. https://doi.org/10.1016/0004-6981(82)90290-6

Satapathy, A., Prasad, S. and Mishra, S.D., 2009. Development of protective coatings using fly ash premixed with metal powder on aluminium substrates. *Waste Management & Research.* 28:660–666. https://doi.org/10.1177/0734242x09348016

Sethi, B., 2007. Physico chemical analysis of clay and flyash in Jhansi area for brick making. M.Phil dissertation, Sirsa University, Haryana.

Shaila, K., Nisha, D., Pralhad, P. and, Deepa, P., 2015. Zeolite synthesis strategies from coal fly ash: A comprehensive review of literature. *International Research Journal of Environmental Sciences.* 4(3):9.

Shemi, A., Ndlovu, S., Sibanda, V. and van Dyk, L.D., 2015. Extraction of alumina from coal fly ash using an acid leach-sinter-acid leach technique. *Hydrometallurgy.* 157:348–355. https://doi.org/10.1016/j.hydromet.2015.08.023

Sombatsompop, N., Thongsang, S., Markpin, T. and Wimolmala, E., 2004. Fly ash particles and precipitated silica as fillers in rubbers. I. Untreated fillers in natural rubber and styrene-butadiene rubber compounds. *Journal of Applied Polymer Science.* 93:2119–2130. https://doi.org/10.1002/app.20693

Surabhi, 2017. Fly ash in India: Generation vis-à-vis utilization and global perspective. *International Journal of Applied Chemistry.* 13(1):29–52.

Tang, Y. and Chen, F.L., 2008. Extracting alumina from fly ash by soda lime sintering method. *Mining and Metallurgical Engineering (in Chinese).* 28(6):73–75.

Thamilselvi, J. and Balamurugan, P., 2018. Extraction of alumina from coal fly ash. *International Research Journal of Engineering and Technology.* 5(4):2677–2681.

Thongsang, S. and Sombatsompop, N., 2006. Effect of NaOH and Si69 treatments on the properties of fly ash/natural rubber composites. *Polymer Composites.* 27:30–40. https://doi.org/10.1002/pc.20163

Tiwari, S.and Saxena, M., 1999. Use of fly ash in high performance industrial coatings. *British Corrosion Journal.* 34:184–191. https://doi.org/10.1179/000705999101500824

Upadhyay, A. and Kamal, M., 2007. Characterisation and utilization of fly ash. B.tech (mining) dissertation, Department of Mining Engineering, NIT, Rourkela, Orissa.

Wang, S. and Wu, H., 2006. Environmental-benign utilisation of fly ash as low-cost adsorbents. *Journal of Hazardous Materials.* 136(3):482–501. https://doi.org/10.1016/j.jhazmat.2006.01.067

Zhang, B.Y. and Zhou, F.L., 2007. The limestone sintering process to produce alumina with fly ash. *Journal of Light Metals (in Chinese).* 27(6):17–18.

Zhuang, X.Y., Chen, L., Komarneni, S., Zhou, C.H., Tong, D.S., Yang, H.M., Yu, W.H. and Wang, H., 2016. Fly ash-based geopolymer: Clean production, properties and applications. *Journal of Cleaner Production.* 125:253–267. https://doi.org/10.1016/j.jclepro.2016.03.019

19 Fugitive Dust Control in Cement Industries

Ashok K. Rathoure

CONTENTS

19.1 Introduction 317
19.2 Necessity for Control of Fugitive Dust 320
19.3 Control Techniques for Fugitive Dust 320
    19.3.1 Environmental Protection Measures 321
    19.3.2 Design Description of Air Pollution Control Equipment 325
        19.3.2.1 Raw Mill/Kiln Bag House 325
        19.3.2.2 Coal Mill Bag House 325
        19.3.2.3 Bypass Bag House 326
        19.3.2.4 Cooler Esp 326
        19.3.2.5 Cement Mill Bag House 326
19.4 Conclusion 326
References 326

19.1 INTRODUCTION

Cement is a mineral, non-metallic material with hydraulic binding properties, and is used as a bonding agent in building materials. It is a fine powder, typically grey in colour, consisting of a mixture of hydraulic cement minerals to which one or more forms of calcium sulphate have been added (Greer et al., 1992). The cement manufacturing process can be divided into the following primary process components:

- Raw materials acquisition and handling
- Kiln feed preparation
- Pyro processing or clinkerization
- Finished cement grinding.

The first step in cement manufacturing is raw materials acquisition. Raw material includes limestone, chalk, seashells and an impure limestone known as natural cement rock obtained from open-face quarries. Clinker and additives like gypsum, fly ash and slag is filled in to the respective hoppers through suitable material handling system. An arrangement for cement mill is considered for clinker storage, additives storage cement storage and packing plant (Watson et al., 2000). During transfer operation involving free fall of material from a higher to a lower level,

317

emissions are generated. These emissions contain primarily the fine dust generated during upstream crushing operation and some fresh fine dust is also generated as a result of breaking of lumps due to impact during the free fall and by breaking due to movement/ conveying of material. Clinker is produced by pyro processing in large kilns. These kiln systems evaporate the inherent water in the raw meal, calcine the carbonate constituents, and form cement minerals. The main pyro processing kiln type used is the rotary kiln. The ground raw material, fed into the top of the kiln, moves down the tube counter current to the flow of gases and toward the flame-end of the rotary kiln, where the raw meal is dried, calcined, and enters into the sintering zone. The cement industry is depicted in Figure 19.1.

In a wet rotary kiln, the raw meal typically covers approximately 36% moisture. These kilns were developed as an upgrade of the novel long dry kiln to progress the chemical uniformity in the raw meal. The water is first evaporated in the kiln in the low temperature zone. The evaporation step makes a long kiln necessary. After cooling, the clinker can be stored in the clinker dome, silos, bins, or outside. Grinding of cement clinker, together with additives can be done in ball mills, ball mills in combination with roller presses, roller mills, or roller presses (Kema, 2005). The manufacturing process of cement (clinkerization) is given in Figure 19.2.

Closed View of Rotary Kiln

Back View of Cement Plant

Aerial View of Raw Material Storage

Conveyor Belt

FIGURE 19.1 Cement Industry

Fugitive Dust Control in Cement Industries

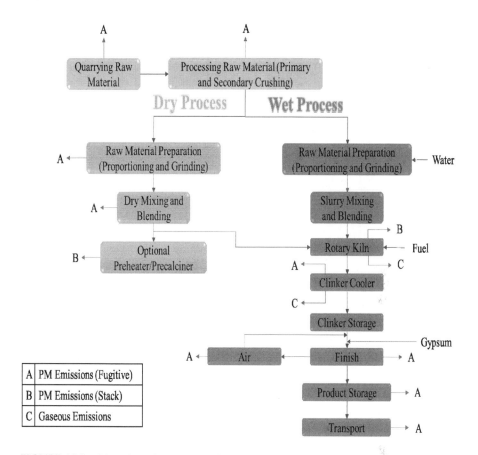

FIGURE 19.2 Manufacturing Process of Cement

The fugitive dust emission in cement manufacturing plant can be broadly divided in two categories: process-related fugitive dust emissions and non-process-related fugitive dust emissions. Process fugitive emission sources include emissions from process related essential operations like material handling and transfer, raw milling operations, and finish milling operations. Since the dust from these units is returned to the process, they are considered to be process units as well as air pollution control devices. Non-process-related fugitive dust emission sources include emissions from activities not essential for process such as vehicle traffic at the cement manufacturing site, raw material storage piles, and clinker storage piles etc. (CPCB, 2007). The impact of a fugitive dust source on air pollution depends on the quantity and drift potential of the dust particles injected in to the atmosphere. In addition to large dust particles that settle down near the source considerable amounts of fine particles also are emitted and dispersed over much greater distances from the source. PM_{10} represents relatively fine particle size range and, as such, is not overly susceptible to gravitational settling. Fugitive dust refers to small particles of geological origin that are suspended into the atmosphere from conducted emitters. Open fields and

Zero Waste

TABLE 19.1
Pollution Control Equipment

S. No.	Stack Attached to	Stack Height	Control Equipment	Parameters to Be Consider
1.	Raw Mill/Kiln	As per SPCB, CPCB, under EP Act 1986	Pulse Jet bag filter	PM, SO_2, NOx
2.	Coal mills		Pulse Jet bag filter	PM
3.	KPD by-pass		Electro-static precipitator	PM
4.	Cooler		Electro-static precipitator	PM
5.	Cement Mills		Pulse Jet bag filter	PM

parking lots, paved and unpaved roads, agricultural fields, construction sites, unenclosed strong piles and material transfer systems are the major sources of fugitive dust (Watson et al., 2000). The pollution control equipment required to be installed in cement industry is listed in Table 19.1.

19.2 NECESSITY FOR CONTROL OF FUGITIVE DUST

The dust generated from cement plant might be classified as inert dust or nuisance dust and can be defined as dust which contains less than 1% quartz. Because of low content of silicates, nuisance dust has a little adverse effect on the lungs. Any reaction which should occur from such dust is potentially reversible. If irritative dust is in extreme concentrations in the workplace, it shall reduce visibility, can cause unpleasant deposits in eyes, ears, and nasal passages, and can cause injury to the skin by chemical or mechanical action. Dust is classified based on an occupational health point of view by size into three primary categories: respirable, inhalable and total. Respirable dust refers to those dust particles that are sufficiently small to penetrate the nose and upper respiratory system and deep into the lungs. Particles are penetrated deep into the respiratory system which is generally beyond the body's natural clearance mechanisms of cilia and mucous and are retained. The inhalable dust is a fraction of dust, which enters the body, but is trapped in the nose, throat, and upper respiratory tract. The median aerodynamic diameter of inhalable dust is about 10 microns. Total dust includes all airborne particles, regardless of their size or composition. Because of these adverse effects of fugitive dust, its control is necessary. If it is reused or recycled, we may reduce its effects and quantity for disposal (CPCB, 2007).

19.3 CONTROL TECHNIQUES FOR FUGITIVE DUST

For the purpose of controlling fugitive emissions from different sections, various cement industries have adopted different combination of control measures with varying degree of effectiveness. The sources and control measures for fugitive dust

Fugitive Dust Control in Cement Industries

emission is given in Table 19.2, and Table 19.3 has aspects and impact analysis for cement manufacturing.

19.3.1 ENVIRONMENTAL PROTECTION MEASURES

- Use of a simple, linear layout for materials handling operations to reduce the need for multiple transfer points.
- Use of enclosed belt conveyors for materials transportation and emission controls at transfer points.
- Cleaning of return belts in the conveyor belt system.
- Storage of pulverized coal in silos.

TABLE 19.2
Source and Control Measures for Fugitive Dust Emission

Area	Control Measures
Unloading Section (Limestone, Coal & Other Relevant Material)	• Enclosure should be provided for all unloading operations, except wet materials like gypsum. • Water should be sprayed on the material prior and during unloading.
Fuel Handling	• High Efficiency Intensive Pulse jet bag filter should be installed as APC. • Regular water sprinkling should be carried out to control the dust generation. • Regular health check-up should be carried out for workers. • Dust masks should be provided to workers. • CEMS is recommended for online monitoring of air pollution. • Fuel should be transported up to the storage bunkers through conveyor belts. The closed belt conveyor system and provision of de-dusting bag filters should be provided. • Airborne dust at all transfer operations/points would be controlled either by spraying water or by extracting to bag filter. • Generated used oil should be collected and utilized as alternative fuel in kilns. • Pucca Floor should be provided at stock yard.
Fly Ash of Limestone/Gypsum Handling	• Regular water sprinkling should be done to avoid dust generation from transportation. • Transportation vehicles should be covered with tarpaulin sheet while transporting materials. • Dust collected from air pollution control equipment should be 100% recycled in system.

(Continued)

TABLE 19.2 (Continued)
Source and Control Measures for Fugitive Dust Emission

Area	Control Measures
	• Generated used oil from plant machinery should be utilized as alternative fuel.
Cement Packing Section	• Provide dust extraction arrangement for packing machines.
	• Provide adequate ventilation for the packing hall.
	• Spillage of cement on floor should be minimized and cleared daily to prevent fugitive emissions.
	• Prevent emissions from the recycling screen by installing appropriate dust extraction system.
Clinker Handling	• Regular water sprinkling should be done to avoid dust generation from transportation.
	• Transportation vehicles should be covered with tarpaulin sheet while transporting materials.
Fugitive Dust due to Transportation	• High efficiency intensive pulse jet bag filters should be considered to arrest the air borne dust at all the locations where transfer of material takes place.
	• The automatic bagging machine with bag filters should be installed for packing plant.
	• Unloading of coal trucks should be carried out with proper care avoiding dropping of the materials from height. It is advisable to moist the material by sprinkling water while unloading.
	• The sprinkling of water should be done along the internal roads in the plant in order to control the dust arising due to the movement of vehicular traffic.
	• Thick green belt should be developed around the plant to arrest the fugitive emissions.
	• The plant should be designed as fully automatic with alarming system, Auto shut down (DCS Control) system and interlinking to plant shut down.
	• Dust extraction system using mist other than regular water sprinkling.
	• Vacuum sweeping system for whole plant should be developed with no dust left behind.
	• All the workers inside the plant should be provided with disposable dust masks and other PPEs as per the requirement.
	• Only valid PUC vehicle should be used for the transportation of materials and equipment.
	• Construction site should be barricaded with sheet to avoid dust emission due to wind from project site.
	• Fuel should be transported up to the storage bunkers through close conveyor belts system.
	• Environmental guidelines for prevention and control of fugitive emissions from cement plants should be followed.

(Continued)

Fugitive Dust Control in Cement Industries

TABLE 19.2 (Continued)

Source and Control Measures for Fugitive Dust Emission

Area	Control Measures
	• The spilled cement from the packing machine should be collected properly and should be sent for recycling.
	• Proper engineering controls to prevent the fugitive emissions may include arrangements like providing guiding plate, scrapper brush for removing adhered dust on cement bag etc.
	• A suitably qualified person should be designated to operate as Dust Control Officer and he should be provided necessary training and should be aware of operational, maintenance aspects. Dust Control Officer (Environmental Officer) should be responsible for proper control of fugitive emissions.
	• Entire fuel storage area should be covered with permanent weather shed roofing and side walls (i.e. in closed shed).
	• Proper drainage system should be provided in all fuel storage area.
Silo Section	• The silo vent should be provided with a bag filter type system to vent out the air borne fines.

TABLE 19.3

Aspects and Impacts Analysis for Cement Plants

Activity	Environmental Attribute	Cause	Impact Characteristics			
			Nature	Duration	Reversibility	Intensity and Significance
Vehicle movement and utilities operation	Air quality	Exhaust emissions *i.e.* NO_x, SO_2 Fugitive emission	Negative	Short term	Reversible	Low, due to movement of vehicle only for loading and unloading of raw material and finished goods.
	Noise levels	Noise generation	Negative	Short term	Reversible	Low, due to limited activity
	Risk & hazards	Accidents, collision of transport vehicles	Negative	Short term	Reversible	Medium due to loss of property and injury to manpower.

(Continued)

324 Zero Waste

TABLE 19.3 (Continued)
Aspects and Impacts Analysis for Cement Plants

Activity	Environmental Attribute	Cause	Impact Characteristics			
			Nature	Duration	Reversibility	Intensity and Significance
Product and raw material handling, storage and processing	Noise levels	Noise generation from utilities, processing machinery	Negative	Long term	Reversible	Low, due to noise reduction measures
	Work zone air quality	Process gas and flue gas generation	Negative	Long term	Reversible	Medium, due to air pollution control devices (APC's)
	Risk & hazards	Catch fire accidents	Negative	Short term	Reversible	High, due to injury to manpower and loss of properties and fatal accidents
Sewage generation	Water quality	Generation of wastewater	Negative	Short term	Reversible	The plant should be operated on 'Zero Discharge Basis'.
Solid/ hazardous waste disposal	Land and soil	Generation of solid waste	Negative	Short term	Reversible	No solid waste generation
Green belt development	Ecology	Planting of trees	Positive	Long term	Reversible	High, Positive Impact
	Air quality	Dust barrier	Positive	Short term	irreversible	Low, Positive Impact
	Noise	Noise barrier	Positive	Short term	irreversible	Low, Positive Impact
	LU/LC	Conservation of land	Positive	Long term	irreversible	Medium, Positive Impact
	Soil	Increase in soil fertility	Positive	Long term	Irreversible	Medium, Positive Impact
	Water	Water consumption	Negative	Long term	Irreversible	Low, due to water circulation

(Continued)

Fugitive Dust Control in Cement Industries

TABLE 19.3 (Continued)
Aspects and Impacts Analysis for Cement Plants

Activity	Environmental Attribute	Cause	Impact Characteristics			
			Nature	Duration	Reversibility	Intensity and Significance
Employment generation	Socio economic status	Direct and indirect employment	Positive	Long term	Irreversible	High, the project should generate employment

- Storage of clinker in covered/closed bays or silos with mechanized extraction.
- Storage of cements in silos with automatic reclaiming and loading of bulk tankers. Implementation of automatic bag filling and handling systems to the extent possible, including: using a rotary bag filling machine with fugitive emission control, using automatic weight control for each bag during discharge.
- Conduct material handling (e.g. crushing operations, raw milling and clinker grinding) in enclosed systems maintained under negative pressure by exhaust fans. Collecting ventilation air and removing dust using cyclones and bag filters.
- Using low NO_X burners and in line calciner to minimize NO_X generation.

19.3.2 DESIGN DESCRIPTION OF AIR POLLUTION CONTROL EQUIPMENT

19.3.2.1 Raw Mill/Kiln Bag House

Gases from kiln/raw mill should pass through a set of cyclones and raw mill fan and should be dedusted in a bag house provided with a bag house vent fan in the compound operation. In direct operation, gases from the PH fan should directly go to the bag house by passing the raw mill. High temperature gases should be diluted with fresh air to bring down the gas temperatures from 308°C to 240°C. Pulsejet-type cleaning of bag house (low pressure cleaning 2.5–3.0 Kg/cm²) has been envisaged. Raw meal dust collected in the bag house hoppers should be extracted through combination of slide gates and rotary airlocks. It should be transported by a set of chain conveyor/air slide to a bucket elevator for transporting the material to blending silo or hot dust bin. The bag house system should be designed considering the process-upset conditions also, such as increase in dust load, rise in gas temperatures etc. Accordingly, mechanical/auxiliary systems should be designed to take care of surges and sudden rise of temperature.

19.3.2.2 Coal Mill Bag House

Product laden gases from coal mill should pass through the pulsejet bag house provided with a mill vent fan. Product should be collected in the bag house hoppers and extracted through slide gates and rotary airlocks. It should be transported by a set of screw conveyors to the fine coal bins. Provisions and recommendations for suitable safety measures should be provided such as purging of inert gases, explosion vents

326

Zero Waste

at suitable locations etc. Moreover, lignite is rich in sulphur; proper attention should be given while designing the expose surfaces that are exposed to gases of bag house, so that they can withstand the erosion/corrosion of the surfaces. Any sharp edges in the metallic parts should be avoided considering material being handled like coal.

19.3.2.3 Bypass Bag House

Kiln bypass gases should be taken into the mixing chamber for cooling and then de-dusted in filter. Pulsejet type baghouse has been envisaged for this application. A cooling fan, a mixing chamber, connecting ducts and control flaps should be installed before baghouse. For the de-dusting filter, rotary gates screw/chain conveyors, samplers etc. should be installed. Dust collected from bypass filter should be conveyed to a storage bin.

19.3.2.4 Cooler ESP

Dedusting of cooler gases should be done through ESP. Material should be discharge directly on the main clinker transport DPC. ESP system should be designed considering the process-upset conditions, increase in dust load, rise in gas temperatures etc. Accordingly, mechanical systems should be designed to take care of surges and sudden rise of temperature.

19.3.2.5 Cement Mill Bag House

Product laden gases from cement mill should pass through the pulsejet bag house provided with a vent fan. Product should be collected in the bag house hoppers and extracted through slide gates and rotary airlocks. It should be transported by a set of screw conveyors/air slides to the silo feeding bucket elevator.

19.4 CONCLUSION

Cement belongs to the most often used building materials and its production is increasing over the world because of rapid urbanization. The cement industry is an energy enormous intensive and produces several emissions, odours and noise. It is a major source of emissions such as CO_2, NO_x, SO_x, particulate matter etc. These can adversely affect human health as well as environment. A current trend in the field of cement production is the focus on low-energy cement utilization of waste in cement production and the associated reduction of CO_2 emissions. Therefore, reuse and recycling are an important part. Main source of pollution from cement industry is fugitive dust. So, fugitive dust control from cement industry is a very important aspect. We can control fugitive dust in different ways like reusing dust, which is generated from pollution control devices.

REFERENCES

CPCB, 2007. Assessment of fugitive emissions & development of environmental guidelines for control of fugitive emissions in cement manufacturing industries. 1–192. Available online at http://cpcb.nic.in/openpdffile.php?id=UHVibGljYXRpb25GaWxlLzQwMl8 xNDU3NDE4NTc3X1B1YmxpY2F0aW9uXzQ4NV9wYWNrYWdlX2NlbWVVudF9m dWdpdGl2ZV9lbWlzc2lvbi5wZGY=

Greer, W. L., Johnson, M. D., Morton, E. L., Raught, E. C., Steuch, H. E. and Trusty Jr., C. B., 1992. Portland Cement. In *Air Pollution Engineering Manual*, Anthony J. Buonicore and Waynte T. Devis. (eds.). New York: Van Nostrand Reinhold.

KEMA, 2005. *Industrial Case Study: The Cement Industry.* Available online at www.calmac.org/publications/IndustrialCementFinalKEMA.pdf

Watson, J., Chow, J. and Pace, T., 2000. Fugitive dust emissions. *Air Pollution Engineering Manual.* 117–135. Available online at www.researchgate.net/publication/235341819_Fugitive_dust_emissions

Index

Note: Page numbers in *italics* indicate figures; page numbers in **bold** indicate tables.

A

acoustic enclosures, 34
activated sludge, industrial wastewater, 126
adsorbent, fly ash as low-cost, 310
aerobic digestion, wastewater, 262
aerosol or pressurized containers, WHO
    category, 80
Agenda 21-Earth Summit, 117
agriculture
    beneficial effects of fly ash on soil, 308–309
    fertilizer value of fly ash, 309
    fly ash in, 308–309
    source of waste, 4
alumina, recovery from fly ash, 309
ambient air quality standards, noise, **33**
anaerobic digestion (AD), 14, *14*
    wastewater, 261, *262*
animal waste, 4
anti-erosion bunds, 9
Apparel Export Promotion Council (AEPC), 52
apparel industry, *see* woven shirt manufacturing
atmospheric noise, 27
autoclave, biomedical waste, 85–86
avenue plantation, 9

B

barricading, 10
bar screen, effluent treatment plant, 19, *20*
biodegradation processes, municipal solid waste
    (MSW), 172
biodiesel production, wastewater, 264–265
biodiversity, importance of wetlands, 204–205
    loss of as vulnerability of wetlands, 210–211
bioelectricity generation, wastewater, *264*, 265
bioethanol production, wastewater, 264
biohydrogen production, wastewater, 263
biological waste treatment
    aerobic and anaerobic digestion of solid
        wastes, 136–138
    bioreactor landfills, 138
    microbes in e-waste management, 138–139
    vermicomposting, 135–136, *136*
biomedical waste, 3
    categories and processes, 161, **162–164**
    classification of, 78–80
    disposal methods of, 85–87

generation, 82–83
hydroclave process treatment, 86
incineration method, 86
Indian scenario of management, 160
landfilling process, 86–87
management, 82–85, *83*
management by Indian legislation (2016),
    160–161
merits of management system, 82
microwave treatment, 86
moist heat sterilization, 85–86
need for management, 82
points of issue, 80–82
segregation of, 83
source of, 5, *77*, 77–78
storage of, 83–84
transportation of, 84
treatment by bag colour, **84**
type of toxicity levels, 78–79
World Health Organization categories, 79–80
biomethane production, wastewater, 263–264
bioremediation of wastewater, 10
    domestic wastewater treatment, 257
    industrial treatment, 257
    mechanisms involved in, 259–260, **260–261**
    microorganisms used in, 257–258, **258–259**
    phenolic wastewater treatment, 256
    treatment of brown, black, yellow and grey
        water, 256
    treatment of oil mill wastewater (OMW),
        256–257
    treatment process of wastewater effluents,
        256–257
biosolids incineration, wastewater, 262–263
black water
    treatment systems for, 256
    wastewater, 250
Bodhjungnagar Industrial Growth Centre, 35
Borosil Mansingh Survismeter (BMS), 231
brown water
    treatment systems for, 256
    wastewater, 250
Bureau of Indian Standards (BIS), 109

C

Canadian Environmental Protection Act (CEPA),
    hazardous waste management, 180

329

canteen wastewater
  design for treatment system, 191–192, *192*,
    192, *193*
  hydroponic treatment for, 189–191; *see also*
    hydroponic treatment
carbon sequestration, importance of wetlands,
    205–206
cardiovascular issues, noise and, 29
cement
  aspects and impacts analysis for cement
    plants, **323–325**
  as building material, 317, 326
  control techniques for fugitive dust, 321,
    325–326
  environmental protection measures, 325
  fugitive dust emission, 319–320
  industry, *318*
  manufacturing process, 317–320, *319*
  necessity for control of fugitive dust, 320–321
  pollution control equipment, **320**, 325–326
  source and control measures for fugitive dust
    emission, **321–323**
cement kilns, co-processing of plastic in, *108*,
    111, 156
Central Pollution Control Board (CPCB), 182
ceramic tiles, fly ash utilization in, 311
chemical dosing, effluent treatment plant, 19, *20*
chemical recycling process, e-waste
    treatment, 159
chemical waste, WHO category, 79
chromium, *see* heavy metal immobilization
cleaner production (CP), zero waste
    management, 7
climate change, vulnerability of wetlands,
    209–210
clinical waste, 77
collection tank, effluent treatment plant, 19, *20*
commercial waste, municipal, 5
common biomedical waste treatment facility
    (CBWTF), 161
communication trouble, noise and, 29
community plantation, 9
composting
  fly ash in agriculture, 309
  municipal solid waste (MSW), 172
construction and demolition (C&D) waste, 92,
    103–104
  applications of, 100
  collection and transportation of, 97
  components of, 93–94
  contributors of, 93, *94*
  disposal of, 100, *166*, 167
  estimation of in India, 101–102
  generation in cities/towns, 165
  handling, 101
  landfill, *166*, 167
  management, 95–100

management rules by Indian legislation
    (2016), 161, 165–167
  overview of construction industry in India,
    102
  recycling, 166, *166*
  recycling of, 99
  reduce of, 97–98
  reuse of, 9, 166, *166*
  segregation of, 97
  storage of, 96
  types of, 93, *93*
  types of generators, 92
  typical composition of Indian, 92, 94–95, **95**
  uncertainty in quantum of generation, 103
construction materials
  fly ash bricks, 307, **307**
  fly ash in, 305–308
  fly ash in portland cement concrete, 306
  fly ash in road construction, 306–307
  fly ash in stabilized base course, 307
  fly ash in structural fills/embankments,
    306–307
  lightweight aggregates, 308
copper, *see* heavy metal immobilization
Coriandrum sativum L., 278; *see also* heavy
    metal immobilization
  chromium (Cr), 287, **288**, 290
  cluster analysis of essential oils, *296*,
    296–297
  copper (Cu), 285, **286**, 287
  correlation between growth parameters and
    constituents of in soil, **293**
  correlation between metals for, in soil, **289**
  GC and GC-MS analysis of essential oil, 290,
    294, 295
  manure-amended contaminated soil and
    essential oil yield, 290, **291**, **292**
  manure-amended metal contaminated soil
    and essential oil composition, 290, 296
  metal accumulation by, 282–283, 285, 287,
    290
  nickel (Ni), 283, **284**, 285, **285**
  zinc (Zn), 282–283, **283**

D

demolition, *see* construction and demolition
    (C&D) waste
Design Build Operate Finance and Transfer
    (DBOFT) model, 102
design for Six Sigma (DFSS), 52
disposal
  biomedical waste, 84–85
  construction and demolition (C&D) waste,
    100
dissolved air flotation (DAF), effluent treatment
    plant, 19, *20*

Index

331

DMADV (define, measure, analyse, design and verify), 52
DMAIC (define, measure, analyze, improve and control)
 analyse phase, 61–68
 cause and effect diagram for defect generation, *61*
 control phase, 69–70, *71*
 define phase, 57
 improve phase, 68–69
 measure phase, 60
 methodology, 53
 Six Sigma-based framework using, 71–72
 study in apparel manufacturing, 53–54, **54**; *see also* woven shirt manufacturing
domestic waste, source of, 3
domestic wastewater, treatment, 257
Drug-Friccohesity Interaction (DFI), 241

E

ear muffs/plugs, 34
economic importance
 hydroponic wastewater treatment, 194, **195**
 of wetlands, 208–209
ecosystem
 wetlands as, 203–204, 213–214; *see also* wetlands
effluent treatment plant (ETP)
 detail of, 19–20
 flow diagram of, *20*
 sample of water jet, *21*
Environment (Protection) Act of 1986, India, 152, 160, 183
environmental issues
 industrial waste, 116
 noise, 27–28
 plastic waste, 107
environmental pollution
 biomedical waste, 81
 noise as, 26
Environmental Protection Agency (USEPA), 119, 178
Environment and Climate Change Canada (ECCC), hazardous waste management, 180
enzymes, *see* microbial enzymes
eutrophication, vulnerability of wetlands, 213
evaporator, effluent treatment plant, 20, *20*
e-waste management
 Indian scenario for, 157–158
 role of microbes in, 138–139
 treatment of, 158–160

F

fennel, *see Foeniculum vulgare*
five R's, waste management, *6*, 6–7

F-List, listed waste, 119, 168, 177
flocculation, effluent treatment plant, 19, *20*
flora
 industrial wastes and, 122
 land use change of wetlands, 210–211
fly ash, 302, 312–313
 in agriculture, 308, 308–309
 alumina recovery from, 309
 bricks, 307, **307**
 characterization of, 302–305
 chemical composition of, 303–304
 in construction materials, 305–308
 in development of protective coatings, 310–311
 fertilizer value of, 309
 in geopolymer, 311–312
 improving soil properties, 308–309
 industrial waste, 302, 312–313
 in lightweight aggregates, 308
 as low-cost adsorbent, 310
 in making ceramic tiles, 311
 mineralogical characteristics of, 304–305
 physical characteristics, 303
 polymer composites as wood substitute, 310
 in portland cement concrete, 306
 qualitative mineralogical analysis of typical sample, **305**
 recovery of value materials, 309–310
 as reinforcement filler, 311
 in road construction, 306–307
 SEM micrograph of, *304*
 silica recovery from, 310
 in synthesis of zeolites, 312
 utilization of, 305–312
Foeniculum vulgare, 220
 catalytic activity of synthesized AgNPs, 222–224
 phytochemical screening, 221–222
 phytochemical screening of aqueous seed extract of, **222**
 preparation of seed extract, 220–221
 synthesis of silver nanoparticles (AgNPs), 221
 UV-vis spectral analysis of nanoparticles of seed extract, 221, 222; *see also* silver nanoparticles
fugitive dust control, *see* cement

G

gaseous waste, 2
gasification, solid waste, 134
general waste, WHO category, 79
geopolymer, fly ash-based, 311–312
Government of the Northwest Territories (GNWT), hazardous waste management, 178, 180
green belt development (GBD), 35–40
 action plan, 40–41

332   Index

case study, 40–46
design considerations, 37–38, *38*
guidelines for plantation, 43, **45**
illustration of plants attenuating sound, *37*
mechanism, 36–37
noise and, 26–27
roadside plantation, 43, **44**, **45**
sample budgetary provision of, **45**
sample map of, *46*
selection of plants for, 41, **42–43**
techniques for, 9–10
three tier plantation management, **45**
trees for noise control, **39–40**
vegetated solid barriers, 38–39
vegetation, 34–35
visual barrier between noise source and
hearer, *36*
zero waste management, 9–10
grey cloth, 18, *18*
grey water
treatment systems for, 256
wastewater, 249
groove plantation, 9

H

hazardous waste
background of management, 175–176
biological treatment, 170, *181*
carriers for transportation of, *179*, 180
characteristics of, *118*
classification of, 168, *177*, 177–178
corrosivity, *118*, 118–119
generators of, 179, *179*
identification of, *176*, 176–177
ignitability, *118*, 119
Indian scenario for, 167–168
on-site/off-site recycling, 182–183
physical and chemical treatment, 168–169, *181*
reactivity, *118*, 119
receivers of, *179*, 180
regulatory groups for management, 178–181
segregation of, 177–178
source reduction strategy, 182
thermal treatment, 169, *181*
toxicity, 117–118, *118*, 119
treatment, *181*, 181–182
treatment of, 168–170
health care waste, 76, 77; *see also* biomedical waste
health issues, noise and, 28
hearing problems
noise and, 28
protectors, 33–34
heavy metal immobilization, 278
chromium in *Coriandrum sativum*, 287,
288, 290

copper in *C. sativum*, 285, **286**, 287
correlation between metals (Zn, Ni, Cu and
Cr) for *C. sativum* in soil, **289**
effect of cow dung amendment, **280**
effect of manure amendment, **280**, 281–282
materials and methods, 279, 281
metal content in aerial parts of coriander
grown in soil, *287*
nickel in *C. sativum*, 283, **284**, 285, **285**
pot experiment, 279
sampling and analysis, 279, 281
zinc in *C. sativum*, 282–283, **283**
high-density polyethylene (HDPE), 106, 108, **154**
hospital waste, 3, 77
household waste, 2
humans, industrial wastes and, 122
hydroclave process treatment, biomedical waste, 86
hydroponic treatment, 10, 187–188
canteen wastewater in Park College of
Technology, 189–191
economics of, 194, **195**
literature review of, 188
overall performance of, *201*, **201**, *202*, **202**
performance of, 193, 195, *196*, *197*, *198*, *199*,
200, *201*
plant layout and design, 191–192, *192*, **192**
quality of raw wastewater, **190**
removal of oil and grease, 191
removal of suspended matter from
wastewater, 190–191
selection of plants, 193, *194*
wastewater quantity and quality for, 189–191
hygiene, waste of, 77

I

incineration, 17
biomedical waste, 86
municipal solid waste (MSW), 172
plastic waste, *108*, 110
solid industrial waste, 127, *127*
solid waste, 134
India
biodiversity of wetlands, 204
bioplastics, 112
carbon sequestration of wetlands, 205
construction and demolition (C&D) waste, 92
construction industry in, 102
Environment (Protection) Act of, 1986, 152
handling C&D waste, 101
noise standards in, 32–33, **33**
plastics in, 105
plastic waste generation in, 153–154
profile of plastic industry in, 108
promoting Environment Protection Act
(1986), 183

Index

State Pollution Control Board, 116
typical composition of C&D waste, 94–95, **95**
Indian legislation
  biomedical waste categories, 161, **162–164**
  biomedical waste management rules (2016),
    160–161
  construction and demolition (C&D) waste
    management rules (2016), 161, 165–167
  e-waste (management and handling) rules
    (2011), 157–160
  hazardous and other wastes rules (2016),
    167–170
  municipal solid wastes rules (2000),
    170–172
  plastic waste (management and handling)
    rules (2011), 153–157
  wastes and their rules, **153**
industrial development, 115–116
industrial waste, 2–3
  based on toxicity of, 117–118
  characteristics of, 120–122
  characteristics of hazardous waste, *118*,
    118–119
  chemical characteristics of, 121–122
  classification of, 117–120
  colour of, **120**
  controlling generation of, 124
  definition, 116
  disposal of liquid, 126
  flow cycle management of, *123*
  fly ash, 302; *see also* fly ash
  improper disposal of, 116–117
  incineration, 127, *127*
  landfilling, 128, *128*
  liquid, 125–126
  listed waste, 119
  management approach, 122–123
  mixed waste, 120
  non-hazardous waste, 120
  physical characteristics of, 120–121
  recycling, 126–127
  reducing sources, 126
  reusing, 126
  segregation of, 125
  solid, 126–128
  source of, 3–4, **121**
  treatment methods, 125–128
  universal waste, 120
  upshots of, 122
infectious diseases, 76
infectious waste, WHO category, 80
inorganic chemicals, industrial wastes, 121
International Civil Aviation Organization
  (ICAO), 180
International Clinical Epidemiology
  Network, 160

International Maritime Dangerous Goods Code
  (IMDG), 180

K

K-List, listed waste, 119, 168, 178
Knop, W., 188

L

landfill/landfilling
  biomedical waste, 86–87
  bioreactor, 138
  construction and demolition (C&D) waste,
    166, 167
  main features of modern, *135*
  municipal solid waste (MSW), 172
  plastic waste, *108*, 110
  solid industrial waste, 128, *128*
  solid waste disposal, 134–135
  waste management hierarchies, *96*
land use change, vulnerability of wetlands, 210–211
liquid fuel, plastic waste conversion to, *108*, 111
liquid solutions, 227–229; *see also*
  Survismeter
  aqueous interaction mechanisms, 228
liquid waste, 2
low-density polyethylene (LDPE), 106, 108, **154**

M

management
  biomedical waste, 82–85
  concept of 3R for waste, 96, *96*
  construction and demolition (C&D) waste,
    95–100
  industrial waste, 122–123, *123*
  plastic waste, 108–111
  waste, 10–11
mangrove plantation, 9
manure; *see also* heavy metal immobilization
  effect on growth parameters of *Coriandrum*
    sativum, **280**, 282
  effect on physicochemical properties of soil,
    280, 281–282
  physicochemical properties of, **279**
  pot experiment with, 279
marine ecosystem, industrial wastes and, 122
mechanical recycling process, e-waste
  treatment, 159
mechanical vapour recompression (MVR), 16
medical waste, *see* biomedical waste
methylene blue dye
  contact time and adsorption of, 224
  preparation of, 221
  silver nanoparticles and, 220, 221

Index

microbes
in aerobic and anaerobic digestion of solid
wastes, 136–138
in e-waste management, 138–139
in vermicomposting, 135–136, **136**
in xenobiotics degradation, 139
microbial enzymes
applications in waste management, **144–145**
cellulases, 142, *143*, **145**
degradation of aromatic compounds, *140, 141*
laccases, 141, **144**
lipases, 142, **145**
oxidoreductases, 140, **144**
oxygenases, 140, **144**
peroxidases, 141–142, **145**
proposed mechanism for hydrolysis of
cellulose, *143*
proposed pathway for protease hydrolysis, *143*
proteases, 142, *143*
role in waste management, 139–143
triolein hydrolysis by *Candida rugosa* lipase,
142
microbial fuel cell (MFC)
hybrid approach of, 267–268
scaling up, 267
wastewater, 265
microorganisms, bioremediation use, 257–258,
258–259
microwave treatment, biomedical waste, 86
Ministry of Urban Development (MOUD),
construction and demolition (C&D)
wastes, 103
moist heat sterilization, biomedical waste, 85–86
multiple effect evaporator (MEE), 15, *16*
municipal solid waste (MSW), 4
biodegradation processes, 172
composting, 172
incineration, 172
Indian scenario of, 170–171
landfill, 172
management, 171–172
plastic waste, 107
source of, 4–5
type and treatment of, **171**

N

nanoparticles, 219–220; *see also* silver
nanoparticles
natural resources, minimal use in zero waste
management, 7
nickel, *see* heavy metal immobilization
noise
atmospheric, 27
barriers for reducing, 26
cardiovascular issues, 29
classes of, 27–28
dosimeter, 31

effects of, 28–29
environmental, 27–28
health issues, 28
hearing problems and, 28
measurement of, 29–32
occupational, 28
recommendations for reducing pollution, 47
sleeping disorders, 29
standards in India, 32–33, **33**
wildlife and, 29; *see also* green belt
development (GBD)
Noise Monitoring System (NMS), 31–32
noise pollution
classes of noise, 27–28
mitigation measures for, 33–35
recommendation for noise reduction of, 47
nuclear wastes, source of, 5
nutrient removal, importance of wetlands,
206–207

O

Occupational Health and Safety (OHS)
Regulations, 181
occupational noise, 28
oil mill wastewater (OMW), treatment of, 256–257
oil skimmer, effluent treatment plant, 19, *20*
open burning, solid waste, 134
organic chemicals, industrial wastes, 121
organic waste composting (OWC), zero waste
management, 8
Organisation for Economic Co-operation and
Development (OECD), 167
ornamental grasses, for noise control, **40**

P

P&U List, listed waste, 119, 168, 178
PARIVESH (Pro-Active and Responsive facilitation
by Interactive and Virtuous Environment
Single-window Hub) Portal, 152
Park College of Technology, *see* hydroponic
treatment
pathological waste WHO category, 79
personal protective equipment (PPE), 33–34
pharmaceutical sciences, 237
Survismeter, 237
pharmaceutical waste WHO category, 80
plants
heavy metal removal from waters in
wetlands, 207–208
ornamental grasses for noise control, **40**
selection for green belts, 41, **42–43**
selection for hydroponic treatment, 193, *194*
shrubs for noise control, **40**
three tier plantation management, **45**
trees for noise control, **39**
plasma pyrolysis technology, *108*, 111

Index

plastic coated bitumen road, *108*, 110–111
plastics
  biodegradable, 111–112
  categories of, 106–107
  degradable, 111–112
  non-recyclable or thermoset, 106–107
  profile of industry in India, 108
  recyclable or thermoplastic, 106
  type and uses, 154, **154–155**
  utilization of, 105–106
plastic waste
  alternative ways to reduce impact of, 112
  conversion into liquid fuel, *108*, 111, 157
  co-processing of plastic in cement kilns, *108*,
    111, 156
  disposal technologies, 155, *156*
  environmental effects, 107
  flow diagram for management of, *156*
  incineration, *108*, 110
  Indian legislation (2011), 153–157
  landfilling, *108*, 110
  management approaches, 108–111
  management by new technologies, 110–111
  management per Indian legislation, 155
  management practices, 108–110
  plasma pyrolysis technology, *108*, 111
  plasma pyrolysis technology (PPT) for
    disposal, 157
  plastic coated bitumen road, *108*, 110–111
  recycling, *108*, 108–109, 159
  sources of, 107
  utilization in road construction, 155–156
pollution, vulnerability of wetlands, 211–213
pollution control equipment, cement
    manufacturing, **320**, 325–326
polyethylene terephthalate (PET), 106, 107, 108,
    109, 110, **154**
polymer composites, fly ash in, 310
polypropylene (PP), 106, **154**
polystyrene (PS), 106, **155**
polyvinyl chloride (PVC), 106, 108, **154**
precious metals recovery, e-waste treatment,
    159–160
processing waste, 4
protective coatings, fly ash in development of,
    310–311
pyrolysis, solid waste, 134

Q

quality control, physicochemical analysis of
    liquid solutions, 237–238

R

radioactive waste, 5
  types of, **80**
  WHO category, 79–80, **80**

rainwater harvesting (RWH), zero waste
    management, 7–8
recover, waste minimization, *6*
recycling/recycle
  construction and demolition (C&D) waste, 99,
    166, 166–167
  hazardous waste, 182–183
  plastic waste, *108*, 108–109
  solid industrial waste, 126
  solid waste, 133
  waste management hierarchies, *96*
  waste minimization, *6*
reduce/reduction
  construction and demolition (C&D) waste,
    97–98, *166*, 166–167
  hazardous waste, 182
  solid industrial waste, 126
  solid waste, 132
  waste management hierarchies, *96*
  waste minimization, *6*
refuse, waste minimization, *6*
regulatory groups, hazardous waste management,
    178–181
reinforcement filler, fly ash as, 311
Resource Conservation and Recovery Act
    (RCRA), 172, 176
reuse
  construction and demolition (C&D) waste, 98
  solid industrial waste, 126
  solid waste, 132
  waste management hierarchies, *96*
  waste minimization, *6*
reverse osmosis (RO), 14–15, *15*
  effluent treatment plant (ETP), 19–20, *20*

S

sanpro waste, 77
segregation
  biomedical waste, 83
  construction and demolition (C&D) waste, 97
semi-insert ear plugs, 34
sequestration, wetlands, 205–206
sewage, municipal, 4–5
sharp material waste, WHO category, 80
shrubs, for noise control, **40**
silica, recovery from fly ash, 310
silver nanoparticles
  catalytic activity of synthesized, 222–224, *223*
  development of methods for synthesizing, 220
  Foeniculum vulgare seeds, 220–221, 224
  methylene blue (MB) dye and, 220, 224
  synthesis of, 221
  UV-vis analysis of, 222, *223*
Singh, Man, 243
Six Sigma, 52; *see also* woven shirt
    manufacturing
  case study using DMAIC approach and, 55–70

design for (DFSS), 52
sleeping disorders, noise and, 29
social forestry, 9
soil; *see also* heavy metal immobilization
    fly ash in agriculture, 308–309
solid waste, 2–3, 131–132
    aerobic and anaerobic digestion of, 136–138
    biomedical, 3
    gasification, 134
    household, 2
    incineration, 134
    industrial, 2–3
    landfills, 134–135, *135*
    management hierarchy, *133*
    open burning, 134
    pyrolysis, 134
    recycling, 133
    reduction of, 132
    reuse of, 132
    schematic of anaerobic digester, *137*
    thermal treatment, 134
    treatment and disposal, 133–135
solvent strippers, 16–17
sound levels
    instrumentation, 30–31
    measurement of, 29–30
    noise dosimeter, 31
    Noise Monitoring System (NMS), 31–32
    plants attenuating, 26–27
    sound level meter (SLM), 30–31
spray dryer absorption (SDA), 17
State Pollution Control Board, India, 116
storage
    biomedical waste, 83–84
    construction and demolition (C&D)
        waste, 96
Survismeter
    advantages of using, **239–240**
    conceptualization and emergence, 230–231
    fluid mechanics models, *236*
    fluid mechanics models operational in η, *235*
    Pendant Drop Number (PDN), 231, 235
    physicochemical sensing, 240–241
    salient studies conducted by, **238–239**
    schematic representation of components, *232*
    science of, 241–242
    significance as cutting edge of routines,
        236–238, 240
    viscous flow time (VFT), 231, 235
    working fascinations, 238, 240
    working mechanism of η and γ measurement,
        231, 235–236
    working snapshots, *233, 234*
synthetic dyes, 219–220; *see also* silver
    nanoparticles

T

Technology Information, Forecasting and
    Assessment Council (TIFAC), 101
terrace gardening, 10
thermal recycling process, e-waste treatment,
    159
transportation
    biomedical waste, 84
    construction and demolition (C&D)
        waste, 97
Transportation of Dangerous Goods Regulations
    (TDGR), 180
Transport Canada, hazardous waste, 180
trees
    green belt development, 35–36
    for noise control, **39**
trickling filters, industrial wastewater, 126

V

vegetation (green belt), 34–35
    mechanism for attenuating sound, 36–37
    selection of plants for noise control, 41,
        42–43
    solid barriers with, 38–39
vermicomposting
    microbiology of, 135–136
    role of microbes in, **136**
vertical gardening, 10
Visionmeter, 238
volatile organic carbons (VOCs), industrial
    wastes, 121
von Sachs, J., 188

W

warp knitting, 17
waste(s), 1–2; *see also* zero waste
    agricultural, 4
    basic terminologies, 77
    biomedical, 3, 5
    domestic, 3
    gaseous, 2
    household, 2
    industrial, 2–3, 3–4
    liquid, 2
    management, 10–11
    municipal, 4–5
    nuclear, 5
    by sharp materials, 80
    solid, 2–3
    sources of, 3–5
    types of, 2–3
Waste Electrical or Electronic Equipment
    (WEEE), 157–158
waste management

Index

application of enzymes in, **144–145**
approach for industrial waste, 122–123, *123*
biomedical, 82–85, **83**
hierarchical relationship of, 124–125, *125*
improper system, *81*
Indian government, 152
objective for industrial waste, 124
treatment processes, *85*
waste minimization, five R's of, *6*, 6–7
waste stabilization ponds, industrial
    wastewater, 126
waste treatment, medical waste, 84–85
wastewater; *see also* bioremediation of
    wastewater
biological treatment, 255–256
black water, 250
brown water, 250
characterization of, 251, **252**
chemical treatment, 255
composition of, 251
energy product recovery from, 267–268
grey water, 249
hybrid approach to treatment, 267–268
methods of treatment, *254*
physical treatment, 254–255
primary treatment, 253
process and mechanism of energy production
    from, 266, **266**
secondary treatment, 253
sources of, 250, **250–251**
spray dryer, 17
tertiary treatment, 253–254
treatment methods, 253–256
types of, 249
yellow water, 249
wastewater digestion
aerobic digestion, 262
anaerobic digestion (AD), 261, *262*
biodiesel production, 264–265
bioelectricity generation, 265
bioethanol production, 264
biohydrogen production, 263
biomethane production, 263–264
biosolids incineration, 262–263
common methods for energy generation,
    261–263
microbial fuel cell (MFC), 265
useful recovered energy products from,
    263–265
wastewater treatment; *see also* hydroponic
    treatment
hydroponics, 188
quantity and quality of, 189–191
water
generation of wastewater, 248–249
global demand for, 187–188
quality in wetlands, 207–208

water jet looms; *see also* weaving industry
effluent treatment plant, 19–20
grey cloth, 18
Indian scenario of ZLD, 22
water jet weaving machine, 18, *18*
water requirement, 20, *21*
zero liquid discharge system for, 17–22
weaving industry
bar screen, 19, *20*
collection tank, 19, *20*
detail of effluent treatment plant (ETP),
    19–20, *20*
dissolved air flotation (DAF), 19, *20*
flocculation and chemical dosing, 19, *20*
grey cloth, 18
oil skimmer, 19, *20*
production process, 18, *18*
reverse osmosis (RO), 19–20, *20*
warp knitting, 17
water balance diagram of, *21*
water jet weaving machine, 18, *18*
weft knitting, 17
weft knitting, 17
wetlands
biodiversity, 204–205
carbon sequestration, 205–206
causes of vulnerability of, 209–213
climate change, 209–210
economic importance of, 208–209
as ecosystem, 203–204, 213–214
eutrophication, 213
importance of, 204–209
land use change and biodiversity loss of,
    210–211
nutrient removal, 206–207
pollution, 211–213
water quality, 207–208
wildlife
conservation, 10
noise and, 29
wood substitute, fly ash-based polymer
    composites as, 310
Worker's Safety and Compensation Commission
    (WSCC), hazardous waste, 181
working fascinations, Survismeter, 238, 240
World Health Organization (WHO), 76,
    79–80, 160
woven shirt manufacturing
analyse phase, 61–68
cause and effect diagram for defect
    generation, *61*
control phase, 69–70, *71*
corrective actions, 69
cumulative defect distribution, **63**
define phase, 57
down stitch defect analysis, 63–65, *64*
improve phase, 68–69

338 Index

list of critical defects, **57–59**
list of operations, **54**
measure phase, 60
operation in final assembly sewing section, *56*
operation in preparatory sewing section, *55*
operator quality performance card, 68
operator self-inspection report (OSIR), 68
raw edge defect analysis, 66–68
signal system, 68–69
Six Sigma and DMAIC application, 55–70, *71*
skip stitch defect analysis, *65*, 65–66

X

xenobiotics, microbes in degradation of, 139

Y

yellow water
treatment systems for, 256
wastewater, 249

Z

zeolites, fly ash in, 312
zero defects, *see* woven shirt manufacturing

zero liquid discharge (ZLD); *see also* water jet
looms; weaving industry
anaerobic digestion (AD), 14, *14*
case study for water jet looms,
17–22
incineration, 17
Indian scenario of, 22
mechanical vapour recompression
(MVR), 16
multiple effect evaporator (MEE), 15, *16*
process of, 13–14
reverse osmosis, 14–15, *15*
solvent strippers, 16–17
wastewater spray dryer, 17
zero waste management, 8–9
zero waste, 5
goal of, 5–6
zero waste management practices
cleaner production (CP), 7
five Rs, *6*, 6–7
green belt development (GBD), 9–10
minimum use of natural resources, 7
organic waste composting (OWC), 8
rainwater harvesting (RWH), 7–8
zero liquid discharge (ZLD), 8–9
zinc, *see* heavy metal immobilization